Vallicelli. Yacht Design

Valerio Paolo Mosco

VALLICELLI

Yacht Design

SKIRA

Cover
Azzurra II, photo by Carlo Borlenghi and below construction plan of *Azzurra*

Art Director
Marcello Francone

Design
Luigi Fiore

Editorial Coordinator
Vincenza Russo

Editing
Maria Conconi
Routre 42 Studio

Layout
Paola Ranzini Pallavicini

Translations
James Taylor with
Adam Victor for Scriptum, Rome
Lucian Comoy for Language Consulting Congressi, Milan

Iconographical Research
Paola Lamanna

Graphic design concept and reworking of the drawings
Jessica Lagatta

Photographic material processing
Luca Mincarini

First published in Italy in 2020 by
Skira editore S.p.A.
Palazzo Casati Stampa
via Torino 61
20123 Milano
Italy

www.skira.net

Printed and bound in Italy.
First edition

ISBN: 978-88-572-4194-4

Distributed in USA, Canada, Central & South America by ARTBOOK I D.A.P. 75 Broad Street Suite 630, New York, NY 10004, USA.

Distributed elsewhere in the world by Thames and Hudson Ld., 181 High Holborn, London WC1V /QX, United Kingdom.

Valerio Paolo Mosco wrote the texts collecting the direct testimony of Andrea Vallicelli, founder of the Studio.

Photo Credits
© François Chevalier e Jacques Taglang: pp. 39, 42 right, 49, 51, 53, 59
© 1982 Ugo Nespolo by SIAE: p.118 right
Archivio A. Vallicelli & C.: pp. 12, 23, 25 bottom, 29, 30-33, 36, 42 left, 45, 47, 48, 50, 55, 56, 58, 60, 61 left, 66, 68, 69 center, 78, 79 bottom, 81 bottom, 83 top, 84 right, 88 bottom, 90, 95 left, 96 first from left, 98, 100 right, 108 bottom, 109, 114 bottom, 120, 121, 124, 125 bottom left, 129 first and second from left, 131 bottom, 134 bottom, 137 right, 139 top left, 140 right, 141 top, 142 bottom, 146, 147 top, 156 bottom, 162, 165 top left, 167 top left, 168 bottom, 170, 171 top right, 176, 178, 184 bottom, 187 top, 204 top, 210, 211 top right, 214 bottom, 215, 219 bottom, 227 right, 228 bottom, 233 bottom, 240, 242, 244, 246, 247 top, 248, 250 first, second and third from the left
Fanatek: pp. 200 e 201
Jonathan Eastland / Ajax News & Feature Service / Alamy Stock Photo: p. 28 left
Francesco Longanesi: pp. 25 top, 38 top
Courtesy of:
A&B Photodesign: p. 249 top
Leonardo Andreoni photographer: p. 251 top
Associazione Italiana Arpege: p. 16 left
Carlo Borlenghi photographer: cover photo, pp. 34, 35, 41 left, 110, 117, 123 top, 125 top, 126, 129 third and fourth from left top, 132 bottom left, 151 top, 153, 155 top, 172, 173 left, 183 left, 197 right, 202, 205 top right
Massimo Bucchi: p. 41 right
Jeff Brown photographer: p. 243 top
Davide Bulli photographer: pp. 230, 231 left
Luca Buttò photographer (Studio Borlenghi): p. 158
Cantiere Sangermani: p. 13 left
Pier Giovanni Carta photographer: pp. 196, 199 top
Andrea De Maria photographer: pp. 69 right, 226, 234, 236 bottom
Forattini: p. 37 bottom
Alessandro Galeani: p. 198 bottom
"Giornale della Vela": pp. 24, 26, 28 bottom, 69 left, 112, 181 bottom, 192, 216 bottom, 220 bottom, 221 bottom, 222, 224 right
Jerome Kelagopian photographer: p. 241 top
Martino Majno photographer: p. 216
Sebastiano Morassutti: pp. 175, 186 top
Ugo Nespolo: p. 118 right
Olimpic Sails: p. 148 right
Franco Pace photographer: pp. 132 bottom right, 188, 190, 191 bottom
Max Ranchi photographer: pp. 185 bottom, 187 bottom right
Bert Richner photographer: p. 167 top right
Studio Sciomachen: p. 15 bottom
Famiglia Spadolini: p. 164 right
Elisabetta Strikland: p. 118 left
Kristina Strobel photographer: p. 245 top
Fabio Taccola photographer: pp. 57, 61 right, 195 top
James Taylor photographer: p. 145 left
"Vela e Motore" (Edisport): pp. 20 bottom, 28 right, 43, 76, 86 last right, 93 top, 96 ssecond and third from left, 102 letf, 104 first from left bottom, 107 top, 125 bottom right, 208
Famiglia Vigliani: pp. 84 left, 86 first and second from top left, 102 right, 104 second, third and fourth from left bottom
Marco Zaoli: p. 52

The Co.Ge.Di.-Uliveto-Rocchetta Group, leader in the mineral water sector, was pleased to support the architect Andrea Vallicelli, leader in the design of what is destined to live in water, to make the publication of this volume possible.

Contents

Introduction

Andrea Vallicelli

A few years ago, when I started thinking about writing an anthology of more than forty years of work at my practice, one thing was very clear to me: I was not interested in publishing just a review of designs; I wanted to share the experiences and the stories behind the projects. This tale would be told primarily through our drawings and more specifically, the drawings sent to the boatbuilders, giving them the information necessary to build our yachts.

Then, as so often happens in design, this project ran off in its direction and the results took on their own, unforeseen shape. This is most definitely true of this book. As work proceeded on selecting and digitizing the many traditional graphics (almost half of our archive), I felt it was necessary to provide a framework for the context in which this story unfurled. At that point, I approached Valerio Paolo Mosco, a man who knows our products and the world they are part of very, very well. Recollections of my experiences here at our practice, in shipyards and during yacht races, have been gathered together and turned into this story by Valerio, supplemented by some thoughts about the technical, sporting, commercial and social sides of yachting over these decades.

So many things – materials, equipment, production systems – have changed over the last forty years. Yachting, or *marina lusoria* as it was called in period prints, has grown both in the sheer quantity and the size of units produced, hand in hand with an expanding market. Sailing sector companies, after many years of operating halfway between craftsmanship and industry, have gradually taken on specific productive, organizational and commercial structures. Something precious has inevitably been lost in this evolutionary transition: some of the shipyards, or rather boat workshops, where highly accomplished shipwrights worked and whom I was lucky enough to meet, are disappearing. Unfortunately, apart from some very rare cases, nobody seems able to carry on the legacy of their experience.

No other country can boast Italy's tradition of excellence in this field. In Italy, the artisan wooden boatbuilding industry is an expression not just of rare aesthetic qualities, but of how these artisans have allowed us to experiment with innovative technical and performance-related solutions, such as the creation of cold-molded structures with high modulus synthetic fibers (carbon, Kevlar, glass). So the reason why so many pages in this book are dedicated to wooden boat designs is not out of nostalgia, but because they bear witness to an emblematic chapter of research and experimentation in Italian design.

Another important development over the last two decades has been the advent of educational offerings dedicated specifically to the boating industry, culminating in specialized degrees and Ph.D. level courses. Compared to the '70s, a decade characterized by enthusiasm and a certain degree of amateurism, today's industry players undoubtedly benefit from greater professionalism.

Changes in the design realm, reflecting the evolution in IT tools, have very much affected us, which is why these changes play such an important role in this book. This radical transformation has been rapid and pervasive in yacht design as indeed it has been in almost all sectors of design, from architecture to industrial design. Indeed, the effects have not yet fully permeated these fields. In just a couple of decades, we have transitioned from graphic techniques whose roots go back for centuries to modern digital techniques that have modified not just how we represent things, but the very process of how we develop ideas, inevitably conditioning the design results themselves.

As part of this framework, we are keen to share our experience of what has changed in the practice of design creativity and presentation, giving a prominent role to technical drawings which, in part for confidentiality-related reasons, are rarely published.

Our Studio's archive, which is miraculously still intact, spans some 150 projects, about thirty of which were conceived for mass production. While the number is impossible for me to quantify, our archive contains a huge number of drawings, perhaps more than ten thousand in all, spanning both traditional and digital graphic work. The main reason why it is impossible to be more precise is that most of the traditional drawings were made on paper in a wide variety of formats and they have never been rigorously cataloged. For example, the formats of *Azzurra*'s approximately 250 shop drawings range from drawings of a size close to A3 (a small handmade sketch recently turned up in the Pesaro shipyard, used for the mast foot), to drawings that measure one meter by four meters for hull models tested in the towing tank.

Traditionally, technical drawings were made in pencil or ink on a variety of media: plain paper, tracing paper, or synthetic sheets. For several years we used a type of paper that wasn't available in Italy and which we had specially sent over from the United States, where it was regularly used for naval and mechanical drawings. It had the texture of tracing paper, the transparency of sketching paper and required the use of not very hard but very sharp graphite leads. To print these out, we used a special heliographic paper that reproduced the lines of the drawing in blue against a light blue background. We thought this color scheme was far more elegant than the traditional one, with its black lines against a grey background; plus, there was a "protective" reason for this choice: these blue line printouts could not be copied in turn via these same processes.

We also used polyester sheets (which were very expensive at the time) to prevent drawings from deforming due to humidity. Any deformity was a considerable problem, given that the geometrical coordinates of a boat's shape (hull, deck, appendages, etc.) subsequently used to develop a full-sized outline were optically measured from the scale drawings using a ruler. A deformed medium inevitably made these measurements more approximate, with easy-to-imagine negative consequences on precision and control.

The need to reduce the degree of approximation, especially on the lines for the boat's architectural forms, prompted us to increase the size of the sheets of paper to reduce differences between the measurements in the design plans and the boat's real-size measurements. Sometimes, when I was in the drafting room where the boat was redrawn in full size, I was reminded of the cartographers in a famous literary passage by Jorge Luis Borges who, to perfect the map of their empire, gradually drew a map that was as big as the empire itself. This paradox was, to a certain extent, resolved with the transition to digital tools: because they don't use in-scale dimensions, digital tools tend to compress the operational and conceptual distance between the various different design process stages. These tools also allow for a dynamic, three-dimensional preview of the boat, revolutionizing our abilities to predict sailing performance through Computational Fluid Dynamics (CFD), as well as how the project itself is visualized, including virtual reality.

We first started making our drawings digitally in the early '80s: though at first only for hull plans, we progressively expanded to all the other areas of design – general plans, sail plans, fittings, structures, renderings, etc. For a while, we printed out these drawings and saved them in paper form, before finally giving way to digital-only storage, for obvious reasons of archive rationalization.

This book is divided up into several sections, each of which presents our most representative projects, grouped by category: racing, cruising, mass-produced and power boats. This approach makes it easier to follow the changes that have swept through these different spheres over time. A summary lists all of the boats we have built, in chronological order.

For the title, in both the Italian and English versions, we chose to use the English word "yacht," a word that does not have an adequate equivalent in Italian. The Italian words *barca*, *imbarcazione*, *panfilo*, or *veliero* seemed inappropriate or sounded outdated. Besides, the word "yacht" has an interesting etymology. It first appeared in English in the second half of the 1500s. As is common knowledge, it derives from the Dutch word *jaghte* or *jaghtschip* (fast boat, hunting boat) and the German word *jagen* (hunting). A line runs all the way back from the old German *jagon* connecting to Akkadian words like *ziaqu* (to move quickly, said of the wind), or *daku* (to slay an animal), and to the Sumerian sag, *ag* (to strike, assault). When hunting, it is necessary to be aggressive, fast, agile and careful. In order not to be seen by the prey, it is essential to pay attention to the wind: all qualities that, by analogy, apply to the finest sailing boats.

I must confess that when it came to selecting the designs for publication, I did have my doubts about including drawings for motoryachts (the puritanical streak that lurks in every sailor is still full of life in me). Then I realized it would be nonsensical to exclude them: when all is considered, with this decision, I found myself in good, indeed a very good company. Masters of the caliber of Herreshoff, Stephens and Giles to name but a few have not worried about this nuance and have given us many interesting (and often beautiful) motoryachts.

A DESIGN HISTORY

The Special Context of My Formative Years

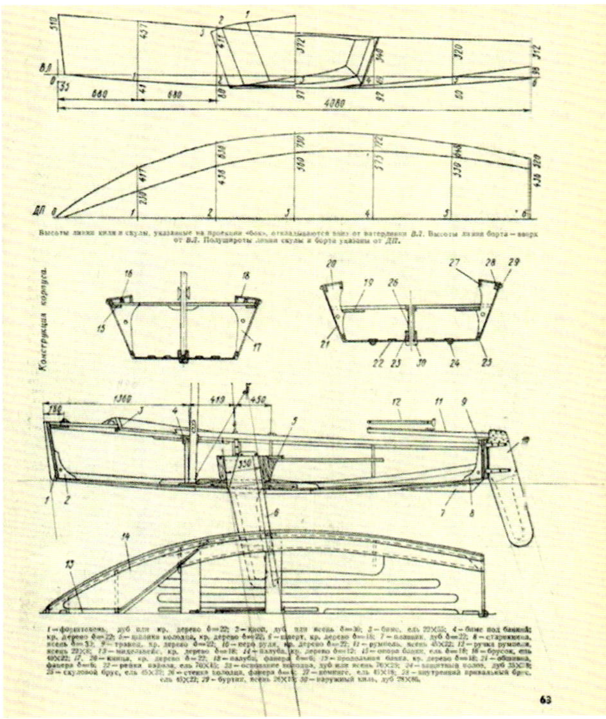

The Vaurien dinghy, an essential 1951 design by Jean Jacques Herbulot.

The years that spanned the period from my first exposure to sailing to when I became a yacht designer were unique in the yachting world. The radical changes introduced from the late '60s onwards had transformed sailing from an elite pursuit to an accessible and democratic pastime. The first signs of these changes appeared in England when competitive events resumed after the Second World War: the majestic J-Class and their kind had almost completely disappeared from the quayside at Cowes, replaced by significantly smaller craft. Yet there were many more of them, marking the birth of a trend that would continue through to the turn of the new millennium, when, with larger and more costly vessels, yachting rediscovered its elite pre-war lineage.[1]

During the two decades that followed World War II, in countries like the United States, Great Britain and France, the rising middle class became truly infatuated with sailing, as if to prove that the suffering and privations of the Great Depression and yet another World War were finally exorcised. But the war was also responsible for the remarkable technological developments that played a vital role in the transformation of pleasure boating.

During the war, the French government invested significant resources in the French Riviera, building nautical camping facilities for young people complete with dinghies and small cabin cruisers. In keeping with the spirit of nautical camping, in 1951 Jean Jacques Herbulot designed and built a revolutionary new boat, the Vaurien, a small single-chine dinghy for two (the name in French is a play on the words *vaut rien*, which means "not worth a thing," perfectly in tune with their goal of increasing the popularity of boating). When introduced, the Vaurien was revolutionary in its simplicity: the structure was pared down to a minimum, clad in just three sheets of pressure treated marine plywood – two for the sides and one for the bottom; the whole boat weighed in at just 85 kg fully rigged.[2]

In 1951 the use of plywood was the latest thing. Made from thin leaves of mahogany glued together, plywood was invented during the war by the U.S. Navy to build small service craft and made possible by a breakthrough in chemistry: phenolic adhesives that were finally waterproof. Marine plywood brought significant advantages to the table, as it is particularly hard-wearing, extremely lightweight and economical – less expensive woods work perfectly well, and plywood is easy to handle, so less specialized labor is required.

For me, the Vaurien was not only the first boat I ever sailed on my own, it was also the first boat for which, in an almost tactile way, I came to understand the underlying reasons for its design and construction. What struck me then (indeed, it still does to this day)

[1] Franco Giorgetti, *Storia ed evoluzione degli yacht da regata*, Edizione White Star, Vercelli 2000, pp. 152-160.

[2] www.vaurien.it; https://sites.google.com/site/vaurien2012/the-vaurien/history.

about the Vaurien is how balanced its components were: although pared back to a bare minimum, the hull, deck, rig and sail plan are all perfectly proportionate to each other. The conviction behind my entire design philosophy – that, first and foremost, the sum of the different components of a yacht must be harmonically proportioned – began to emerge with the Vaurien, this brilliant "worth nothing" boat.

The other dinghy I sailed back then was the Flying Dutchman, a radical boat with a genoa and trapeze, capable of planing even when sailing upwind. In those formative years, if the Vaurien was an ideal template for a properly proportioned minimalist boat, the Flying Dutchman was an ideal template for a very extreme boat. Interestingly enough, both the Vaurien and the Flying Dutchman were designed in the same year, 1951.

Let us return to marine plywood. As the Vaurien showed, it was, without doubt, an ingenious material. It did, however, have one critical flaw that could hardly be ignored: it could only be used to build chined hulls. From a fluid dynamics point of view, a chined hull is only slightly slower than a rounded hull, but a chine does nothing for esthetics, a necessary requirement for sailboats: we all unconsciously project expectations of lightness and smoothness onto these crafts. Phenolic glue came to the rescue of this critical issue. By the early '50s, the molded wood (*bois moulé*) technique circumvented the problem by overlapping thin layers of wood and gluing them together over the curved forms of mold. But unlike marine plywood, molded wood required highly-skilled craftsmanship. It was, therefore, no coincidence that the first boatyards to use molded wood were top-tier boatyards, with which Italy was well-endowed.

The famous Sangermani boatyard is a fine example of this. Founded in Lavagna back in 1896, in the 1950s and '60s the Sangermani yard became a benchmark for shipbuilding, not only in Italy, with yachts that have gone down in history – *Chiaro di Luna*, *Artica II*,

The Flying Dutchman, the acrobatic dinghy designed by Conrad Gülcher and Uus van Essen in 1951.

Gitana IV, *Mania* and *Mabelle*. These yachts were designed by several of the finest yacht designers of their times – Laurent Giles, Olin Stephens and Peter Nicholson. Yet Sangermani's success was primarily based on the fact that the yard could draw on the highly skilled craftsman, able to adapt to any new technique that came to market rapidly. In this, Sangermani was no exception. Since the 1950s, Italy had a consolidated cultural tradition of wooden boat building, and at the Studio Vallicelli, we were lucky enough to be able to cull the best of this culture and follow it's lively transformation from wood to composites.

The transfer of technology in the materials and processing methods developed during wartime went well beyond wood and included metal alloys. By the early '50s, U.S. boatyards had begun adapting duralumin, used during the war to build fighters and bombers, to marine construction. Duralumin was an alloy with notable mechanical qualities that considerably lightened structures, while at the same time guaranteeing improved resistance to the electrolytic corrosion that had assured such a short life for the great J-Class yachts

[3] A perfect example of this is SAI (Società Aereonautica Italiana), probably Italy's foremost aluminum boatyard, founded by engineer Angelo Ambrosini and based in Passignano sul Trasimeno (the nearby lake was perfect for seaplanes). After the war, the plant converted to civilian use and worked on commissions for Breda Ferroviaria, Oto Melara and NATO. Its craftsmen were renowned for their skill, and were among the first in Italy to perform TIG sheet welding. From Danilo Fabbroni, *La Vela (e l'amore) ai tempi dello IOR* Youcanprint, 2012, p. 174.

[4] Fiberglass first appeared on cruising yachts, before making it to racing yachts. In 1960, the year when it became an Olympic class, the first *Flying Dutchman* made out of fiberglass went on to win the Olympics. Fifteen years later, the first ever advanced composite FD (in Kevlar) was manufactured. Dufour of France was among the first companies to make fiberglass boats. In Italy, early production fiberglass boats included the 1965 Classic 26, designed by Epaminonda Ceccarelli.

[5] These were located near ports or even directly on the beach, usually with a bare minimum of covered space. Newer boatyards tended to eschew port basins, as the new wave of customers came not just from the local area and as production increased, the yards required more roofed space. Therefore new boatyards sprung up on the outskirts of port towns, close to major road routes.

Gitana IV, designed and built by Cesare Sangermani: an archetypical example of the Italian artisan culture of those years.

Sagittario, a successful cutter designed in 1971 by the forerunner of Italian yacht designers, Carlo Sciarrelli. Built by the Craglietto yard in Trieste.

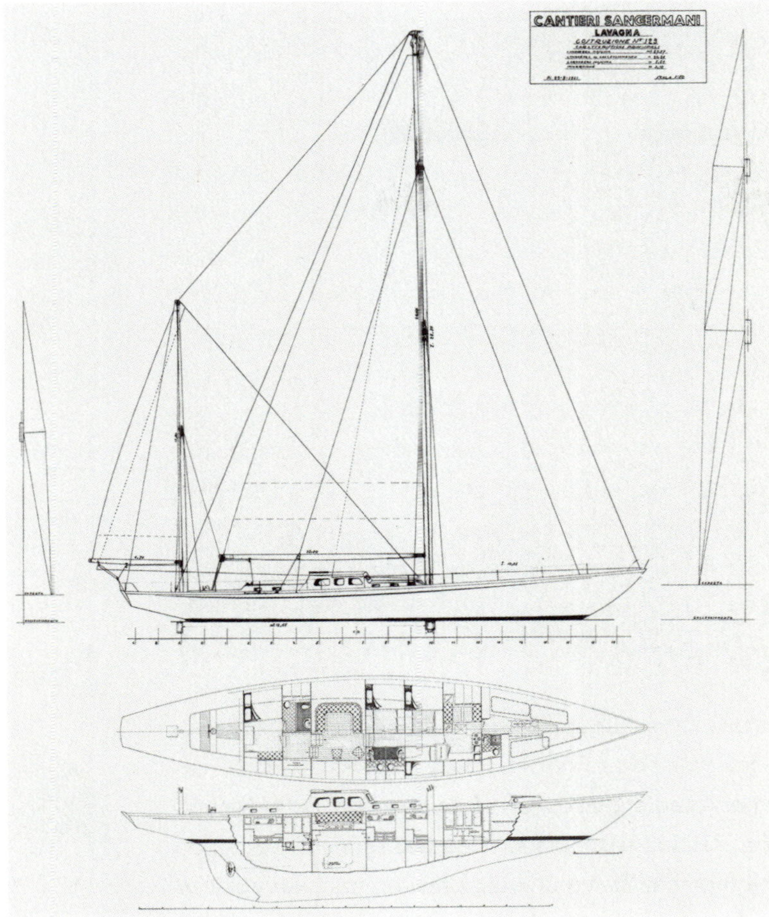

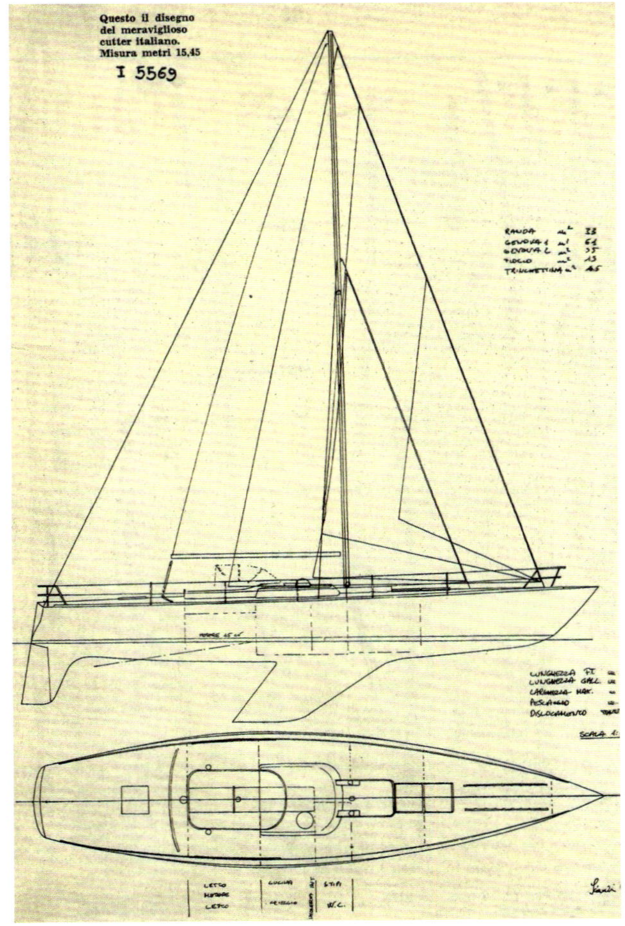

of the '20s and '30s. Further improvements ushered in by duralumin included the replacement of riveted joints with welded joints, making it possible to significantly reduce the quantity of filler needed to prepare the hull.[3]

Yet there can be no question that the material most responsible for the worldwide development of pleasure boat sailing is fiberglass. This versatile technology, which has various interpretations, is based on impregnating sheets of ordinary glass fibers with polyester resin. This can be achieved either by impregnating a mat (a piece of felt made up of glass-fiber threads arranged uniformly but in random directions) with resin, or else by using what is known as the chopped strand method, in which fiber, chopped up into small pieces, is sprayed onto fresh resin and then spread onto a mold .[4] All these techniques were faster than building with wood and could also be performed by less-skilled workers.

The very first fiberglass vessel was built in the '40s for the U.S. war effort. However, fiberglass did not take off in the years immediately after the war; there was so little demand that boatyards were unable to write off the cost of molds. But by the late '50s, with rapidly increasing demand for pleasure boats, mold costs ceased to be an issue. Within a very short space of time, the first industrial – or at least semi-industrial – boatyards began to spring up, a far different proposition from their wholly crafts-based forerunners.[5]

Unlike wood construction, in which it was already specialized, Italy had to import the new fiberglass techniques. Once they arrived, they spread quickly as skilled artisan boatbuilders, of whom Italy boasted so many, were eager to learn the new techniques. The artisans' agility would have a vital impact on the emergence of industrial districts, whose importance in Italian shipbuilding came to the fore in the '80s, as it did in other industries.

The history of Fiumicino, Rome's port at the mouth of the Tiber, is a perfect example of how this process took place. By the '70s, the town was teeming with artisans and boatyards capable of building excellent fiberglass boats. This could be ascribed to the fact that soon after the war, the Americans brought the Chris Craft company to Fiumicino. Specialized in manufacturing fiberglass powerboats, Chris Craft employed the local workforce that promptly learned the technology from the Americans. The ever-increasing demand for small boats then encouraged a number of local craftsmen to go into business and start up small boatyards, usually tied together as a network. From the early '70s onwards, fiberglass boatbuilding in Italy made quite a name for itself, particularly in the center and north of the country. Since then, this type of boatbuilding has generally followed the ups and downs of the Italian shipbuilding industry.

If there is one date when pleasure boating can be said to have taken off in Italy, it would be 1962, the year of the first *Salone Nautico di Genova* (Genoa Boat Show).[6] The first boat shows were small affairs, trade shows for local and national boatbuilding. It did not take long, however, for the *Salone* to really take off, as shipbuilding became one of the industries that drove what was known as the Italian economic boom.

By the late '60s, French mass-produced fiberglass boats were being shown in the Genoa Boat Show. They were soon followed by Italian emulations that, from the very start, were competitive technically and on-price. Three boatyards, in particular, stand out from that period: Alpa, operating out of Fiesco (near Cremona) between 1969 and 1972, whose elegant polyester resin boats designed by Buchanan, Illingworth and later Sparkman & Stephens – the Alpa 11.50, Alpa 9.50 and Alpa 12.70 – remain remarkable and elegant to this day. Next was Cima in Anagni, southeast of Rome, which began production in 1972 of another noteworthy design in Italian boatbuilding: the Brigand 7.50 (followed by the Brigand 9.50), designed by Camper & Nicholson. Last but not least, we have Sipla (later known as Comar) in Forlì on the Adriatic coast. This company reaped success in the early '70s with their diminutive Meteor, a six meter sloop designed by Van de Stadt in 1968, and then with the Comet 910, designed by Jean-Marie Finot and Van de Stadt. The Meteor was highly innovative for the time, as well as being fast, it was built using the counter-molding method that had made Dufour's boats so famous. Thanks to this method, for the first time boats, could be marketed as meticulous and smooth design objects: no longer did their superstructures and interiors appear to be simply assembled parts, they looked like integral parts of a whole. Interestingly, the molding principle of a shell or a fiberglass skin made headway in both airline interiors (so-called full body) and yachting at around about the same time.[7] The Comet 910 deliberately played up its design innovations: the cooker was a foldaway unit on runners, as was the chart table; instead of having the usual portholes, the coachroof featured an unbroken wraparound sheath of smoked perspex, typical of that period. With the 910, Comar perfected its first industrial production line, or at least, a semi-industrial line.

With our office's help, this process reached full fruition in the following decade. We must remember that back then, the output of Italy's three largest boatbuilders were boats designed by non-Italians. The few Italian designers out there, Carlo Sciarrelli or Giulio Cesare Carcano, were commissioned to design one-offs rather than production boats.

In the early '70s, cracks began to appear in the dominance of non-Italian designers. The first act of emancipation may be traced back to one boat, *Arlecchino*. Designed by Franco Sciomachen, the architect from Bologna worked with his brother Ernesto and son

[6] Antonio Fulci, Fabio Pozzo, Franco Belloni, *La storia della nautica in 50 edizioni del Salone Internazionale di Genova*, Sagep, Genoa, 2010. See also: http://www.nauticareport.it/dettnews.php?idx=6&pg=3996.
[7] The Meteor was built by Sipla, which a few years later was renamed Comar. The design was manufactured between 1968 and 1984. Initially, Sipla was a company that assembled items after sub-contracting production to craftsmen in and around the Italian town of Forlì. Subsequently, thanks to the efforts of Renzo Zavatta, a man with significant insight into the development of Italy's industrial sailing business, production was shifted to a boatyard that became one of Italy's first industrial boatyards. The Comet 910, managed and sailed by Cino Ricci and his team, won the Middle Sea Race in 1973, and that was just one of an impressive haul of races. The boat was manufactured from 1971 to 1977.

Comet 910, a 1971 design by Van de Stadt and Jean-Marie Finot built by Comar (ex Sipla); this may be considered the first Italian boat to be a product of industrial design.

Arlecchino, design by Franco Sciomachen, was one of the first Italian boats to attain international success: a second place in the 1971 Quarter Ton Cup.

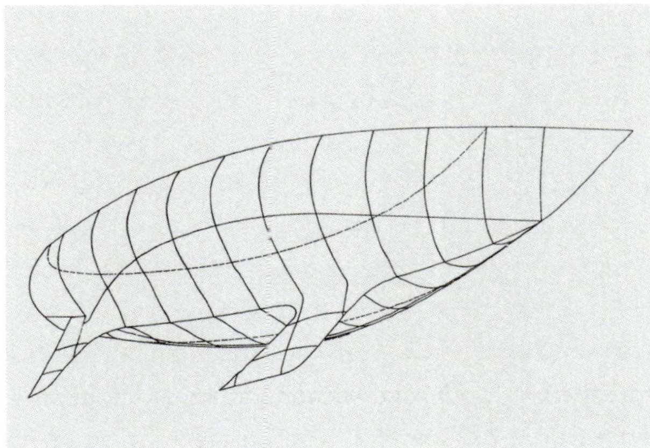

Aldo on a number of projects that were to go down in Italian yacht design history. *Arlecchino* was a graceful wooden quarter-tonner built by Cantiere Sartini in Cervia. The boat was exceptionally fast, especially upwind in light-to-medium conditions. *Arlecchino* was the first successful Italian offshore sailing yacht, taking second place in the 1971 Quarter Ton Cup in La Rochelle. Cino Ricci, who was at *Arlecchino*'s helm, would become Italy's leading exponent of offshore sailing in the '80s.[8]

As new boatyards opened up, a new generation of designers emerged, the first sailboat hardware companies appeared (Barbarossa for winches, Canclini for masts and spars), and sail lofts like Carozzo and Peer expanded their operations, riding the wave of success of Italian offshore sailing in the early '70s.

Demand was principally for boats of around nine to ten meters, fast sailers with full interiors, capable of doing well in the ever more numerous club races that were being held along Italy's coast. Boat owners hailed from the new middle-class that was driving Italy's economic boom. While the early profits generated by the '50s economic boom were sunk into real estate, the next wave went on luxury goods, vacations, cars and in the end, crowning it all, pleasure craft. At the same time, the boats were changing, catering to the requirements of a new age. One sign of this was in the interior arrangements, where cabins for paid-hands began to disappear or were no longer separated from the rest of the living space, and owners' cabins became less prominent, even disappearing completely from racing boats, while the dinettes and shared spaces became increasingly important. Unlike today, interior space quality was measured in terms of the number of available berths; indeed, it was not unusual to find as many as seven berths on an eight meter boat.

Italy's harbors were also affected by the ever increasing demand that fueled the revolution in boating. Following France's example, marinas began to spring up along the coasts. These brand new structures, wholly dedicated to pleasure boating, spawned boatyards, ship chandlers and stores offering a variety of services.

A flurry of new yachting magazines crowned this state of euphoria. In the early '70s, the long-standing *Nautica* magazine was joined by *Vela e motore*, *Uomo mare*, *Forza 7* (where I collaborated before I became a designer) and the *Giornale della Vela*, a magazine dedicated almost exclusively to offshore racing, whose inaugural issue appeared in 1975, the year of my first design. The first photographers to work exclusively in yacht racing were supported by these magazines: men like Bert Richner,[9] Franco Pace and later Carlo Borlenghi.

Another driving force in the expansion of Italian pleasure boating should not be underestimated: the '70s were known in Italy as the Years of Lead, where a perverse combination of what was defined as a strategy of tension, based on terrorist attacks, rendered

[8] At La Rochelle, *Arlecchino* barely lost the World Championship, finishing half a point behind another important design, the Philippe Harlé-designed *Tequila*. *Arlecchino* was manufactured by Sartini in Rexilon, a marine plywood laminate with perpendicular grain. It is worth remembering that before *Arlecchino*, the Sartini boatyard had built *Passatore*. This Quarter-Tonnerdesigned by Jean Marie Finot, has his signature lines: a very beamy stern, large volumes and an almostflat bottom. *Passatore* was built on two half-moulds that were subsequently joined lengthwise.

[9] One of the first illustrated offshore sailing books was *Mediterranean Yachts* by the American photographer Bert Richner, Ripala edizioni, Milan 1976. Browsing those pages, one is struck by how the size of yachts has grown over the last forty years.

The Arpège, a half tonner designed by Michel Dufour in 1967 and built by his yard of the same name; a cutting-edge boat both in design and in its fiberglass construction.

Scampi, the half tonner designed by Peter Norlin in 1969, winner of three Half Ton Cups.

The C&C 37, a 1971 Cuthbertson & Cassian studio design built by the Benello yard in Livorno in a fiberglass sandwich with Airex core.

the atmosphere in the country asphyxiating, particularly in its major cities. In this bleak situation, the sea provided an escape from increasingly trying tensions ashore. But the tragic nature of those years, culminating in the kidnapping and murder of former Prime Minister Aldo Moro, would soon end. The extent of *Azzurra's* media success – something that surprised us as much as anyone – is an example of how, in just the five years between 1978 and 1983, the country's climate completely changed: what had been an escape to the open sea became a celebration of hedonism, fueled by the return of economic wellbeing – the Years of Lead had been consigned to history.

After the Vaurien, my sailboat apprenticeship spanned three boats owned by my family, all of which were typical of the design and construction developments of the day. This first was the Arpège, a Michel Dufour design from 1967, a fast nine meter boat with graceful lines built in a boatyard owned by the designer himself, a leader in fiberglass construction.[10] The boat's design was groundbreaking for its unusual maximum beam (one third of its length), which not only guaranteed a great deal of form stability (meaning less ballast), but it also made the interior extremely spacious, particularly in and around the dinette. The Arpège was also a trailblazer with its construction technique; the use of counter-molds made it possible to assemble two fiberglass shells, one inside the other. This gave the boat a refined, fully-finished look inside and out; indeed, the Arpège may lay claim to being the first industrial design product in pleasure boating. The boat made its debut in 1967, four years before the Comet 910 that we mentioned earlier as the first Italian industrial design product. These four years show how

[10] Michel Dufour, who designed *Arpège* and owned his own boatyard, was one of the most interesting characters in sailing back then. In the early '60s, Dufour worked as an engineer at a railway carriage factory, where he was in charge of the fiberglass department – in its infancy, the technique was used for many parts of locomotives. Drawing on this experience, Dufour was one of the first to realize the potential of using fiberglass and polyester resin in pleasure craft. In 1965, he designed a Quarter-Tonner boat, *Sylphe*. The following year he developed this into his design for *Arpège*. After designing these initial one-design boats, Dufour set up his own boatyard for mass production. In 1967, they made 19 boats, the following year 58, the year after that 158. By the early '70s, Dufour was turning out a boat a day. http://www.arpege.it/it/arpege-barca-epoca.

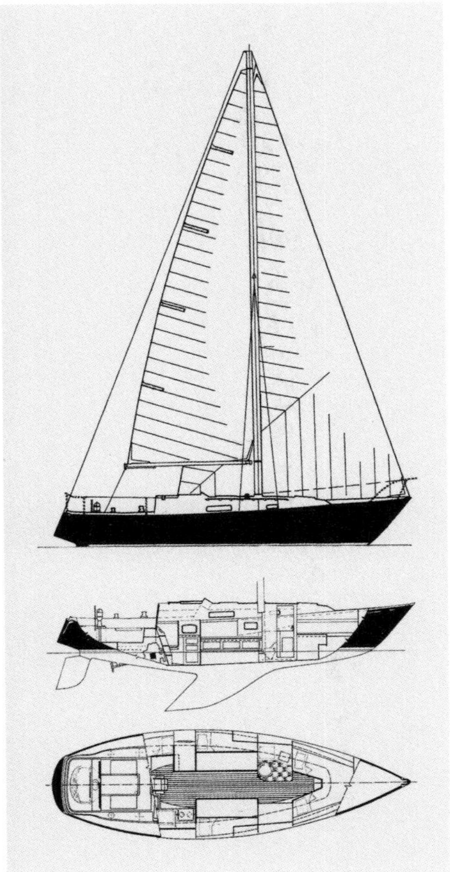

quickly Italy was closing the gap with France and England, where modern boatbuilding had begun. This gap was fully closed by the late '70s when the Studio Vallicelli was in its infancy.

Next came *Scampi*, a Peter Norlin design the same length as the Arpège but from a successive generation. Unlike the Dufour boat, *Scampi* was designed to comply with the newly created IOR (International Offshore Rule, 1970) rating rule, and to fully exploit its characteristics. It is no coincidence that this yacht won three Half Ton Cups starting in 1969. *Scampi*'s lines maintained the Arpège's generous maximum beam, but set it further forward than the Arpège and merged it a bit more sharply towards *Scampi's* longer overhangs. Overall appearance was a step ahead compared with the old spindle-shaped effect. The result was a very well-balanced boat suited to all conditions. *Scampi* was indeed a true all-rounder, perhaps even the first of its kind.

Our third boat was a C&C 37, the rather unique design by Robert W. Ball for Cuthbertson & Cassian. Longer than the Arpège and *Scampi*, with her short overhangs, a maximum beam that was very far aft, an unusual rounded bottom (for those days) coupled to one of the earliest lead keel bulbs, she didn't adhere very closely to the by then consolidated IOR. The C&C 37 was conceived for the waters of Canadian and North American lakes, with their calmer sea states, making her well-suited to summer conditions in the Mediterranean and offsetting her otherwise high rating.

The C&C 37 was a somewhat eccentric boat. Experimental rather than bizarre, it was skillfully built by the Benello boatyard in Livorno. Giacomo Benello's story is an interesting one: after seeing combat with the British navy in the war, he moved to the U.S., where he met the Stephens brothers, and at the same time, absorbed the techniques of fiberglass construction. Benello returned to Livorno after the war with the technical know-how he had picked up in the States. He joined forces with the Romoli e Mazzantini boatyard, at that time specialized in wooden construction, and persuaded the owners to take on fiberglass. Subsequently, Benello asked Olin Stephens to design a boat to mark the start of his new project. However, Stephens apparently sold the same design to both Nautor and Romoli e Mazzantini. This boat, the Swan 36, went on to become a myth in yacht design, bringing a great deal of success to Nautor. Benello cut off all links with Stephens, who was not keen on the boat having a sandwich construction, and approached Cuthbertson & Cassian in Canada. The result of this contact was the C&C 37, a boat that, among other things, adopted an innovative construction of fiberglass sandwich with Airex core construction.

Benello's story is a fine example of the trend in Italian boatbuilding at the time, as the country gradually moved beyond imported designs and, within a short space of time, developed and consolidated into the latest offshoot of the Italian economic boom.[11]

[11] https://mariacristinacc37.com/history/

The Early Years of the IOR: Different Design Philosophies Compared

Throughout the '60s, issues with measurement and rating rule problems slowed the development of offshore racing. The U.S. raced using a rating rule approved by the CCA (Cruising Club of America); the Europeans followed another set of rules released by the RORC (Royal Ocean Racing Club). Conceptually and technically conceived from different criteria, these two sets of measurements and rating rules inevitably resulted in boats that were extremely hard to compare against one another.

In 1970, after four years of negotiations, the Americans and Europeans reached agreement on a rating rule based on new criteria: the IOR (International Offshore Rule). I began designing during the early years of the IOR,[12] a time of great ferment for the industry as American and European designers focused on how to increase hull performance, while at the same time how to best interpret the new measurement rule.

The biggest showcases for the latest developments were the Ton Cups, world championships organized for fixed-rating classes, where it was possible to compete in real-time without handicaps. In addition to these championships, another showcase was the prestigious Admiral's Cup, a team race in the tricky waters of the Solent, in which each nation competed with three boats, culminating in the challenging Fastnet race.

In the '60s, Olin Stephens was the undisputed champion of yacht design and had been for the best part of thirty years. It was no coincidence that he and Paul Spens were the men behind the IOR rating rule. Stephens' clean, elegant, medium-heavy displacement boats with V-shaped hulls, finned appendages faired with the hull and reduced maximum beam, dominated the scene.

And yet the adoption of the IOR rating rule that he had conceived marked the beginning of the end for Stephens' dominance. The first to inflict a blow on his dominance was fellow American Dick Carter, who may be considered the first designer to fully exploit the new measurement rule. Carter first appeared on the scene between 1965 and 1967 with three boats: *Rabbit* (restored and brought back to life in 2018) and, most importantly, *Tina* (winner of the 1966 One Ton Cup) and *Tina II*. These boats were markedly wider than Stephens', giving them greater form stability; they were also lighter, their rudder and keel increasingly separate from the hull. Although *Rabbit*, the two *Tinas* and then *Optimist*, which won the One Ton Cup in 1967 and 1968, were boats developed under the CCA rating rule, the solutions they adopted foreshadowed solutions adopted in the early years of the IOR: first and foremost, the idea of increasingly rounded hulls down to flat bottoms, then separating appendages from the hull. Carter was the first to use aerodynamic profiles on trapezoidal keels and rudders, with swept-back leading edges and

[12] The newly-founded International Technical Committee (I.T.C.) on which designers Olin Stephens, Dick Carter, Ricas Van der Stadt and Gustav Plym sat, began working on the new measurement rule in 1968.

The plans for *Ydra*, Dick Carter's design, winner of the 1973 One Ton Cup with Agostino Straulino at the helm: the first great Italian success in a world level-class championship.

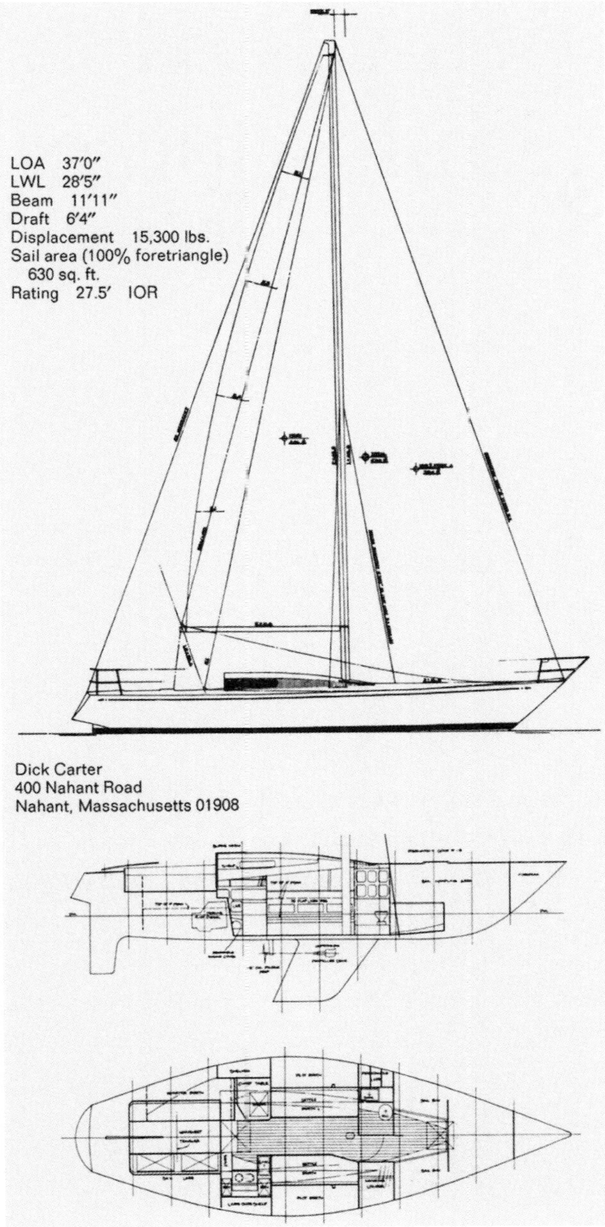

vertical edges aft. He used profiles created by the NACA, the National Advisory Committee on Aeronautics (which would later become NASA), designing shapes that were characteristic of IOR boats until the late '80s. In Carter's second season, increasingly concentrated on interpreting the IOR, he designed *Gitana V*, *Red Rooster* and *Ydra*, the last of which would win the 1973 One Ton Cup under Italian colors.

Other American designers like Gary Mull came up with their own personal interpretations of Carter's approach: unlike Carter, Mull opted for extended overhangs when he designed the extremely competitive *Improbable* in 1971. New Yorker Scott Kaufman, who had trained at Sparkman & Stephens, designed *America Jane III*, another boat that in the early '70s had a major impact on design. The inventive Britton Chance moved away from Carter's approach and won the One Ton Cup in 1976 with the radical center-boarder *Resolute Salmon*: extremely fast, this short, medium-heavy displacement boat was able to win despite a penalty for centerboards, recently adapted by the ITC. Another eccentric and experimental designer, the Frenchman Jean Marie Finot, followed yet another design philosophy. His were both the aforementioned Comet 910 and the cutting-edge *Revolution*: extremely wide and rounded, with high freeboard, a flush deck and with maximum beam set very far aft.

The One Ton Cup was the most prestigious world championship level-class trophy of the day. Two One Ton Cups proved to be turning points in IOR history: in 1973, off Porto Cervo in Italy, and the following year at Torquay, in England. *Ydra*, helmed by Tino Straulino, a hero of Italian sailing with a gold medal in the renowned Star class at the 1952 Helsinki Olympics to his credit, won at Porto Cervo. Carter's *Ydra* was an aluminum design, a versatile boat particularly well-suited to strong wind and waves. She was commissioned by an Italian owner, Marina Spaccarelli Bulgari, who liked to name her boats after Greek islands (her previous boats were named *Paxos* and *Kerkyra*). Before the One Ton Cup, in an unusual move for the time as yacht racing was still an amateur pursuit, Ms. Spaccarelli Bulgari decided to run trails for the two boats (*Paxos* and *Ydra*) at Marina di Carrara before deciding which one to race at Porto Cervo. *Ydra* proved to be considerably faster and went on to become the first Italian boat to win the prestigious race.

However, the surprise at Porto Cervo was not the quick *Ydra*, but *Ganbare*, designed by a young Califor-

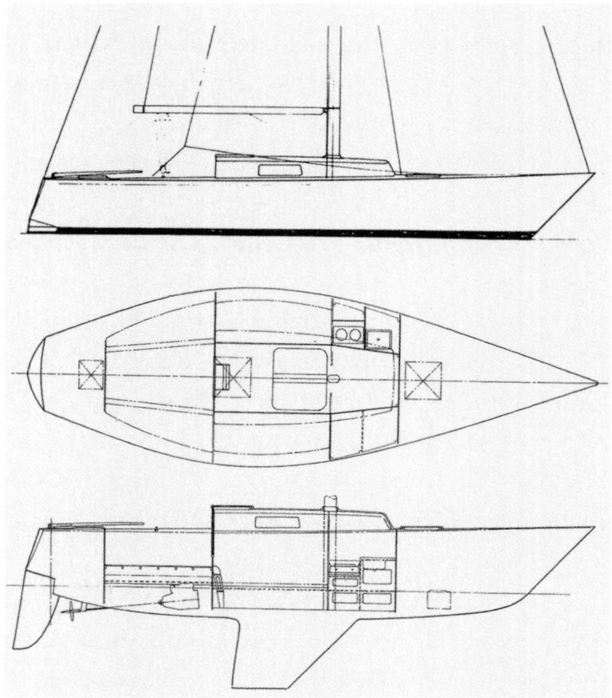

Ganbare: designed by Doug Peterson, she lost the 1973 One Ton Cup by a whisker but nevertheless revolutionized the design of racing boats.

nian named Doug Peterson. *Ganbare* was the moral victor in that competition, forfeiting the title only after being disqualified for rounding a mark the wrong way. *Ganbare* was clearly a descendant of Carter's designs with her medium displacement, pronounced volumes, flat bottom and trapezoidal keel, but unlike Carter's boats, she was shorter and narrower to more smoothly reconcile the ends of the boat with the maximum beam. *Ganbare* also had more volume aft and bustle, a highly pronounced blister between the hull and skeg to exploit the IOR girth measurements. Although built from extremely lightweight cedar-strip planking, *Ganbare* looked so pared back compared to the other yachts competing in the 1973 One Ton Cup that she seemed almost naked. With just 4 winches, all control lines (also reduced to a minimum) were run to the cockpit; moreover, to lighten the ends of the boat to reduce pitching, all weight belowdecks was concentrated amidships, leaving nothing literally forward of the mast.

While not extreme like the boats that Finot or Mull designed, *Ganbare* was a strikingly innovative boat. Looking at her when moored immediately gave one the impression that Stephens' all-round philosophy had found a new proponent in Peterson. *Ganbare*'s crew was also noteworthy for appearing like the embodiment of American counterculture at that time: long haired, cordially laid-back and almost no sign of any hierarchy aboard: the *Ganbare* spirit was as seductive as the boat's performance. While Admiral Straulino's elegant *Ydra*, with her equally elegant owner Marina Spaccarelli Bulgari and a highly-disciplined crew, represented a vision of offshore sailing that was gradually disappearing, *Ganbare* marked the emergence of a new style that, consciously and unconsciously, would fascinate Vittorio Mariani, Nicola Sironi, Patrizia Ferri and myself, the founding members of our design office. The roll call on board Peterson's *Ganbare* was stellar: in addition to the boat's designer, the crew included New Zealander Ron Holland, who would, not too long into the future, become Peterson's fiercest rival in yacht design; Lowell North, founder of what remains to this day the most advanced sail maker in the world; Bill Green, who at his boatyard in Lymington was to go on to build most of the world's fastest yachts; Jim Pugh, who after working with Peterson would also become a famous designer; and Gary Weisman, the future chairman of North Sails, often present on board our yachts in later years.

The revolution forecast by *Ganbare* would fully show its effects during the following year's One Ton

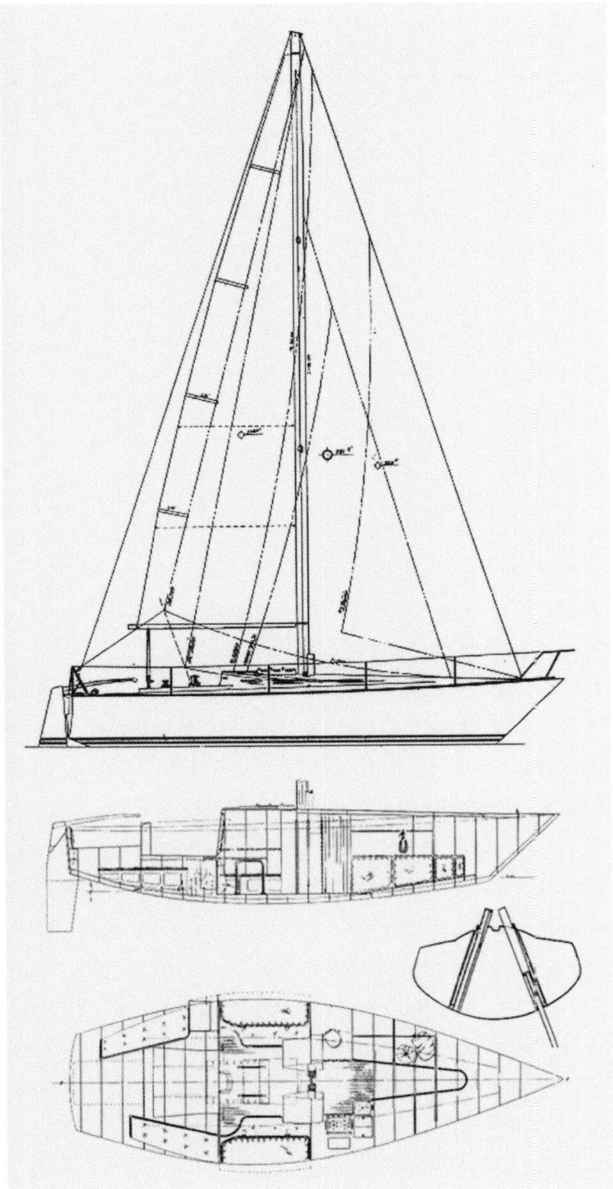

Terrorist, designed by Bruce King, at the 1974 One Ton Cup. The boat had an innovative system of double retractable keels, and an unusual rudder hung from the transom.

[13] Actually, my first design wasn't for a boat but for a sail: a radial spinnaker that I cut and sewed myself, in two pieces, one for the head, the other for the lower horizontal panels.

Cup in Torquay, in which I personally took part. The quay in Torquay was a veritable smorgasbord of design inventions. To start with, there were the new developments adopted by Bruce King's very fast design, *Terrorist*. Short, with a rounded bottom, she had twin centerboards, angled at 15° off the boat's vertical axis. Upwind, *Terrorist* lowered the leeward centerboard (the most efficient solution, given that it was more perpendicular to the water plane) and pulled up the windward one; when running both boards were brought up to reduce resistance. The result was an extremely fast, variable configuration boat, though not easy to handle. Like Bob Miller's boats at Torquay, *Temibile* and the elegant *Celtic*, *Terrorist*'s deck was without a coachroof, giving her an aggressive, determined feel.

Golden Apple, designed by New Zealander Ron Holland (who the year before had raced in Sardinia on *Ganbare*), was another very fast boat at Torquay. *Golden Apple*'s elegant varnished wood hull revealed in the color and grain of the various woods used in its construction. Holland followed the design choices espoused by Carter and Peterson while taking a different approach with the bustle, which in his case was less extreme. *Golden Apple's* mast had a recessed channel for the mainsail luff to reduce windage, and a complex system of swept-back spreaders with diamond shrouds, allowing for a notable reduction in spar section. After these episodes, experiments like these ceased for quite a while, to return twenty years later in the Open Classes and the generation of America's Cup yachts that followed the 12 Meters. At Torquay, there were also designs from Bob Miller, Britton Chance, Gary Mull, Van de Stadt and Doug Peterson. Peterson's *Gumboots* (a composite construction from the then well-known Jeremy Rogers boatyard) gave him his revenge for the One Ton Cup, which he had just missed winning the previous year. In Torquay, *Ydra* performed admirably. The writing was on the wall, however: the new IOR generation was ousting the previous generation of yachts.

It was in Torquay, where I had the chance to see the best that sailing had to offer, that I took the decision to go into designing racing yachts. I also found my guiding light: the Californian school of medium displacement yachts, whose leading exponent was Doug Peterson. What I liked about Peterson's yachts was that they were not extreme. I believed that a medium displacement boat with a long waterline length and a generous sail plan, its volumes harmoniously distributed and ends properly faired to a moderate maximum beam (as opposed to the very narrow beam in Bob Miller's designs or very wide as in Finot's boats) would work in the Mediterranean.

The U.S. design approach also ensured adequate living space belowdecks, something that was out of the question on such an extreme boat as *Terrorist*. By following the design approach initiated by Carter and developed by Peterson, it would be possible to satisfy the demand for series produced yachts that was beginning to burgeon in Italy. *Ganbare* was proof of this theory: built in aluminum at a Dutch boatyard and in fiberglass at the Puccinelli yard at Castiglione della Pescaia in Italy, the series built design continued to win on the racecourse.

This was in 1975, the year I designed my first two yachts. They were launched the following year, to race in two major Ton Cups in the Med: the One Ton Cup in Marseille and the Half Ton Cup in Trieste.[13]

Our First Two Yachts

Gattone was the name we gave to the one-tonner I designed with Carlo Alberto Tiberio, Professor of Physics at the University of Rome and co-owner of Artmare, the Fiumicino boatyard where the boat was built and then series produced until 1983 under the moniker V-Cat 38.

The influence of Doug Peterson and Gary Mull on *Gattone* was abundantly clear, even if, on this first boat, we explored variations to the U.S. approach. We decided to shift the maximum beam a little further aft and to reduce volume forward. The deck arrangement, with a double cockpit, one for the helmsman and the other for the crew, differed too from U.S. standards. I had also come up with an aesthetic for *Gattone* that foreshadowed future Studio Vallicelli yachts. It began with a low, unbroken, tapering coachroof, joined smoothly to the deck and without portholes, making it as visually integrated with the hull as possible. Following the example of Bob Miller's yachts, I chose to characterize the hull with a large, single-color stripe (sky blue for *Gattone*, amaranth for my next design, *Ziggurat*) that, close to the sheer and the waterline, brought out the white hull.

Gattone introduced a number of unique solutions in her construction. The yacht was built out of fretted sandwich laminate, a construction patent that Artmare held, where the fiberglass sandwich's core was set with laminate fretting containing polyurethane foam, which is lighter than PVC foam (Airex). The rudder had a skeg, which Peterson did not employ on his yachts, while the keel was made out of cast iron, with a hollow to pour the lead into.

On her maiden outings, *Gattone* performed admirably well, winning the Giraglia and the Mediterranean Championship. It went less well at the One Ton Cup in Marseilles, where we barely managed a twentieth overall. This taught us how difficult it would be to win on the international circuit, and how very different Ton Cups were from club races. We came up against new adversaries in Marseille too, not just well-known medium-displacement Peterson and Britton Chance designs, but brand new light-displacement boats from Bruce Farr, which after this One Ton Cup would become a bone of contention in the IOR's evolution.

My second design for that year was *Ziggurat*, a half-tonner to which I owe my first international victories. As was true three years earlier with *Ganbare*, *Ziggurat* was more than just a boat: she was a team project, created by a group of friends who shared the desire to excel in Italian sailing's event of the year, the Half Ton Cup scheduled for early September in Trieste. That 1976 would be the year level-class racing exploded was confirmed by the incredible number of entries in that year's Half Ton Cup. To choose the yachts for the Italian team, not just selection rounds but pre-se-

Gattone, one tonner designed by Carlo Alberto Tiberio and Andrea Vallicelli and mass-produced under the name V-Cat 38. She participated in the One Ton Cup of 1976, winning the Mediterranean Championship in the same year.

lection rounds were needed, in both the Tyrrhenian and Adriatic.

Ziggurat was a boat that, in the day's parlance, was self-managed. Every single one of us – and there were more than ten of us involved – had a stake in the ownership; every single one of us contributed what they could to implementing and completing the project. The yacht's hull was made out of a fiberglass sandwich with a balsa core in Verona, at the well-regarded Sigovich boatyard. The shell was then transported to Fiumicino, where we all worked on the finishing out.

The team that coalesced around *Ziggurat* at the newly-founded Studio Vallicelli included Vittorio Mariani (her future helmsman), responsible for the deck layout, rigging and boat management; Nicola Sironi, yacht structures and fine-tuning; Patrizia Ferri (my wife) interiors and esthetic choices; and then our friends Gigio Russo (who would later work for North Sails Italia), Gary Redfern (who at that time was a deck hardware distributor), Mario Vallicelli (my father, involved in developing and racing *Gattone*), plus a cast of others who would continue to revolve around the studio and help handle the many commissions that were to come in after *Ziggurat*.

The *Ziggurat* project pressed ahead, reaffirming our variations to the U.S. approach undertaken with *Gattone*. The hull's maximum beam was set well aft, and volume astern was increased. Above all, the hull's forward entry was slimmed down to better adapt to the light-to-medium winds of the Mediterranean, even with

Ziggurat, the first boat designed by Andrea Vallicelli, finished third in the 1976 Half Ton Cup in Trieste and was subsequently mass-produced by the CPR yard in Fiumicino.

her less pronounced bow overhang. Given the results, all of these solutions would be adopted in our following designs.

Notably, as had been true of *Gattone*, almost all of the boat's rigging was Italian: the mast came from Canclini, the sails from Carozzo e Peer and so on. The only exceptions were the winches and the headstay foil, which back then were not manufactured in Italy at all. Although our rigging was not yet at the same level as that made in the U.S., it was plain to see what great strides Italian boatbuilding had made in the space of just a few years.

When we launched *Ziggurat* in the spring of '76, the yacht immediately proved fast, especially in the light-to-medium wind conditions for which she was conceived. That June, she won the pre-selection trials in Castiglione della Pescaia, beating helmsmen of the caliber of Paul Elvestrom and yachts designed by Gary Mull, Britton Chance and Ron Holland. The Trieste selection races that September did not go quite as well: *Ziggurat* barely scraped into the top ten, and almost suffered the ignominy of being knocked off the team.

The 1976 Trieste Half Ton Cup was an enormous success, disputed by more than fifty brand new yachts, of which at least a third were considered podium material. *Ziggurat* did not start off particularly well: though sailing with the lead pack, she lacked the winning spurt that would only appear in the final races. Even we were surprised as she began to perform better and better, but leaving a somewhat bitter taste in our mouths for all the early races when, if we had sailed with greater confidence, we could have won. *Ziggurat* wound up third overall, behind Harold Cudmore helming a Ron Holland design, and Oscar Tonoli, who was at the helm of a cold-molded Gary Mull design built by the Galetti boatyard in Italy.

After the racing was over, Elio de Sabata, the son of a famous orchestra conductor and soon to become a dear friend, bought *Ziggurat* on the spot. The money we earned was enough to cover not just the boat's costs and six months of racing, we even had a little leftover to distribute among the team. Back in Fiumicino, the boatyard where we had fitted out the boat, CPR purchased the design and put it into series production. Over the following years, production *Ziggurat* yachts would go on to win two Italian Championships and any number of other yacht races.

All of us at the newly-founded Studio Vallicelli were under twenty-five at the time. The chance to design yachts as a profession suddenly seemed possible. Indeed, given the investments that rolled in after Trieste, it became a necessity.

The Vallicelli Practice – The Early Years

In Rome in the early 1980s, from left to right: Vittorio Mariani, Andrea Vallicelli, Patrizia Ferri and Nicola Sironi.

The Vallicelli studio in the early 1980s. From left to right: Antonio di Stefano, Andrea Vallicelli, Paola Catella, Alessandro Nazareth, Vittorio Mariani, Nicola Sironi and Patrizia Ferri.

At the quay in Trieste, a local boat owner asked us to design a one-tonner. We designed *El Cid*, built from wood by Crisman & Giraldi, also known as Sergio & Nico, soon to be the winner of two editions of the Barcolana. A logical evolution of *Gattone*, she sported finer lines. Also, on the quay, a Milanese couple, Piergiorgio Vigliani and Riccardo Trionfi came over to talk to us. They were keen to compete in the Three Quarter Ton Cup, the world championship scheduled to take place at La Rochelle in France the following year. Before talking to us, Vigliani and Trionfi had been in contact with other designers – big names like Kaufman, Holland and Finot – but such was their flair for taking risks and their enthusiasm to help raise the profile of Italian offshore sailing, that they chose to gamble on our Studio.[14]

Argento vivo, the name we gave this successful design, proved to be a defining moment in the history of our practice. For this yacht, the owners insisted on a level of a professional organization that was new to us. The project was to build an extremely moderate yacht without forcing anything: a medium displacement hull with 50% of the ballast in the lead keel and a generous beam amidships, faired smoothly with a slightly finer bow and a wider stern than *Ziggurat*. Our idea was to come up with another all-round boat that would be at its best during the reaching and running conditions that make up the better part of offshore races. Due to their high scoring coefficient, it was important to do well in these races to place overall in the Ton Cups

We commissioned De Cesari of Cervia, a solid and reliable boatyard with considerable experience in wooden racing yachts, to build the craft. Like *Ziggurat* and *Gattone* before her (with the exception of some fittings and the North sails, at that time the top sail loft), *Argento vivo* was almost entirely Italian.

This yacht was, in all respects, our Studio's first true departure from the American approach. Immediately after her launch in spring 1977, she won the Italian Championships. However, *Argento vivo*'s performance revealed some shortcomings, particularly in light winds, which were the conditions expected that August in La Rochelle. Vigliani and Trionfi were in favor of modifying the yacht to be at her best for France. At our newly-founded Studio, we decided to remove weight from the keel to reduce wetted surface and to change the propeller, replacing the folding type with one of the earliest pitch propellers (from MaxProp, whose design would soon become a common feature on all racing yachts), lowering our rating. This lets us increase the sail area of the spinnaker.

At La Rochelle after these modifications, *Argento vivo* began to perform significantly better, considerably helped by fine-tuning against another very fast Italian boat that had given us a hard time at the Italian Cham-

[14] Riccardo Trionfi, Piergiorgio Vigliani, *Argento vivo story*, "Giornale della vela", no. 2, 1978, pp. 22-23.

pionships. *Regolo* was designed by Mario Violati, another Italian designer who had recently appeared on the scene.[15] In addition to the owners and their group aboard *Argento vivo*, Vittorio Mariani was at the helm as he had been on *Ziggurat,* along with Nicola Sironi, who had led the recent fine-tuning.

The yacht performed consistently at the World Championships, growing in the latter part of the series and ultimately finishing with an impressive second place. *Argento vivo* was beaten by *Joe Louis*, designed by rising star Bruce Farr of New Zealand. Good luck favored us at La Rochelle. We were racing against boats from the young French designers Briand, Berret and Joubert-Nivelt, who, as they were all from the same country, tended to give each other a hand. This approach wound up costing them dearly: during the offshore race, the group of French boats failed to round one of the marks, disqualifying the whole group.

The following year, 1978, *Argento vivo* won a number of races and the Italian Championship once again; the owners put her up for sale and moved on to another new design. This, too they commissioned from Studio Vallicelli. The new boat was to take part in the 1980 One Ton Cup, in the Gulf of Naples.

El Cid, *Almic* (another one-tonner, built by Morri e Para) and *Argento vivo* were our first yachts built of cold-molded laminated wood. To perfect the executive drawings, we called on marine engineer Carlo Bertorello of Naples, who introduced us to the very refined world of Italian wooden boatbuilding. *Argento vivo*'s hull was made by laying up to four strips of 4mm wood in diagonal, cold gluing them together. The inner ply was spruce; the other three, okumé mahogany. Together with the owners, we had decided against using cedar, a lighter wood that we had used with *Ganbare*, as it was not strong enough. The layering of the hull's skin was done with a fast and reliable method developed by De Cesari: layers of wood were assembled onto a throwaway mold with longitudinal stringers closely spaced over frames; the whole was held together by the longitudinal keel beam. Once the shell laminate was finished, the hull was taken off the mold and turned right-side up to finish framing. To give rigidity to the whole, the keel, chainplates and mast step were secured to the centerline floor by means of a steel structure. On *Argento vivo* we allowed ourselves a bit of styling: a thin layer of seamless, wide strake teak was laid over the 8mm plywood deck.

De Cesari's elegant construction technique was not uncommon at the time. Italy had – and continues to have – a long-consolidated tradition in wooden vessel construction, with a variety of different approaches, each of which uses its particular construction technique. In the Adriatic, the Trieste school was trailblazed by Craglietto, Petronio and Pecarich, Crisman and Giraldi and Sergio e Nico (who built *El Cid*). Then

Argento vivo, a medium displacement design built of laminated wood by the De Cesari yard. She finished second in the 1977 Three Quarter Ton Cup and was also the winner of two Italian Championships.

[15] This is what the panorama of Italian designers looked like at the end of the '70s. The country's top post-war designer was Carlo Sciarelli, who designed magnificent fast cruising yachts like *Sagittario* (1972), which were very quick but paid little heed to the new IOR measurement rule. There was also Epaminonda Ceccarelli, who designed Italy's first mass-produced fiberglass boat, the Classic 26 (1965), and later the EC series. Giulio Cesare Carcano, who was also a mechanical engineer (he designed the Moto Guzzi V-twin engines), designed eccentric yachts throughout the '70s that had very light, rounded, planing hulls, often with transom hung rudders and trim tabs, and all had a name that began with a "V" and ended in an "A", such as *Villanella* (winner of the Two-Ton Cup at Sanremo in 1971, the first Ton Cup ever won by an Italian design), *Vihuela*, *Vanessa*, *Vinca*, etc. The next generation of designers was led by Aldo and Ernesto Sciomachen, who designed *Arlecchino* (which came second at the 1972 Quarter-Ton Cup), and introduced his very fast *Balanzone* at Trieste in 1976, which was in the fight for victory right to the end; he won the 1983 One-Ton Cup with *Linda*. Then there was Mino Simeone of Naples, who designed and built the Minaldo yachts, and Mario Violati, who before *Regolo* had designed a One-Tonner, *My Lochness*, before going on to design the highly successful One-Tonner *Ragtime band*. The first boats designed by the Fontana-Maletto-Navone trio appeared in the late '70s, winning the Mini-Ton Cup in 1978 with *Rivetto*, and immediately becoming the go-to designers for Mini and Quarter-Ton yachts. In 1980, they designed *Buonalena* for that year's One-Ton Cup. Flying the U.S. flag, she took second place, just a few points behind *Filo da torcere*. In 1983, they designed *Botta dritta*, the deserving winner of the Three-Quarter Ton Cup in Trieste.

Imp, a two tonner designed by Ron Holland, a very fast protagonist of the 1977 and 1979 Admiral's Cup. The composite hull was reinforced with an reticular aluminum structure.

there was the Venetian school, spearheaded by Crosato and Favero and, moving further south down the Adriatic, in Rimini the approach espoused by Carlini with Morri e Para, two characters for whom wooden boat building was regarded as something mystical.[16] De Cesari was located in Cervia, where he built not just *Argento vivo*, but Vigliani and Trionfi's next yacht, *Filo da torcere*. In the Tyrrhenian Sea, moving north to south, we had the Maccianti school in Chiavari, then Pezzini in Viareggio, Bani, at Santo Stefano on the Argentario and then Gallinari in Anzio. Unlike De Cesari, Gallinari used a throwaway male mold to create a shell where it was easier to glue (and nail during the temporary layering process) the various layers of wooden laminate together.

Towards the late '70s, for a short but intense period, wood was the most popular material for building racing yachts. American designers like Gary Mull and Doug Peterson were stunned by how skilled Italians were at handling a material that required enormous craftsmanship, not to mention their national propensity for adapting to new technologies such as the West System, (a special epoxy-resin preparation) or, like Morri e Para and De Cesari were among the first to do, to Kevlar and carbon fibers, which had only recently appeared, using them between the various layers of wooden laminate.

Between 1977 and 1979, we designed and built a number of yachts: *Nostra signora dei turchi*, which Biasi built in Fiumicino from fiberglass, and which was probably the fastest of our first-generation one-tonners; *Caccia alla volpe*, a chine built light-displacement boat Bani built from marine plywood on the Argentario, which wound up in Antigua and is still racing; *Sogno bagnato*, built out of fiberglass by Perotto; *Botta dritta*, a Gallinari built cold-molded ¾ tonner for Luigi Carpaneda; *Manua*, a cold-molded second class yacht built by Petronio e Pecarich. *Koala* V, a cold-molded cruiser/racer, was built by De Cesari for the Florentine architect Pierluigi Spadolini – he contributed to the design of the interiors and soon became one of our Studio's closest friends; we also built his next yacht, *Koala VI*. Among the smaller boats were the Ziggurat Junior, a ¼ tonner, and the mini-tonner Avventura 703, both of which were mass produced.

In our early years, Studio Vallicelli witnessed the emergence of a new star in the small universe of boat designers. This was in 1976 at the One Ton Cup in Marseille, where our *Gattone* was competing. Two boats from the young New Zealand designer Bruce Farr were also in the fray: *45 South* and *Jiminy Cricket*. Both were very light and wide; they looked a lot like huge dinghies, with their undersized and at the time unusual 7/8 fractional rig with a big main and a small foretriangle. Running backstays gave control to sail shape, shifting the center of an effort to adapt to changing conditions.

The Farr boats had considerable wetted surface and high form stability, which meant they were boats with a low block coefficient. As their hull shape became asymmetrical when heeled, constant fore and aft trimming was called for, given the hull's completely different behavior in light and heavy winds. Their rigs were reduced to a bare minimum, even the normal genoa sheet tracks were replaced by a system of blocks and lines (barber haulers)[17] that adjusted the genoa sheet points both fore and aft and inboard and outboard. They were planing hulls; the following year, Farr gave them centerboards, moving all the ballast into the bilges.

Also racing against our *Ziggurat* at the 1976 Half Ton Cup in Trieste was a new design by another young New Zealander, Paul Whiting. His design followed the same concepts as Farr had, in an even more extreme guise, with a completely open transom that looked like the open maw of a landing craft.

Farr's yachts were seductive. Marina Spaccarelli Bulgari, who owned *Ydra*, commissioned *Kea* from him for the One Ton Cup in Marseille. However, this time round the results were anything but impressive: *Kea* proved to be slow, out of its element in the Mediterranean's summer breeze.

[16] To find out more, see: Flavia Segnan, *Arrigo Petronio "dotor in segadura"*, Luglio editore, Trieste 2017.
[17] As is generally true of wide, planing boats with high form stability, in light wind their greater wetted area increases hull resistance when compared with narrower boats. It must be remembered that most resistance at low speed is directly proportional to the wetted area, dominating any residual resistance (waves and induced resistance).

Gerontius, from 1975, was one of Bruce Farr's first boats: light, beamy, with great form stability, she inaugurated the season of light displacement hulls.

The innovative lines of Giulio Cesare Carcano's boats foreshadowed the success of light displacement hulls with their great form of stability. In the photo, one of his most successful boats, *Villanella*.

The New Zealanders' yachts (designed by Bruce Farr, Laurie Davidson, and Paul Whiting) weren't all-rounders like the American boats: they may have been very fast – practically uncatchable – in plenty of wind, but they more or less stopped moving when the wind dropped. The stiff winds had helped them out in Marseille, but even then, they failed to win the World Championships. They were pipped not by a light-displacement but by a medium-heavy displacement hull, albeit one equipped with a lifting centerboard: Britton Chance's *Resolute Salmon*.

The following year, 1977, both the One Ton Cup and the Half Ton Cup took place in and around New Zealand and Australia. Competing on home turf with brand new boats, evolutions of his previous designs, Farr went on to win both competitions, as well as beating *Argento Vivo* in the 3/4 Ton Cup at La Rochelle. Farr's unprecedented success put his extreme boats, which were not at all suited for conversion to cruising, ahead of the by-then classic yachts designed by Peterson and Holland. But in the space of just a year, Farr's boats improved considerably in light-to-medium wind conditions. The astute New Zealand-born designer had realized that to reduce the resistance generated by their ample beam and considerable wetted surface (Farr's yachts had a very elevated metacentric height), it was necessary to work on the appendages. Here he designed keels and rudders with a high aspect ratio that reduced resistance and offset the higher wetted surface of the hull.

Farr was happy to admit that eclectic Italian designer Giulio Cesare Carcano was an inspiration for him. Carcano, creator of the famous Moto Guzzi V-engined motorcycles, designed racing yachts for his enjoyment. A Farr boat – *Gerontius* – made her appearance at the 1975 Admiral's Cup, foreshadowing his later designs.

The interior of one of the boats designed by Studio Vallicelli between the end of the '70s and the early '80s: this was taken during the years of the maximum development of wooden construction, soon to be superseded by advanced composites.

Its shape, light displacement, and flat, extremely wide bottom were very similar to a Carcano boat, *Vihuela*, that sailed in the same event, and was a direct evolution of Carcano's *Villanella*, which had won the Two Ton Cup in Sanremo a few years before. All of the yachts Carcano designed in the '70s sported a woman's name, beginning with a V and ending with an A: *Vihuela*, *Villanella*, *Vanessa*, *Vinca*, *Veronica*. They were difficult boats to sail, very fast but ill-suited to the IOR formula, which benefited medium displacement shapes.

By the end of the 1977 season, after Farr had made a clean sweep of the Ton Cups, it was clear that we had to reckon with 3/4 or 7/8 rigged, light displacement centerboard designs from down under. The debate began in the yachting world as to whether such extreme boats were acceptable, or if they were a breach of the spirit of the IOR, a rating rule that had come into existence to promote convergence between racing and pleasure boat designs, with which Peterson- and Holland-style medium displacement boats managed to comply.[18]

At Studio Vallicelli, we had begun drawing light displacement designs with the intention of creating boats capable of interpreting the lessons learned from the New Zealander, while still maintaining our house approach of all-rounders. Our first semi-planing light displacement project with a centerboard was *Asabranca*, a half-tonner launched just in time to compete in the 1978 Half Ton Cup in Poole, England. Affectionately nicknamed "little rocket," the yacht performed well in competition, although a lack of experience with such extreme craft meant that it was hard to find that winning spurt in tricky English waters. Nevertheless, *Asabranca* taught us that it was possible to combine the New Zealand-inspired revolution with what, by the end of the '70s, had become the American medium-displacement tradition.

[18] Claudio Maletto, *Neanche l'ORC dice bene cosa fare*, "Giornale della vela", no. 2, 1978, p. 7; Mino Simeone, *Architetto hai chiuso con i "buchi"!*, "Giornale della vela", no. 2, 1979, p. 48.

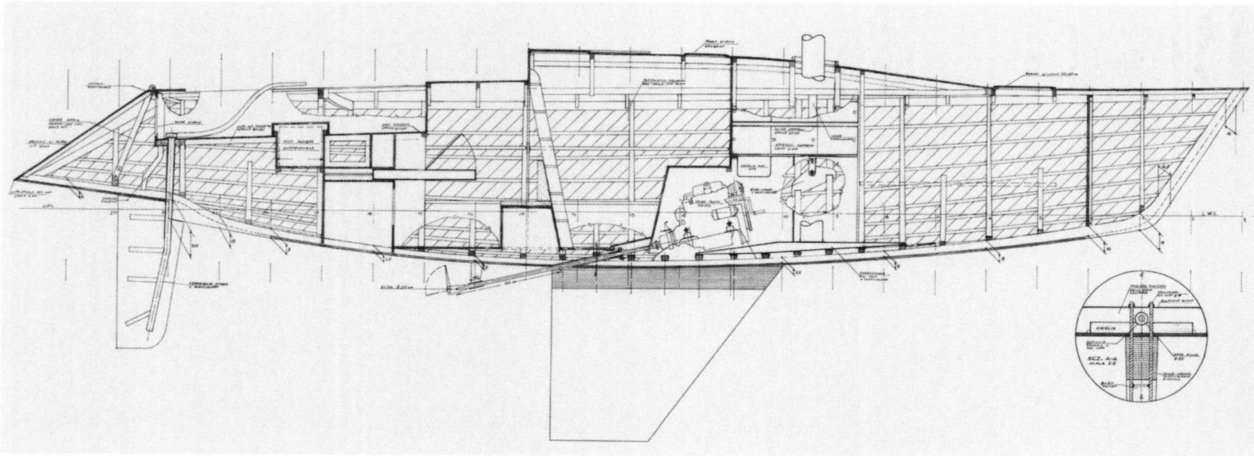

With *Nonno Gigi* and her predecessor *Asabranca* Studio Vallicelli directed its research towards a hybrid position, between American medium displacement hulls and the New Zealand light displacement school.

This compromise was necessary as, by the end of 1978, light displacements and centerboards were seriously penalized under the new IOR rules. Indignant at this penalization, Bruce Farr floated the possibility of New Zealand setting up its own alternative rating rule, irked by what he scornfully referred to as the "beamy heavies," boats that to his mind were very unsafe, given that their weight was a severe test for both hull and rig.

Nonno Gigi, a two-tonner in cold-molded wood built by Bani, 7/8 rigged like *Asabranca* but with a fixed keel, was the design we came up with to mediate between the two approaches. *Nonno Gigi* went to Porto Cervo for the 1979 Admiral's Cup team selection, crewed by Vittorio Mariani and Nicola Sironi, Carlo Croce, Gigi Audizio, Franco Zamorani, Ciri della Porta and at the helm the great Paul Elvestrom. *Nonno Gigi* barely failed to qualify, coming in fourth, just a few points out of third. The early summer calms were unkind to the yacht, and that was a pity, because at that year's sadly infamous Admiral's Cup, with the tragic Fastnet race characterized by strong winds, high seas, and extreme conditions, *Nonno Gigi* would have lived up to its full potential.

The yacht that caught everyone's attention at the Admiral's Cup was an Ed Dubois design, *Police Car*, a design very similar to *Nonno Gigi,* that adopted the same design approach: taking the middle road between the U.S. and New Zealand schools. Together, *Police Car* and *Nonno Gigi* showed that the attempt by the ORC's ITC (International Technical Committee) to penalize the entire New Zealand-inspired design approach had been only partially successful. The next trend was to design hybrid medium- and light-displacements. After the Farr revolution, both Peterson and Holland (and indeed, everyone else) would start to shift maximum beam further aft, widen the stern sections, flatten bottoms even more, and use fractional rigs.

For the following season, 1980, our Studio won a number of commissions to design yachts for the two biggest events of the year: the One Ton Cup, scheduled to be run in the Gulf of Naples, and the second Sardinia Cup, a Mediterranean answer to the Admiral's Cup, which had been held for the first time in 1978, two years earlier.

Two Successful Boats

Filo da torcere, the one tonner winner of the World Championship held in Naples in 1980: a medium-light displacement design built in laminated wood by the De Cesari yard.

We designed three yachts for the One Ton Cup, all variations on the theme of catering as much as possible to the wind and sea conditions expected for the Bay of Naples in early summer. The first, *Excelsior*, was built out of composites by Canados in Ostia and helmed by Paul Elvstrom. The second was *LSD* for Carlo Bixio, built by Galetti in cold-molded wood and helmed by Neapolitan Gianluca Lamaro. The third was *Filo da torcere*, built for *Argento vivo* owners Vigliani and Trionfi by De Cesari from cold-moulded wood and helmed by Stefano Roberti from Rimini.

Filo da torcere's lines were without excesses, creating a sober and linear medium-light displacement boat with the maximum beam set far aft, a long waterline, and a 7/8 fractional rig. With *Filo da torcere*, we aimed for an even more radical slimming down of the bow sections to ensure that the yacht was as competitive as possible when off the wind in light-to-medium conditions. It proved to be a winning combination. The One Ton races off Naples attracted an extremely wide variety of yachts that rotated in the lead during the series: two designs from Ron Holland, *Sharkey*, and *Toda Hesa*, with two superstar sailors on board, Lowell North and John Kolius; *Buonalena*, a design by the Fontana-Maletto-Navone trio for Benetton, flying the U.S. flag; Mino Simeone of Naples' *Cuordileone*, helmed by Picchio Milone, as well as our designs *Excelsior* and *Filo da torcere*. The final standings in the last race, the long double-coefficient run down to Stromboli and back, were decided in a lull off Punta Campanella, at the entrance to the Gulf of Naples. With a light following breeze (ideal conditions for the yacht's design), *Filo da torcere* managed to catch and overhaul her adversaries one by one: her fourth place finish won the One Ton Cup.

This was the first ever victory for a boat designed, built, and sailed by a wholly Italian team. Over a period of three years, *Argento vivo* and *Filo da torcere* owners Vigliani and Trionfi had amassed a One Ton Cup victory, a second place in the 3/4 Ton Cup while winning three Italian Championships and a number of other races. Yet out of 24 races, they had won only five outright. Not only did this give us food for thought, it also demonstrated the effectiveness of the all-round boats on which the Studio Vallicelli had focused since its beginning.

For that year's Sardinia Cup we designed *Blue Show*, a small two-tonner built by the Pezzini boatyard, later produced in series by Barberis. With the valiant Dick Deaver at the helm, *Blue Show* helped Italy take second place behind the Americans. Her specifications were similar to *Filo da torcere*'s, but with a masthead rig and accentuated flatter sections below the counter.

1980 was also the year of *Brava*, a 43' sloop commissioned by Pasquale Landolfi for the Admiral's and

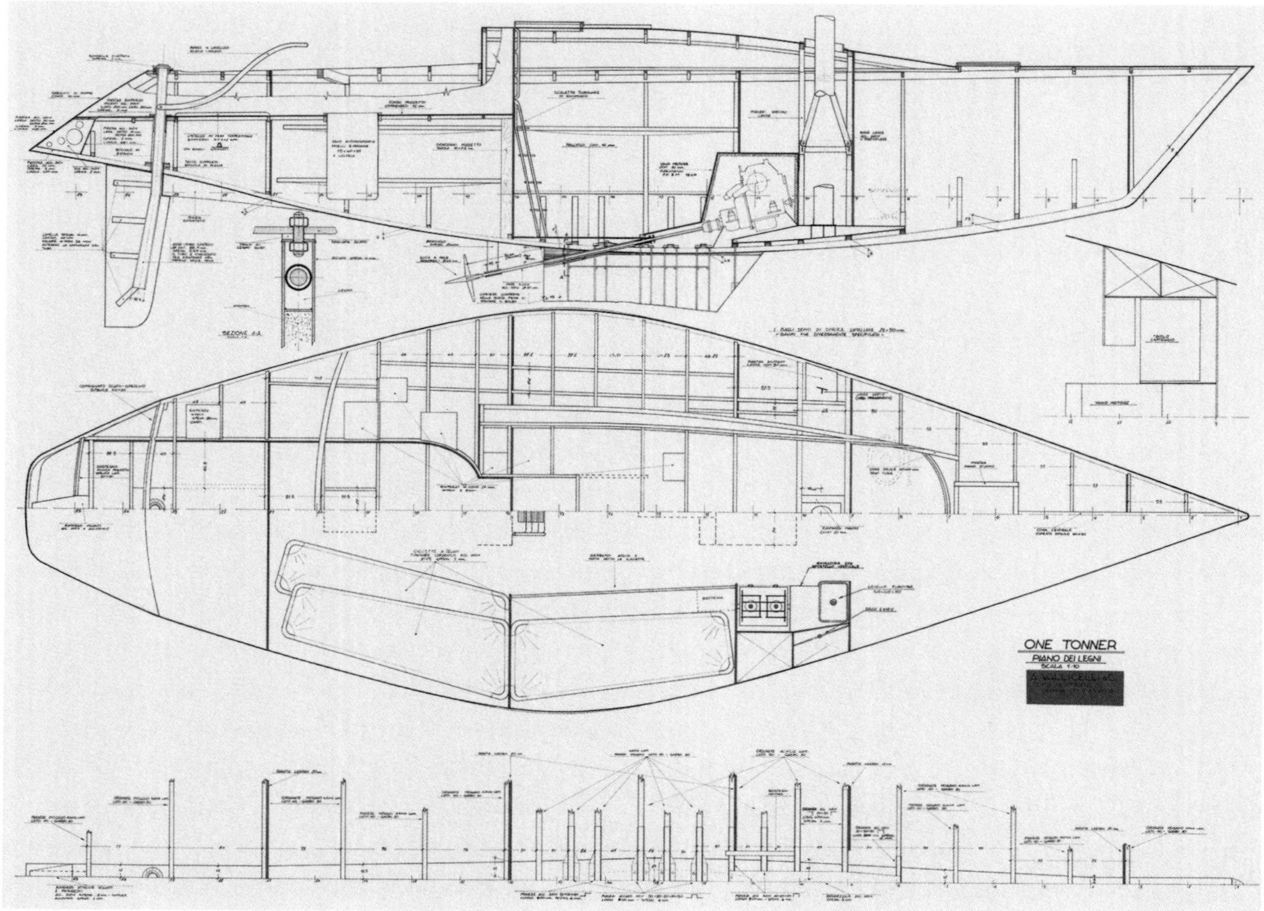

LSD, one of the one tonners designed by Studio Vallicelli for the 1980 One Ton Cup in Naples, built in cold molded wood by the Galetti yard of Lake Garda.

Brava just launched at the US shipyard in Minneford, at the time considered one of the most prestigious shipyards in the world. With the Brava, the Vallicelli studio has successfully entered the international scene.

Sardinia Cups. In the late '70s, Landolfi owned a series produced Canados of our design. He subsequently came to us for a 45' boat; his goal was to be at the top of the top. A successful manager, with *Brava* Landolfi imposed his strict organization to the benefit of our management approach. A few years after we started up, this proved to be a much-needed step-up for us. Following our advice, for outfitting *Brava*, he chose the best the market could offer, no expense barred: spars from Stearn, sails from North, and for her construction the top aluminum boatyard in the U.S.

Located on City Island, not far from Manhattan, Minneford had built mythical America's Cup yachts and some of the best-known yachts in the country.

The *Brava* project had a number of unique aspects, making it something of a gamble. Very much a design specifically for the IOR rule, *Brava* had a medium-heavy displacement hull that was almost a rhomboid in shape, with a very wide maximum beam compared with its adversaries, a high block coefficient below the waterline, and a narrow bow entry angle.

The whole Studio was amazed by how the Minneford people went about construction. On our trips to the U.S., we entered into the good graces of the boatyard's technical director Chuck Sadler, the yard owner's son-in-law, who initiated us to the art of aluminum construction. At Minneford, the sheets used for the hull never exceeded 3mm; welding was so well executed that very little filler was needed.

We were also amazed by the boatyard's modesty. It was located on City Island, a place that had been home to many large shipyards during the war but which

had later been left semi-derelict. The landscape definitely had a look of desolation about it. The boatyard didn't just build racing yachts, it also produced pilot craft and rather ordinary workboats. To us, all this was unexpected to say the least. Those trips to the boatyard, following Chuck Sadler around as the boat was being built, proved to be a learning experience of immense value when the time came to build *Azzurra*, the first Italian America's Cup boat, back in Italy.

Brava was launched during a bitterly cold American winter. On that first trip, I remember the sea spray freezing into ice mid-air. After shipping the yacht back to Italy in early 1980, we entered her in the races at Alassio which, back then, opened the racing season. Another American yachting star, John Marshall, a leading light in fine-tuning yachts, took the helm. One image became famous and immortalized those races: in her first ever race, *Brava* broached, laying flat on the water as Marshall tried unsuccessfully to right her. This spectacular sequence was captured by photographer Carlo Borlenghi, who had the good fortune to be in the right place at the right time, landing a set of images that launched his career.

The boat was spectacular to look at in any case: cobalt blue, the name picked out in gold leaf, graceful yet aggressive lines, with a quality aluminum construction that was hard to find in Europe at that time. Despite failing to qualify for the Italian team, *Brava* competed in that year's Sardinia's Cup under the Canadian flag, where she came in a very impressive fourth.

Although fast, we all had the impression she could improve. For the next season, we decided to make changes to the keel and other areas, so that when the 1981 Admiral's Cup selection races came round, *Brava* would be in tip-top shape. The yacht breezed through the selections with another yachting legend of the time, Star world champion Tom Blackaller, at the helm. In Cowes, for the first time, not just *Brava* but the whole Italian team (which also featured two new Doug Peterson designs, *Almagores* and *Mandrake*) was one of the favorites to win. But the script turned out to be different. Along with the rest of the team, *Brava* plummeted down the leaderboard as a result of a disastrous Fastnet (the final 600 miles, triple coefficient race). That 1981 setback turned out to be a beneficial defeat. We thought long and hard about where we went wrong, and that became our launchpad for victory two years later, this time with Vittorio Mariani at the helm, assisted by Gary Weiseman, for very same Fastnet that had been such a humiliation the year before.

During the 1981 Admiral's Cup, we met Alessandro Nazareth, an engineering student and top-notch sailor who had crewed on *Brava*. Nazareth would soon become one of the pillars of our practice, and before the decade was out, he became a partner. We still work together today.

Brava, the first class designed by Studio Vallicelli and built in aluminum by the Minneford yard in the United States: a successful boat, shown here heeling spectacularly at her debut at the 1980 Alassio regattas.

In 1982 and 1983, *Brava* won not just the Fastnet but everything that was up for grabs: two Italian Championships and the 1982 Sardinia Cup. As time went by, through careful and intense management, the boat just kept getting better and better. Thanks to *Brava*'s wins, commissions began flooding into our Studio for big boats, belonging to what in Italy were known as the first and second classes IOR. *Gemini*, for instance, which the Italian Navy commissioned from us, won plenty of races in Italy, and took part in prestigious SORC races in the US; *Fantasque*, a 45-footer very similar to *Brava*, was built for a French owner; and of course the 50-foot *Entreprise*, with an aluminum hull and a composite deck, commissioned by another French owner who also owned the boatyard where the boat was built.

As well as these first classes, in 1986, we built *Sirena* and *Springbok* out of composites, respectively, in Turkey and the United States. These boats were less extreme than *Brava*, more streamlined, and with a less extreme bow entry angle. Another notable early '80s yacht was our two-tonner *Primadonna*. An evolution of *Filo da torcere*, after her win in category at the 1983 Italian Championships, *Primadonna* was part of the Italian team that finished second in the Admiral's Cup.

Azzurra in Her First America's Cup

Azzurra in training before America's Cup. As soon as the boat was launched, she immediately proved faster than her sparring partner *Enterprise*.

For a long time, there had been talk about an Italian challenge for America's Cup. The idea dated back to 1962, when Gianni Agnelli floated the idea of launching FIAT in the U.S. through an Italian participation in the Cup. Beppe Croce, president of the Italian Yacht Club and a future president of the Federazione Italiana della Vela and later of the ISAF (International Sailing Federation), was to coordinate the effort. The designer was to be Giulio Cesare Carcano, the eccentric and genial designer of the light-displacement yachts that would later provide a template for the New Zealand-born designers. The project fell apart not long afterwards, partly because of difficulties in preparing the entry, and partly because FIAT's market research reported that the U.S. market was not a great match for small Italian cars. Baron Bich, who had lost his way in a Newport Bay fog during the 1970 challenge, also helped dissuade the Italians. But twenty years after that first tentative approach, Italian offshore sailing had made so much progress that the idea of competing in America's Cup was no longer considered a long-shot.

In a little over a decade, yacht racing had become vital to the development of pleasure boating; now cruisers were nothing else but emulations of racing yachts. At Studio Vallicelli, we took the lines from *Gattone*, *Ziggurat*, *Argento vivo*, *Excelsior*, *Blue Show* and *Ziggurat Junior* to create cruising derivates, entering series production in several boatyards. The climate of optimism in the early '80s was buoyed by the results Italian boats brought home: between 1979 and 1983, Italians won the Mini Ton Cup, the Quarter Ton Cup, the 3/4 Ton Cup, the One Ton Cup and the Sardinia Cup; in the 1979 Admiral's Cup, the Italian team placed a brilliant third, and took second in 1983, when *Brava* won the Fastnet.

The time was right to start gearing up for America's Cup. This was in no small part because, as the numbers at the Genoa Boat Show proved, the pleasure boat market was aligning itself with the economic growth that characterized the rest of the Italian economy. As early as 1980, the year *Freedom* won the America's Cup – *Freedom* was the last 12m IC designed by Olin Stephens to do so – many across the country were thinking about an Italian challenge for the 1983 edition. That year, Gianni Agnelli and the Aga Khan discussed this very subject. Names began to circulate after that conversation: Beppe Croce, Mario Violati and Riccardo Bonadeo. There was plenty of talk about America's Cup at the FIV and Yacht Club Costa Smeralda, and it was decided to follow a funding model common among other Cup challenges – finance a club boat to take part in the Admiral's Cup selections and Two Ton Cup scheduled to be held in Porto Cervo the following year. The boat was the (not particularly fast) *Smeralda Prima*.

In Porto Cervo, the Aga Khan hosted early meetings of what would later become the *Azzurra* executive group. However, the project didn't gain traction until Nicola Sironi, who happened to be in the U.S. for a meeting of the ITC, heard that *Enterprise*, the Olin Stephens designed Twelve (built for the 1977 Cup by the syndicate led by Lowell North), was up for sale.

Enterprise had quite a reputation as an excellent boat. She had barely failed to qualify as a defender during her racing days. In 1981, *Enterprise* was still one of the finest sparring partners around, and a sparring partner would be vital to do well in Newport. After Nicola passed this information on to us, we contacted Pasquale Landolfi, the enterprising owner of *Brava*. Without seeking any external funding, he immediately took out an option on *Enterprise*. That was the moment the Italian challenge truly got off the ground.

Although Landolfi made the first tangible move, owing to the internal dynamics of the organizing committee, he was forced to pull out and sell *Enterprise* to the Yacht Club Costa Smeralda.[19] On July 13, 1981, Gianni Agnelli, the Aga Khan and Gianfranco Alberini (from the Yacht Club Costa Smeralda) signed a memorandum of understanding for Italy's first challenge for America's Cup. The triumvirate leading the challenge was augmented by Beppe Croce, Riccardo Bonadeo and Luca Cordero di Montezemolo. While there was no doubt about the appointment of Cino Ricci, who had attended all the meetings, as Technical and Sports Director, who the yacht designer would be had yet to be decided.[20] Three names were in the ring: Mario Violati, Studio Vallicelli and Mario Tarabocchia, an Italian of Croatian origin who, after being forced to expatriate after the war, had joined Sparkman & Stephens.

To resolve the impasse, the hypothesis was that we should all join forces and work together as a group. We at the Studio immediately dissented, afraid that final responsibilities would not be clear, or that timid choices and compromise would result. We had good reason to express our disagreement. At that time, yacht design was still an authorial discipline. The selection of data, which was, for the most part, empirical, and essential for the final design, was up to designer intuition. It was the designer alone who made this call. In later years, with increasing specialization and scientific (therefore less intuitive) numerical modeling, design teams replaced the figure of a sole designer responsible for making operational choices. As we shall have reason to see, later on, this first occurred in Italy in the early '90s with *Moro di Venezia*.

Mr. Agnelli's direct intervention, strong pressure from our champion Pasquale Landolfi and, last but certainly not least, the excellent results of our yachts in recent years, meant that in 1981 we were awarded the commission to design Italy's America's Cup yacht.

Azzurra was designed in double-quick time. The boat's fortuitous name was chosen by a team of Roman advertising executives, along with the graphics, the latter mostly by Riccardo Corato. It has been said that our design for the yacht was merely a slightly updated copy of the already fast *Enterprise*. This hypothesis is simple to refute. Certainly, we began with an analysis of *Enterprise*, but then we made substantial modifications to this template, based on our own offshore yachting experience. Like *Ziggurat*, *Argento vivo*, *Filo da torcere*, and especially *Brava*, *Azzurra* had a finer bow than *Enterprise* and, more importantly, a wider and flatter stern: this was a daring choice, particularly as this was heavily penalized under the 12 Meter rating rule. We also innovated a great deal with *Azzurra*'s keel: after consultations with Attilio Salvetti, Director of the Department of Aerospace Engineering in Pisa, and his assistant Nanni Lombardi, we decided against classic aeronautical profiles and opted instead for profiles developed by NASA for supersonic aircraft with low aspect ratio delta wings. The resulting fin had a sharply profiled leading edge (from aviation strakes), and a highly compressed profile in the middle. Our fin was much leaner than those of other 12 Meters, very similar to an IOR fin.

Enterprise, the 12 Meter yacht designed by Olin Stephens for the 1977 America's Cup and purchased by the Azzurra Consortium as a sparring partner.

[19] http://www.giornaledellavela.com/news/2013/12/27/best-of-2013-il-dietro-le-quinte-di-azzurra/; for the official *Azzurra* film, see: https://www.youtube.com/watch?v=P5fDdAmQs4Q.
[20] Piero Ottone, *Dall'America all'Azzurra: 1851-1983.* Gruppo Editoriale Fabbri, Milan 1983.

The summer of 1983 cover of the weekly "L'Espresso" dedicated to *Azzurra*.

The cartoon by Forattini, which appeared in the "La Repubblica" mocking a well-known politician of the time well summarises *Azzurra* as a popular phenomenon.

We refined the project using the CNR/Rome Italian Navy towing tank, with 1:4 scale models to fine-tune the design. This was the first time that the Navy's Rome tank had been used to test a racing yacht. Working with engineers from the INSEAN Institute (Istituto Nazionale per Studi ed Esperienze di Architettura Navale), we adapted the tank to its new purpose and fine-tuned the model's mechanical drive systems. The test results were encouraging, and yet… In what I can only define as an act of intuition, one of those acts a designer cannot and must not live without, in the final design I decided to make a number of modifications to the model that had emerged from testing, drastically reducing the bulb's section (indeed, it was the thinnest bulb at Newport) and reducing the fin to hull junction at the root.

We commissioned Cantieri di Pesaro to build *Azzurra*. The supervising engineer was Marco Cobau, who would continue to work on subsequent models of *Azzurra*. We were concerned about the building phase of the yacht, or, to be more precise, we wanted to take every possible precaution to ensure an excellent result. At that time, a central rule of America's Cup, one which had made a crucial contribution to the United State's unchallenged supremacy, was that challengers could only utilize people, materials, and technologies from their own country. But that rule began to waver during our first challenge in 1983. We had built *Brava* in Minneford, where we developed an excellent relationship with Chuck Sadler, the boatyard manager. However, for economic reasons, the consortium failed to reach an agreement with him, so we decided to hire Bill Mavroiannis to be our supervisor. Although Mavroiannis had worked both at Sparkman & Stephens and Minneford, he did not turn out to be such a good fallback choice. Indeed, he was more interested in the beauty of the countryside surrounding Pesaro (he was an excellent photographer), than in building the yacht. However, everything seemed to be progressing smoothly at the boatyard: with Cobau's help, we managed to put together a team comparable to Minneford's.

On deadline and satisfying all the America's Cup construction requirements, the boat was launched in Pesaro on July 19, 1982 at a packed popular celebration that filled the entire quay. From that day on to the surprise of all, *Azzurra* became an object of glamour. There were several reasons for this, as over the last few years, Italy had changed a great deal: emerging from the period of political and social tension, Italy had come to embrace the hedonism of renewed economic wellbeing. What became known as the "Milano da bere" period of the '80s dawned, and the country began to discover its patriotic pride. Then, a week before *Azzurra's* launch, on July 11, 1982, Italy's soccer team pulled off an unexpected triumph at the World Cup in Spain. Italy's piazzas and streets overflowed with spontaneous euphoria, filled with seas of tricolor flags for the first time in living memory. The astute affinity between the yacht's name and the nickname for Italy's national soccer team, known as the *azzurri*, lent the boat an undeniable and direct appeal. When America's Cup began the following year, the swell of public interest was as strong as during the famous '50's cycling duel between Coppi and Bartali.

So little racing was available for TV that news bulletins tended to use photos, giving yachts a rather exotic and imaginary aura. Even people who had never set foot on a sailboat could dream. The mere mention of the name *Azzurra* made people feel like they were part of something, even if they didn't understand it, and we could not have been happier with that.

Commercially, the *Azzurra* brand name had attracted a large group of sponsors (twenty in all) who had put up a limited amount of money to use the yacht's logo on their products, expecting to sell within narrow market sectors. Nobody expected the huge media resonance that followed, least of all the sponsors, who had underestimated the pent-up demand in Italy for escapism through national sporting pride and glamour. Rather than the way it works today, with cars or boats completely covered in randomly-placed bits of writing, back then, it was the logo that attracted the sponsors, also because, in the early '80s, overt sponsorship was prohibited in sailing. The expedient of associating companies with the *Azzurra* brand name made it possible to comply with the sponsorship rule, a reminder of the aristocratic and exotic aura sailing had had in the past. Picture essays of the '80s invariably feature *Azzurra*, a trailblazer in publicizing the concept of *Made in Italy*.

The July 1982 launch gave us a year to train before relocating to Newport. The Aga Khan was keen for training to be based in Porto Cervo, but weather conditions in Sardinia were quite windy, too different from those we were likely to encounter in the U.S. It was therefore decided to remain in the Adriatic, in Ravenna, and then in Formia, where conditions were comparable to what we expected to find in Newport Bay. By early 1982, Cino Ricci had already selected the crew, with Mauro Pellaschier at the helm (in preference to Star sailor Flavio Scala); Tiziano Nava as tactician; Cino himself as navigator, while Stefano Roberti, who had helmed *Filo da torcere* when they won the One Ton Cup race in Naples, was on the mainsail. For the first time, Italy had a full-fledged national sailing team.

Once on the water, our first impressions of the yacht were excellent: *Azzurra* seemed to be faster than *Enterprise*, although a bit slow coming out of a tack, so important for the direct upwind one-on-one duels typical of America's Cup. We worked for ages on improving the yacht, trying hard to remedy this weakness. And then something unexpected happened during the yacht's fine-tuning phase: the Americans lodged a protest against our mast for having been made in the United States under Nicola Sironi's supervision, claiming that it qualified as an American rather than an Italian product. Fortunately, Beppe Croce's diplomatic offices rapidly solved the problem. We left for Newport in the summer of 1983 with the intention of acquitting ourselves at least with dignity (or rather, as FIAT boss Agnelli so colorfully said, "let's not melt like a bunch of chocolate-makers").

FORZA AZZURRA!

Lunedì 19 luglio, alle sette del pomeriggio, *Azzurra*, il 12 metri S.I. italiano per la sfida di Coppa America, ha toccato per la prima volta l'acqua nel porto di Pesaro. Oltre alle circa settemila persone che hanno assistito al varo, uno spettacolo davvero inusuale per un avvenimento nautico, erano presenti anche Gianni Agnelli e S.A. il principe Karim Aga Khan promotori dell'impresa, il presidente del CONI Franco Carraro, l'onorevole Forlani e il ministro Signorello. La cerimonia si è svolta secondo le migliori tradizioni: la principessa Begum Salimah, consorte dell'Aga Khan, ha fatto da madrina rompendo la bottiglia sulla fiancata dello scafo. La giornata si è conclusa con una cena offerta dall'Ente soggiorno di Pesaro al Palazzo Ducale (Foto F. Pace).

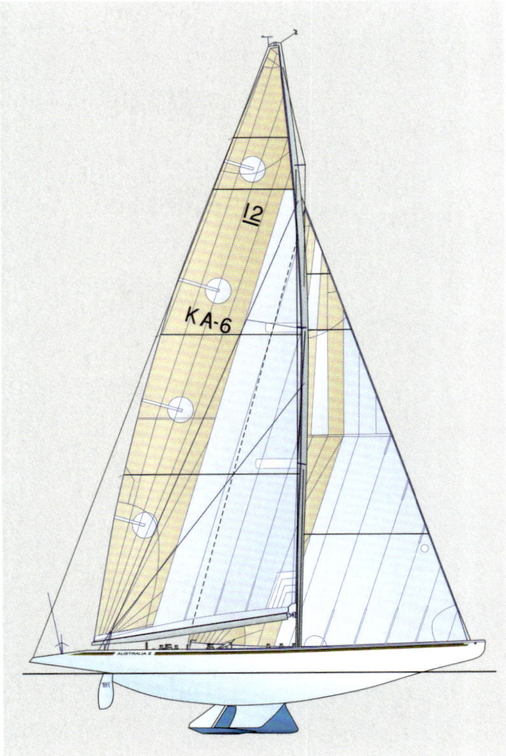

An image of the trials with a 1:4 scale model of *Azzurra* in the Navy's testing tank in Rome.

The launch of *Azzurra* in Pesaro on July 19th, 1982: a few days earlier, the "Azzurri" had won the soccer World Cup.

Australia II, the 12-metre boat designed by Ben Lexcen: an innovative boat with a revolutionary bulb that will enable him to snatch the America's Cup in 1983 from the Americans who had never lost it.

The 1983 America's Cup featured one big new development: the Louis Vuitton Cup, three grueling series of races to choose the challenging yacht. This tough series turned out to be a fabulous opportunity to fine-tune the yachts, in theory allowing the challenger to arrive at the Cup final with a better work-up than the defender: indeed, this factor was to be highly influential in *Australia II*'s success.

Each race left the *Azzurra* team better acquainted not just with the boat but with match racing, which back then was more or less completely unknown in Italy. After a rather muted start, the day came when *Azzurra* managed to beat the great revelation, *Australia II*, in a race. The news spread like wildfire in Italy. During the summer news lull, the national papers leapt at the chance to plaster their front pages and no-end of inside pages with pictures of a yacht that was upholding the nation's pride and duking it out on the high seas. Almost unbelievably, *Azzurra* picked up the nickname "Italy's sweetheart."

During the round robin stage, we made it into the semi-final, where we were pitted against the British consortium with *Victory*. Not only did the consortium have an America's Cup pedigree we could only dream of, the team had built two yachts for that year's challenge, one designed by Ed Dubois, the other (the one that got the nod) by Ian Howlett. *Victory* and *Azzurra* fought it out in a close battle. Alas, we lost in the final race: a jumper spreader fitting had been imprudently replaced with one of inferior grade steel, and we risked losing the mast. *Azzurra* chose to retire, but shortly after, *Victory* broke her rudder. The English boat, however, sailed on using its trim tab (the flap on the trailing edge of the keel) and managed to cross the finishing line just a few seconds before the time limit was up.

This hard luck story did nothing to diminish the fact that the crew and yacht had proven themselves to be extremely competitive. In the late September final, *Australia II* overtook *Liberty*, the American defender designed by Johan Valentjn, in the epic last run of the race; this despite the fact that *Australia II* had proven slower than *Liberty* when off the wind and that Dennis Conner had rounded the previous mark with a full 50 second advantage.

By then, *Azzurra* had returned to Italy to a warm welcome, and we had already started thinking about the next Cup, which would be held outside the U.S. for the first time: in Fremantle, a suburb of Perth and home town to *Australia II's* owner, Alan Bond.

Azzurra in Her Second America's Cup

Australia II won by a small margin, but that took nothing away from the fact that she was a truly fast, ground-breaking boat. Short, light with a lot of sail area, she had a revolutionary keel which had been kept under canvas throughout the race period, arousing the curiosity of media outlets around the world. After the final winning race, *Australia II*, at last, dropped what the press had dubbed "the skirt" to reveal a keel with an inverted profile, with the tip longer than the root, equipped with two winglets sprouting from the after the end of the bulb, their design a complex scored geometry with a slight outward twist.

It was common knowledge on the Newport docks that *Australia II*'s keel had not been designed by Ben Lexcen (a new name that designer Bob Miller had mysteriously chosen to use), but was the result of scientific research whose paternity was not Australian, but Dutch. The new keel was based on research carried out by Joop Sloff, which in turn was based on NASA research to improve wingtip efficiency by using winglets to exploit flow diversion and tame turbulence. Another marine engineer, Peter Van Oossanen, subsequently perfected these studies at the tank in Wageningen, Holland.

This type of keel offered a number of benefits. Tapering off towards the root, it significantly lowered the center of gravity, increasing stability, while at the same time ensuring excellent maneuverability in the key moments of America's Cup races, the tacking duels, and pre-start circling. What's more, when sailing upwind, boat heel caused the winglets to increase the relative draft of the keel, while remaining inside the strict parameters of the 12 Meter Rule. They also reduced tip vortex, considerably increasing performance.[21]

As a result, *Australia II* was extremely fast upwind and coming out of a tack; and yet it was downwind, where *Liberty* had proven faster, that she beat the Americans. *Australia II* hadn't won just because of the Dutch appendages, but because the challenger's technological development had closed the gap with the Americans. Her sails (mainsail and genoa), for example, had an original tri-radial design that was adopted throughout the yacht racing world the following season. This invention may, for the most part, be ascribed to Tom Schnackemberg, a New Zealander who was part of the Australian crew. *Australia II* heralded the start of a new era for sailing, where research into physical and digital models would gradually take over from empirical research.

By the end of the racing, enthusiasm for the Cup wasn't limited to *Australia II* but had spread to the *Azzurra* team too: before we even left to return home, the Costa Smeralda Yacht Club announced it would be launching a second challenge in Australia, and that the Studio Vallicelli would once again be commissioned to

[21] Peter Van Oossanen, who engineered the keel at the MARIN (Maritime Research Institute Netherlands) tank in Wageningen, also invented an extremely sophisticated and efficient towing method for tank testing, by towing the models from the sail plan's center of effort. He also experimented with lead keels to simulate the position of the boat's true centre of gravity, making it possible to record how models behaved in real time regarding resistance and leeway.

The launch of *Azzurra II* on July 25th, 1985, in Venice: note the covers that hide the underwater form of the hull. After *Australia II*, these design intuitions had proven fundamental to the performance of the various boats.

The cartoon by Massimo Bucchi published in "La Repubblica" in 1985 mocking Gianni Agnelli.

design the new yacht or – given that the consortium would be better funded this time round – yachts.

Thanks to the permanent addition of Alessandro Nazareth, who took over a significant portion of our engineering research, we immediately began to focus on model-based research for the design of the new Twelves. Our first move was to involve Van Oossanen, one of the leading lights during the testing and fine-tuning of *Australia II*'s keel. When we began tank testing in Wageningen, we had three different hull shapes in mind. We relied a great deal on VPPs (Velocity Prediction Programs) to design the hulls. The processing power available on modern computers gave us more than encouraging results, at a time when we were making significant investments in digital equipment. We had taken our first step in this direction in 1981 when we acquired a computer for hydrostatic calculations. However, we began wide-scale computerization in 1984 and 1985 with CIRCE, a program developed from the more complex CATIA, which harnessed splines and Bézier curves to make three-dimensional drawings of complex surfaces. We drew up the shapes of *Azzurra II* and *Azzurra III* on CIRCE and then acquired a plotter so we could print out large-scale drawings in real time, right up to the 1:1 hull plans.

Work in our offices on Via Vittoria changed radically with computerization. And although digital technologies were developing at breakneck speed, there was still one critical issue that, at the time, we all underestimated: how reliable these technologies were. The VPPs and CFD (Computational Fluid Dynamics) programs derived from aeronautics had significant processing shortcomings at the time due to limited data sets, considerably less than what would be required to assess complex fluid dynamics fully. It is worth noting that fluid dynamics are far more complex on the water's surface, where two fluids, water and air, come into play, than in the air, for which the programs had been calibrated.

In the meantime, we had begun tank testing in Wageningen using larger models than the ones we had tested in Rome for the first challenge (in 1:3 rather than 1:4 scale). We also had a model built of the first *Azzurra* for comparison purposes. To our great surprise, the results of the new models, whose radical new designs even featured an option for variable geometry keels, performed very poorly compared with the old *Azzurra*.

In hindsight, we realized that the entire data set we'd used to develop the VPPs with Van Oossanen needed to be revised, but by then, our allotted time in the Dutch tanks was drawing to an end. We had to trash a year's work that we had wrongly thought was excellent, and start designing the boat again practically from scratch.

But the design of *Azzurra II* that evolved out of the first *Azzurra* incorporated some highly positive insights. The rounded fin profile was a step in the right direction. It was no coincidence that *Stars & Stripes*, the very fast American winner of the upcoming Cup, designed by Britton Chance Jr., Bruce Nelson and David Pedrick, and helmed by Dennis Conner, would adopt a similar fin. This excellent insight was rendered null, as our research into its proportions was incorrect. But the weaving of the boat's structure in a predominantly longitudinal direction, an excellent new approach we developed with the Finzi practice in Milan and Ruggiero in Genoa and had had preemptively approved by the Rules Committee, was present in many of the yachts that then made it to Perth.[22]

And yet *Azzurra II*'s performance continued to disappoint us, as we found out in Australia at the 1986 Twelve Meter World Championships, the year before America's Cup, in a very different climate than in 1983: it seemed as if *Azzurra II* was broken. The gap between the sporting side of things, for which Cino Ricci was again responsible, and the design side for which we were responsible, was growing wider by the day, and everything that was happening was exacerbated by a lack of political direction.

For this edition of America's Cup, the Costa Smeralda Yacht Club had become the Challenger of Record, responsible for organizing the Louis Vuitton Cup, the series held to select the challenger. This huge commitment kept Alberini from us, in retrospect, a key figure in resolving problems "from the top" during our first challenge. A general climate of mistrust began to trigger a whole series of errors. Our consortium bought *Freedom*, the American yacht that had won the Cup back in 1980, but *Freedom* proved to be slower than *Azzurra I*, making her a useless sparring partner; in addition, several key members of the 1983 crew decided not to take part in the next challenge.

All of these problems came to a head during the 1986 World Championships: *Azzurra II* proved to be

slower than *Australia III* and *French Kiss*, Philippe Briand's French astonishingly fast yacht; we were also slower than *New Zealand*, created by that country's three great designers, Farr, Holland and Davidson (who later produced the first ever fiberglass 12 Meter yacht). Back in the office, we recognized our mistakes, starting with our reliance on theoretical research rather than on-the-water experience. A more logical approach would have been to start by swapping *Azzurra I*'s traditional fin with one that had winglets, and then to start testing in Fremantle: a unique place to race, with a stiff wind that whipped up steep, extremely choppy waves that favored long, heavy yachts without too much sail – just like the eventual winner, *Stars and Stripes*.

After these initial disappointments, we attempted to sort things out with the consortium but were unable to overcome the disorganization and lack of a coherent

[22] Making calculations for the finished elements the Finzi Studio checked structural sizing by comparing predominantly transversal to predominantly longitudinal classic structures.

Azzurra II was a problematic boat, conceived in conformity with theoretical parameters that later proved unreliable. She raced the 12 Meter World Championships in the difficult waters off Perth.

Stars & Stripes, the 12 Meter yacht designed by Britton Chance Jr., Bruce Nelson and David Pedrick. Dennis Conner won the 1987 America's Cup in Australia with her.

Azzurra I and *Azzurra II* in training.

plan. A year out from the racing, we began building two new yachts, one that we designed, *Azzurra III*, the other designed by Studio Sciomachen, *Azzurra IV*. Those were the boats that we took to Perth. *Azzurra IV* turned out to be very slow indeed; *Azzurra III* also struggled. In the meantime, arguments continued to rage. Cino Ricci was dismissed once and for all; Lorenzo Bortolotti took his place. *Azzurra III* began the Louis Vuitton Cup badly and ended badly, even though, after the enormous effort and numerous trips back to the boatyard to make changes, the yacht did improve… just not enough to get on the leaderboard.

When I think back to that painful experience, all I can say is that the main reason why the *Azzurra* mechanism broke was that we were too filled with pride, all of us. Blinded by the sporting and media success of our first challenge, to be able to do without even that top-down governance that becomes fundamental to achieving appreciable results. If the 1983 America's Cup marked the end of a cycle, the 1987 edition marked the beginning of a new, ever more technical cycle, in which the process of engineering and tuning a yacht and it's crew could only be managed by complex organizations working in perfect coordination.

The IOR: From Development to Crisis

As the '80s drew to a close, the IOR, having led the worldwide development of offshore sailing, went into something of a tailspin. Towards the end of the '70s, to preserve the philosophy of a measurement rule designed to foster the development of medium-heavy displacement yachts that could be adapted for use in pleasure boating, the light-displacement shallow planing hulls with centerboards favored by the New Zealanders (first and foremost Bruce Farr) were hit by heavy penalties.[23] Despite this, a decade later, designers began adopting some of the characteristics of the New Zealanders' approach. Attempts were made in the early '80s to strike a compromise between the two approaches, changing the rating for one-tonners from 27.5 to 30.5: now they could compete in the Admiral's and Sardinia Cups, and their hulls and rigs were the most conspicuous examples of cutting-edge design. There would be no revolutionary vessels like *Terrorist*, *Ganbare* or *45 South* as there had been a decade before, but from the mid-'80s onwards the new one-tonners had a similar look: sleek lines and none of the bumps they'd had before; longer stern girths and very flat bottoms; small volume aft sections and higher prismatic coefficients.

The top designers of the '80s were Joubert and Nivelt of France (who designed *Diva*, the one-tonner that won the 1983 Admiral's Cup), Jean Berret (who designed *Phoenix*, the 1985 Admiral's Cup winner), and Philippe Briand. They were joined by Judel and Vrolijk of Germany and Niels Jeppesen of Denmark, who designed and built the incredibly successful X-Yachts series.

New developments appeared in the late '80s for keels, with their evolution from trapezoidal shapes to elliptical shapes, increasingly tapered towards the root to reduce less efficient surface area and redistribute it to more efficient areas. Rudders also became elliptical and propeller shaped, their surface area diminishing significantly as they lengthened downwards to reduce tip vortex and induced drag. This trend also spread to keels, driven in part by new developments in the rating rule that decreasingly penalized draft.[24] As they sought to find the ideal proportions, designers began to use terms like aspect ratio, the relationship between a wing's depth and its average width. Meanwhile, those famous winglets that had appeared on *Australia II* vanished, never to be seen again.

In 1983, we designed the second *Brava*, a 30.5-footer masterfully built by Luigi Morri and Dino Para in Rimini, with a cold-molded hull incorporating unidirectional carbon fiber to stiffen the yacht's frames and floors. To lighten the yacht, we decided to make the deck out of PVC foam-cored sandwich with carbon and Kevlar laminates. In 1984, *Brava 30.5* made her debut.

[23] It is worth noting that despite the IOR's critical issues, in the late '80s, as light displacements became more common, yachts were less extreme than before. Their lines were more harmonious and, especially when it came to new one-tonners, their lines were conceived to give boat speed precedence over exploiting loopholes in the rating. Bustles were reduced to a minimum, while bumps – the artificial swellings at measurement stations that had marred clean lines until then – were dispensed with altogether. Bow and stern overhangs diminished, while maximum beam tended to blend more harmoniously with the rest of the boat.

[24] According to classical wing theory (known as the Slender Body Theory), the point where a wing was attached to the fuselage should be elongated because it is precisely in this area that they achieved their greatest efficiency and produced most lift. On sailboat hulls, things were different: the relationship between the appendage and the adjacent hull was completely different to the relationship between a wing and the fuselage, as indeed is the flow: heel ensured that it is precisely the point of attachment between keel and boat bottom that is the least efficient and most turbulent.

Brava 30.5, a 1983 one tonner by Studio Vallicelli with harmonized lines, less extreme than her antecedents and without the typically forced shapes of first-generation IOR boats.

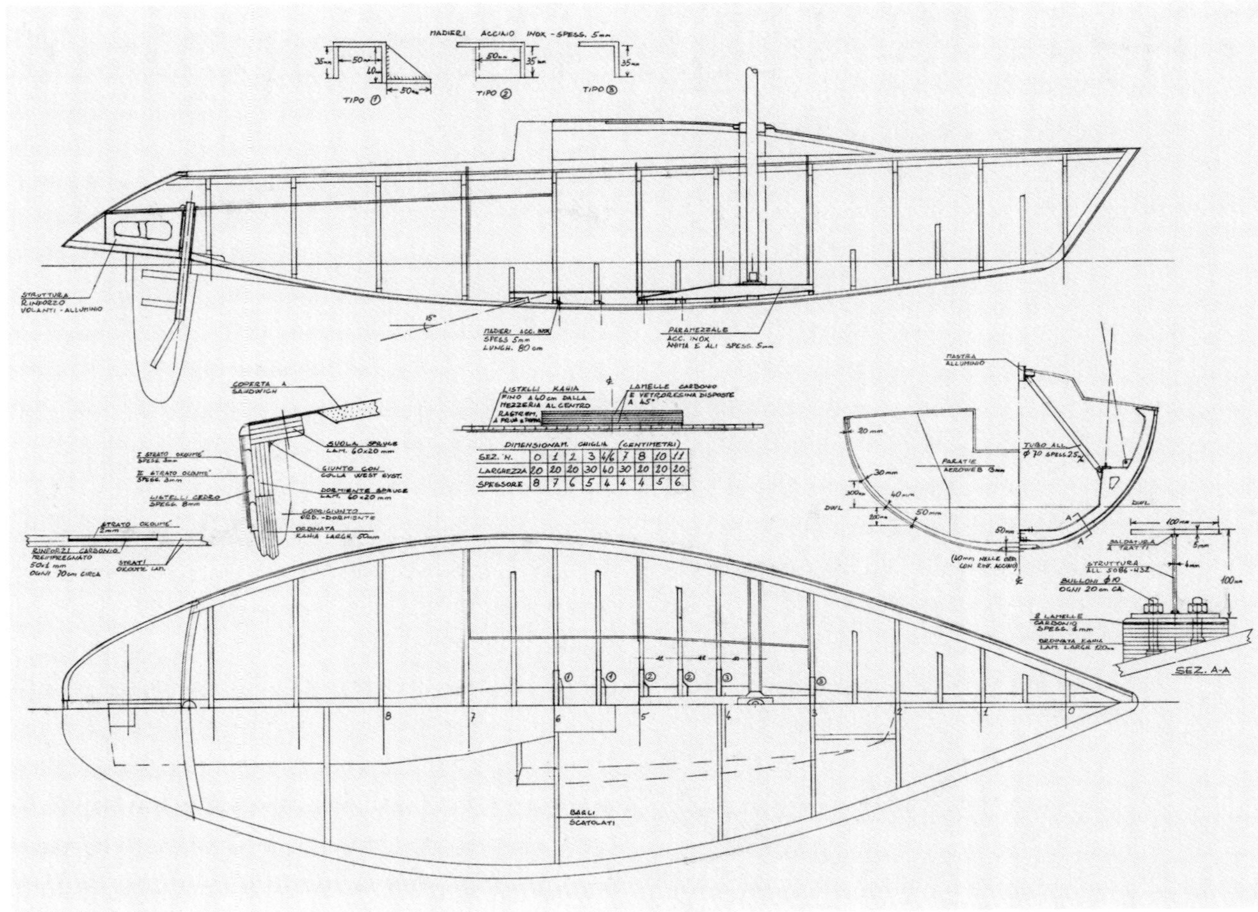

This was, to all intents and purposes, the last-ever generation of wooden boats. At the 1987 Admiral's Cup, with no cold-molded or aluminum yachts, it was clear that offshore sailing was in the midst of radical upheaval. Composite boats dominated, with Kevlar and carbon sandwich and PVC foam or Nomex (a honeycomb structure with an aramid core) cores.[25] For the first time, many of the yachts adopted pre-preg construction, using fabric previously impregnated with resin and stored at low temperatures to prevent catalyzing. This pre-impregnated fabric was applied to molds and then, following oven-controlled programmed heating cycles, catalyzed until fully hardened.

Another first that year in Cowes (where our design *Merope* was competing, composite built with the new techniques[26]) was carbon hardware, now allowed under the IOR, increasingly permissive and open to change.[27] In an attempt to rein in costs, which had started going through the roof, the regulations had prohibited the use of carbon. But as the evidence shows, progress cannot be stopped, and this attempt was destined to failure.

Building on experience gained in America's Cup, and foreshadowing an innovation that would radically change navigation – the GPS, onboard electronics had made their debut and were becoming increasingly sophisticated.

Sails went through a revolution after the 1983 America's Cup when Tom Schnackenberg had come up with his extremely efficient tri-radial genoas for *Australia II*. In the space of just a few years, sail performance significantly improved. After years of (heavy) Dacron as the only fabric used for sailmaking, it disappeared in the

[25] *Imp*, a boat designed in 1977 by Ron Holland and built by Kiwi boats in Florida, was the precursor to high technology composite construction. She was composite built with balsa sandwiched between layersof fiberglass and Kevlar. The hull was stiffened with an unusual internal 3D load-bearing structure made up of aluminum tubes attached to the hull solely at the points where stress was most concentrated (keel, rudder, chainplates and mast step). See Bill Burton, *The Legend of Imp: The Magical Yacht That Rocked The Sailing World*, Direct Response Imaging, San Francisco 2010.

[26] The construction of *Merope*, engineered by SP System which had opened a branch in Italy, employed leading-edge techniques for the day: vacuum lamination (eliminating excess resin by vacuum pump and separating fabric by the use of peel ply), and post-curing the yacht in an oven to bolster the laminate's strength. The rudder had a carbon shaft with a blade of Kevlar and carbon sandwich filled with polyester, tipping the scales at a total of just 12 kg. SP System was subsequently taken over by the Montedison Group, as was its affiliate SP Technology, which specialized in designing composite structures.

[27] At that time, the boatyards with the finest reputation for making composite yachts were the historic Jeremy Rogers boatyard (now Green Marine) in the U.K. and the Cookson boatyard in New Zealand: in the late '80s Cookson went through an extraordinary period of development after being one of the first to adopt the vacuum bag technique; in addition, we must not forget the U.S. boatyard Kiwi boats, which built the ground-breaking *Imp*. And as well as the first carbon spinnaker poles, new Sparcraft masts also came out around that time, assembled from two riveted profiles and manufactured out of the new 7000 alloys that significantly reduced weight.

early '80s, replaced by Mylar and Kevlar. The way sails were cut continued to change radically: the tri-radial was followed first by warp-oriented sails, their panels positioned depending on the direction of strain. These were subsequently supplanted by laminated sails developed by the Sobstad sail loft, where film panels were layered with a dense network of load bearing yarns to create a veritable structure for the sail. In the early '90s, North Sails went one step further, with a technique still in use today: molded sails. These single-panel 3DL sails were shaped on a mold, interwoven with Kevlar and carbon reinforcement yarns.

It is interesting to note how hull and sail development advanced in parallel. For both, there was an escalating use of analytical research into materials to ensure they perform their allotted tasks as well as possible, orienting the fiber reinforcements depending upon tension direction. This new approach to design was predicated on acquiring and managing larger and larger data sets, made possible thanks to the development of IT, and the resultant increased ease of access to the data.

Therefore, the greatest new development – one whose effect is being felt to this day – is the significant amount of weight shed by both hulls and sails. However, this weight reduction and the general enhancement of hull, sail, and component performance came at the cost of massive price hikes. This critical issue, combined with a yacht's decreasing competitiveness over time, would prompt a downwards slide in the IOR's prestige starting from the late '80s.

Alternatives began to appear on the market. The first was one-designs, specifically, the J-24.[28] To give some idea of this phenomenon, which began in Italy around 1983, within five years, there were so many J-24s in Italy that the 1987 World Championship held off Capri had more than 80 boats. Won by rising star Francesco de Angelis, with a fleet including the finest international sailors, the series seemed like Olympic-level racing.

Alongside the development of one-designs, production boat racing also gained in popularity. In 1986, around fifty boats entered the Italian Championship dedicated solely to production boats. That number continued to increase over the following years. The Swan Cup was also held that year, followed two years later by the very successful Grand Soleil Cup. Then there was the *Giro d'Italia* for sailing, an event won the first time by a Russian team and run with Comet 333s designed by our Studio.[29]

To complete the picture of an ever more varied yachting scene was the growing phenomenon of classic yacht rallies, proof that the eternal beauty of classic yachts continues to characterize contemporary taste for vintage charm. In hindsight, it seems that this passion for classic yachts was the harbinger for the extremely glamorous fast cruisers of the '90s.

This interest in vintage yachts had an unforeseen yet fortunate side effect. In the '80s, the boatyards that had so highly excelled in traditional wooden boatbuilding fell upon hard times. The yards that survived found a new market in the '90s, restoring classic yachts. However, this required a high level of care and attention that only a few craftsmen could provide, so it's no exaggeration to say that these restorations revived the traditions of wooden boatbuilding, which had appeared to be on the way out.[30]

As the numbers clearly show, an increasingly broad and varied range of available boats slowly put the IOR on the ropes. The 1989 One Ton Cup in Naples (won by Landolfi's *Brava*, a Bruce Farr design, with De Angelis at the helm and Paul Cayard as tactician), was fought out by 28 yachts, a similar number to the 1980 One Ton Cup that we won with *Filo da torcere*. The following year, 1990, just 10 yachts registered to race, almost all of which were designed either by Bruce Farr or Judel & Vrolijk, the architects who were monopolizing this shrinking market.

It is instructive to compare a one-tonner like *Brava*, which won in Naples, with a one-tonner from a decade

[28] A number of monotypes were in circulation in the '90s: J-24s, as well as the J-22, the Este 24 designed by Germán Frers, the Tripp 26 and the Mumm 36, which was to become the one-design for the Admiral's Cup. Interesting developments included the Melges 24, which came out in the early '90s and seemed like a miniaturized version of the America's Cup IACC, equipped with carbon masts and retractable bowsprits.

[29] It is interesting to note that in the Italian Championships of 1993 under the new IMS measurement rule, one-offs accounted for just 30% of competitors; the rest were cruisers (25% X-Yachts, 15% Grand Soleil, 5% Comar and others). See the "Giornale della vela", no. 7, July 1993.

[30] To give some idea of the extent of this phenomenon, suffice it to say that by the end of the '90s, more than a hundred vessels, some dating as far back as the nineteenth and early twentieth century, attended vintage yacht rallies.

Brava 30.5 was masterfully constructed in carbon fiber reinforced laminate with a composite deck by the Morri and Para yard in Rimini.

Dioscura is a fast cruising boat designed in the 1980s; her lines avoided the extremes that characterized the IOR racing boats of the time and later brought about their decline.

earlier, built from cold-moulded wood, for example, our own *Filo da torcere*. Whereas the older boat's hull planking tipped the scales at around 5 kg/sq.m., by the late '80s one-tonners were being built out of increasingly exotic composites, and the planking's weight was down to some 2 kg/sq.m.: less than half the weight, but at twice the cost.

The crisis rippled through to the ICR's flagship event, the Admiral's Cup. Problems had begun in the late '80s when to counteract the one-tonners' predominance, organizers decided to set limits on the overall rating of the three yachts each nation was entitled to present. The formula that had remained unchanged for years and years (as you would expect from British respect for tradition) was modified and, with an ongoing series of problematic changes, were the first signals of the race's declining importance.

To make races more spectacular, a few years later, the organizers decided to get rid of handicaps altogether, with real time races between three classes: two-tonners, one-tonners and IOR 50-footers. Then, to try to staunch the hemorrhage of participating nations and to put a cap on costs, the Admiral's Cup was suddenly opened up to One-Designs.[31] One of the three boats for every team was to be a Bruce Farr-designed Mumm 36. This choice, however, created a dilemma: was the Admiral's Cup to be a showcase for the latest racing one-offs, as in the past, or a kind of compendium of different One-Designs?

This ambiguous solution failed to pay dividends. The same was true of the rash decision in 1999 to do away with the prestigious Fastnet Race, the tough 600 mile course that until then had been not just the most prestigious race in English waters, but one of IOR's top races *tout court*. After cutting participation from three boats to two, the Admiral's Cup was on the way out. In 2001, the Dutch won against a field of perhaps a quarter the number of boats that had raced during the IOR's golden years and the 2003 edition ended up not even being run.

[31] In 1995, the Italian team won the Admiral's Cup with three boats: *Capricorno* (Big boats category, owned by Rinaldo del Bono); *Brava Q8* (ILC 40 category, owned by Pasquale Lando fi), and *Mummamia* (Mumm 36 category, owned by Paolo Gaia).

New Boats for a New Era and the New IMS Measurement Rule

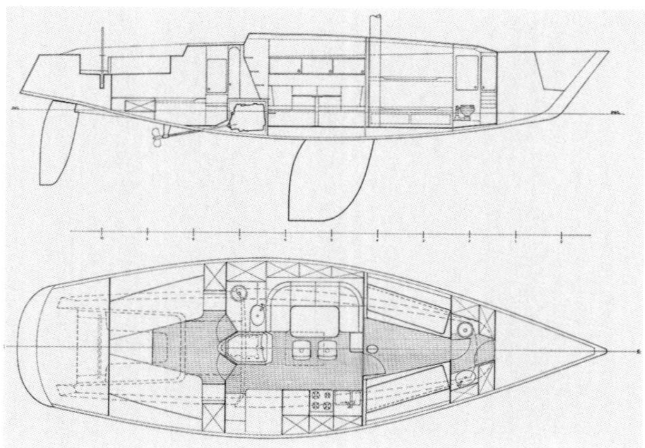

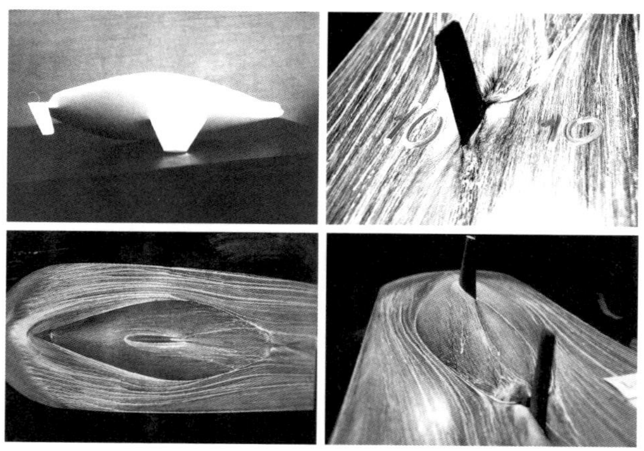

General plans of the *Fenicia*, built in glulam in '90 by the Studio Altura shipyard in Venice, whose drift plan was studied, using scale models, in the wind tunnel of the University of Pisa.

During the IOR crisis years, Maxi yachts enjoyed a period of success. In the second half of the '80s, these 70/80-footers captured the public imagination for their size, which was unusual at the time, and for the incredibly large crew they required. The formula adopted for Maxi yachts was the international circuit. This formula, capable of attracting large numbers of people seduced by the yachting jet-set to a selection of different, glamorous locations, had its genesis in the days of *Azzurra*.

To get an idea of how much prestige Maxi yachts enjoyed at the end of the '80s, everyone in Italy remembers the challenge that pitted *Moro di Venezia* against *Matador*, Maxi yachts owned respectively by Raoul Gardini and Bill Koch, the same owners who would fight it out for the 1992 America's Cup.

Throughout the '80s, Maxi yachts looked like scaled-up IOR boats. Changes were to come at the start of the following decade: two Bruce Farr yachts raced at the 1992 Maxi Worlds after concluding (and in the case of *Steinlager*, winning) the prestigious Whitbread round-the-world race. Soon afterwards, *Steinlager* and *Fisher & Paykel* were bought by Italian owners Giorgio Falck and Gianni Ravasi. Although they were IOR yachts, they looked very different from standard IOR craft of the day, as they had been designed specifically for the Whitbread. They had less wetted surface, low freeboard and were narrower too, almost canoe-like, with a reduced maximum beam that gradually blent into the bow and a truncated stern. They also featured an unusual ketch rig with a surprisingly large mizzen. This rig not only lowered the center of effort, making the yacht stiffer and less prone to rolling, it fully exploited the measurement parameters of the IOR rule.

In the early '90s, we designed a rather special Maxi, conceived as a training ship for the Italian Navy. *Orsa Maggiore* had a sleek and elongated hull. Maximum beam was most certainly less pronounced than on IOR yachts of the '80s, albeit not quite as narrow as on Farr's yachts. Ketch rigged, the big mizzen mast was well-separated from the main mast, and everything was conceived to cater to the requirements of a vessel crewed by officers and students, constantly sailing around the globe. This is why we designed *Orsa Maggiore* with an indoor wheelhouse and lodging in the deck for the tender when hoisted aboard. The distribution of hull volume throughout her length made *Orsa Maggiore* very well-balanced, easing her passage through the waves and minimizing pitching.

In addition to Maxis, another interesting development in the early '90s were the 50-footers. However, like the Maxis, these crafts were exceptions to the IOR's slew of setbacks.

Taking stock of the decade that was ending and looking into the future of sailing, in 1989, the Giornale

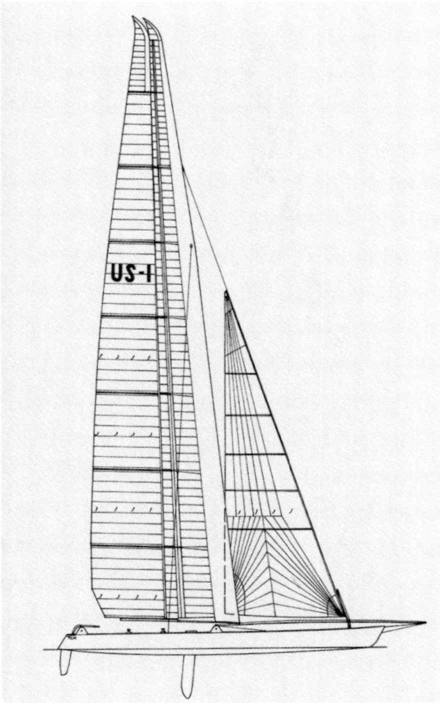

Stars & Stripes, the catamaran designed by Gino Morelli and Britton Chance Jr., winner of the 1988 America's Cup against the immense New Zealand monohull. She featured an innovative rigid wing-sail divided longitudinally into three flaps.

della Vela published a long article listing alternatives to the IOR; the first was the Open Classes.[32] After first appearing in the '50s on German lakes, they were highly developed on Lake Garda in Italy, where the key race for the Open Classes was held since 1951: the Centomiglia. Since the late '60s increasingly extreme designs, similar to the *Flying Dutchman*, competed in the Centomiglia: long, very low in the water, ultra-light and planing, their stability depended on positioning the weight of the crew outboard on hiking racks, often using trapezes like on dinghies.[33]

The rating rule formula for these boats was as unrestrictive as possible, setting only a few parameters: maximum overall length for each of the different categories (the Centomiglia had three: 12.70, 10.70 and 8.70 meters), maximum beam and a stability factor, empirically determined with a capsize test. Provided that they satisfied these parameters, designers could freely express themselves with displacement, draft and sail plan.

The Open Classes were developing an approach to measurements and rating rules that would later be known as the "box rule," whereby a design had to fit inside a solid, and once that was the case, utmost liberty then applied. The box rule approach successfully dovetailed with the liberalization of sailing that began in the early '90s, prompted more than anything else by the need to make races more spectacular for TV coverage, a demand which has increased ever since.[34]

Just ten years after *Azzurra*, the days of a few extremely short visual clips, conveying a distant, slow and ethereal aura, were coming to an end. By the '90s, sailing was becoming more and more spectacular and was shown without filters, becoming (and this was not without its pitfalls) an acrobatic sport for the use and delight of the mass media.

That aforementioned article not only featured the Open Classes, it also included the 70' ocean-going catamaran and trimarans, which were becoming more and more extreme, ever wider and therefore more structurally daring, with longer and more complex outriggers. In the late '80s, one catamaran, in particular, caught the public's attention: *Stars & Stripes*, designed by Gino Morrelli, Britton Chance and others, which in the 1988 America's Cup humiliated the New Zealand challenger, a huge Open Class monohull (actually, a Centomiglia design blown out of all proportion) designed by Bruce Farr, who's speed was not even in the same league as the American catamaran.

Stars & Stripes adopted a stunning wing-sail, made up of three parts: the first, encompassing the leading edge, was followed by a central portion and then an exit flap.[35] Although this was hardly the first time, a rigid wing-sail had been seen, with *Stars & Stripes*, the solution was engineered on a grand scale, leveraging newly-developed, leading edge technology.

In the late '80s, notable development also took place in the Mini-Transat class, small six and a half meter boats designed for their own exhausting solo Trans-Atlantic race – first run in 1977, the race is still going strong. Mini-Transats were (and still are) extremely wide boats with amazing form stability, a huge amount of sail and a planing hull. Right from their inception, radical solutions were developed for these extreme racers: water ballast-tanks that offered a vital contribution to stability, along with twin rudders and canting keels. Since then, the Mini-Transat (and other designs for solo races like the Vendée Globe) acted as templates for putting increasingly extreme design solutions to the test, becoming the precursors of today's acrobatic sailing.

Interestingly, the continuing transformation of yacht design did not occur through a succession of revolutions, each time starting all anew. As naval architecture has evolved, if anything, it has followed the original meaning of the Latin verb *invenire*: to find, or come upon. Progress in this sector is a succession of interrelated improvements leading to an ongoing transfer of knowledge, generating never-ending adaptations of old solutions to new technologies. This is true of all fields in which technology plays a key role. We all know that Leonardo da Vinci invented the helicopter, but it

[32] Stefano Roberti, *La vela di domani*, "Giornale della Vela", no. 4, April 1989, p. 37.
[33] In the early '80s, the Open Class boats designed for the Centomiglia by Bruce Farr, *Grifo* and *Farrneticante*, caused quite a stir. When designing them, the New Zealander drew inspiration from the 18-footers with which he had begun his career.
[34] It is worth remembering how successful Ultimate 30s – the small acrobatic catamarans that were the forerunners of today's acrobatic sailing – were in the U.S. in the late '80s.
[35] The first rigid wing sail would be adopted by the Australian Class C catamaran *Miss Nylex*, which won the Little America's Cup in 1972. Subsequent to this, Tony di Mauro and Dave Hubbard's team perfected the system in *Patient Lady*, which won five Little America's Cups. See Franco Pivoli, *Viva la rigid wing sail*, "Giornale della Vela", no. 4, April 1989.

couldn't be built until the right technologies and materials came along to make it possible for something heavier than air to fly.

A similar principle applies to sailing craft. Returning to *Stars & Stripes*'s 1998 wing-sail, this was hardly a new idea: decades before Australians had made small wing-sailed multihulls. The new development was to make the rigid sails large enough, and this was possible primarily because of materials like carbon and Kevlar, not to mention the huge budget the *Stars & Stripes* consortium had at its disposal.

Mainly for media-related reasons, what may be termed a "relative" innovation was glorified as an "absolute" innovation. The same is happening again today with foils: the earliest patents and models for foils date back to vessels that plied the seas nearly a hundred years ago.[36]

Another key factor in development is economics. Since time immemorial, yacht design has striven for the greatest lightness allied to the greatest strength. This inevitably raises costs: every kilo saved comes at an ever increasing price. But production costs begin to fall only when demand reaches a certain threshold. For example, carbon-fiber is more frequently used today because rising demand has, over time, reduced production costs.

One may legitimately wonder whether sailing as a whole generates its independent technological development, or whether it simply borrows from other sectors. Put another way, is this transfer of techniques and technologies merely incoming, or does sailing also export development to other sectors? The answer is somewhat complicated. The direction of technology transfer is inwards, as may be seen if we consider the development of the underwater appendages in naval architecture, which *de facto* evolved out of aeronautics research. It is worth remembering that initial research for *Australia II*'s famous keel was done in the US by NASA, then developed by a Dutch engineer, to be fine-tuned in the towing tank at Wageningen.

Yacht design has, nevertheless, offered opportunities to experiment more freely than aeronautical research. Aircraft designers must stay within very strict safety coefficients when dimensioning their structures. Far lighter, less complex wings could be manufactured than those in use today, but for a good reason they are prohibited by regulations.

Consequently, in the freer environment of boatbuilding, particularly sailboat construction, it has been possible to experiment with tangible solutions that aeronautics could only investigate theoretically. These solutions, some of which have rebounded from boatbuilding to other sectors, have triggered an ever more integrated and complex interaction of importing and exporting experiences. One such example is composite materials, for which, perhaps for the first time, boatbuilding was the trailblazing sector.

Let's now return to the '90s. The sailing world had, from the end of the previous decade, embarked on a process of liberalization – much like the rest of the economy – and the IOR had gone into something of a tailspin. So back in 1985, the ORC had responded to the early stages of this crisis by introducing a new rating rule: the IMS (International Measurement System).

The IMS was based on a system developed by researchers at MIT in Boston that made it possible for boats of different dimensions to race against one another. This new system for measuring boats replaced the individual point system as used by the IOR with an overall measurement of the hull and its appendages. Special instruments (later replaced by laser scanners) determined this measurement by recreating the construction plan section by section.

The second new development brought in by the new rating rule was the adoption of VPPs (Velocity Prediction Programs). These computer programs, simulating a boat's performance, had been available for some time by then; we used a VPP when designing the very first *Azzurra*, and the programs had become more reliable since then. Under the IMS, the VPPs became the

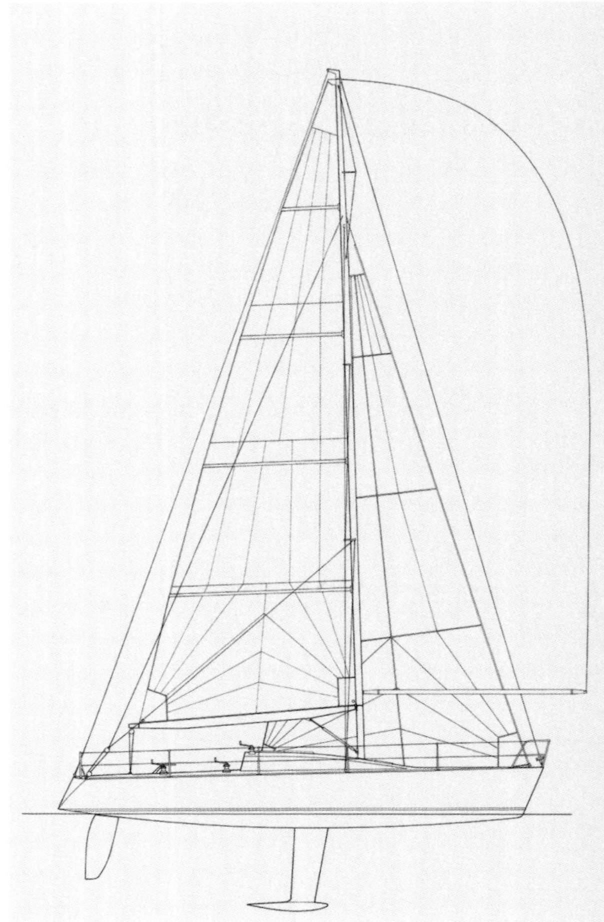

Brancaleone Open was one of the first Open classes designed by Studio Vallicelli: a boat with wide stern sections, a discreetly straight form, reduced wetted surface and relatively slender overall.

[36] To get an idea of the importance of foils applied to sailboats, it is worth recalling the experiments run on *Nancy*, a model catamaran with a 1.5 meter foil built by H.G. Bell and Casey Baldwin in 1913; in 1920, Malcolm McIntyre's patent for a hull with lateral foils; and in 1938, the first experiments carried out for the NACA on small boats by Robert Rowe Gilruth and Bill Carl in the U.S. These experiments would resume after World War II, paving the way for hydrofoils.

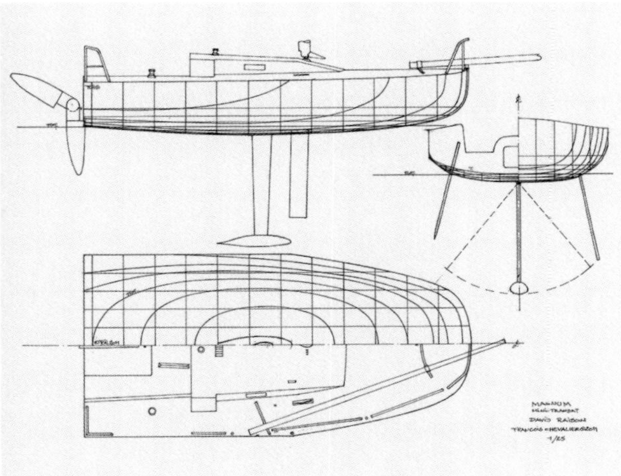

A Mini-Transat from the '90s. The Mini 6.50s are six and a half meters in length and were created for the solo race of the same name, ongoing since 1977. They have always been extreme boats: very beamy, with a huge sail plan and easy to bring onto a plane, they were among the first boats to be equipped with water ballast and canting keels.

basis for working out handicaps and relative allowances. To get some idea of how complex the IMS was, even wind and sea conditions during a race were input factors in computing the handicaps.

The IMS was a decidedly more ambitious measurement rule than the IOR. Its philosophy, based on VPPs calculating the theoretical speeds of boats in different operating conditions, at varying wind, and with real wind angles, has continued to be refined ever since.

Thirty years after it was initially drafted, the IMS has since changed its name to ORC International. It is under permanent review by the International Committee (ITC), chaired by Nicola Sironi from 1986 to 1996, and since 2007 with Alessandro Nazareth at its helm. The numbers show that, with 10,000 certificates issued each year, the ORC is the most wide-spread rating rule in the world.

It is worth mentioning that the greater complexity and structural depth of the IMS has, over the years, resulted in better boats than under the IOR: less extreme at their measurement points, more ergonomic, easier to live with and safer too. This does not, however, mean that the IMS has been a paragon of perfection. After all, measurement rules are like electoral laws, they can always be perfected and improved.

Returning to the period when the new measurement rule came in to use, the arguments were so heated that a split took place that has never fully healed. The RORC of England and UNCL of France responded to the IMS by bringing out their own new measurement rule, the CHS (Channel Handicap System, which in 2000 changed its name to IRC, International Rating Certificate), a much simpler rule whose formulation is secret. With the IMS and CHS being adopted at practically the same time, it seemed as if the yacht racing world had returned to the status quo that preceded the IOR rating rule, when there were two sets of measurements and rating rules, one for the US and one for Europe.[37]

After the IMS was adopted, the first designs strived to optimally interpret the new formula. In general, this meant a less extreme ratio between overall length and width, and a higher prismatic coefficient. These designs also sported reduced bow and stern overhangs and increasingly vertical sides, creating what tended to be U-shaped transverse sections. Without doubt, the new boats were more seaworthy: they heeled and pitched less, in part thanks to the weight-loss effect of technological developments in materials, and part to better distributed volumes.

Over time, as has always been the case, designers began to find loopholes in the new measurement rule. By the late '90s, these loopholes led to hulls with exaggerated freeboard whose sides were so straight they seemed practically vertical. In consequence, the "balance of form" sought by adopting a new formula once again fell by the wayside.

However, the biggest critical issue of all with the IMS was its huge and abstract complexity: racers no longer had the "physical" sensation about how they were doing in the standings – a sensation that relied on a single handicap number, that the IOR, despite its shortcomings, had managed to retain.

For America's Cup, the long season of the 12's came to an end in 1987, when the Americans took the Cup back to their country. After the spurious 1988 challenge between the immense Kiwi monohull (36.57 meters long) and the agile American catamaran (18 meters), which the catamaran won with ease, in catering to media demand for greater entertainment and to incorporate recent technological developments, for 1992 the decision was taken to adopt a new measurement rule, the IACC (International America's Cup Class). This rule remained in place until 2007.[38]

The 1992 America's Cup was simply the *Moro di Venezia* for Italians, the yacht of Raoul Gardini's Montedison company; a watershed in the evolution of national interest in sailing. Prior to the '90s, Gardini had owned a number of successful yachts. After his first boat, a series Orca 43 designed by Dick Carter, in 1973,

[37] To simplify the system and make it more ergonomic, in 1995 the ORC decided to redefine the Level Classes, reducing them from eight to four: 30, 40, 46-footers and Maxi's.

[38] The new measurement rule was adopted in 1988 in Southampton, at a plenary meeting attended by thirty or so designers, rating rule officials and event organizers.

Gardini commissioned another Carter-designed boat, *Naif*, a 45-footer built from wood by Carlini, which raced in the Admiral's Cup that year. In 1976, he commissioned Italy's first Maxi yacht, *Moro di Venezia*, designed by Germán Frers. This was followed in 1980 by *Moro Blu*, an ill-starred Doug Peterson two-tonner, followed in 1983 and 1987 by two more Maxis designed by Frers, winners of a slew of circuit races.[39]

In 1989, when the IACC measurement rule was confirmed, Gardini was ready to try his hand at America's Cup. He organized everything through the Montedison company (at that time, before its ill-fated deal with ENI, in rapid expansion). The company's first move was to set up Tencara, its in-house, leading-edge boatyard at Porto Marghera. Constructed by Technimont, another major Italian engineering firm in the Montedison stable, Tencara went on to build two large-sized yachts to our designs, *Orsa Maggiore* and *Virtuelle*.[40]

Tencara fully exploited Montedison's knowledge and equipment, and no expense was spared. It was one of the first boatyards in Italy to use the recently-developed pre-preg system on a massive scale; it was the first to use numerically-controlled machines, and the first to install a variable dimension oven, essential for post-curing yachts after lamination with pre-preg, considerably improving their lightweight credentials and robustness.

All five of the *Moro di Venezia* built for America's Cup were therefore made by a boatyard that had transformed artisan yacht construction into a high-tech, semi-industrial technology. Gardini's ambitions extended to international sailing too: he didn't just want to win the Cup, or just make it into the finals; he wanted Italy to become a leader in the yacht racing world.

Once Tencara was up and running, Gardini's next move was to launch a takeover bid for IMI (International Marine Industries), a US multi-asset company that in 1990 owned the finest brands in world sailing: mast-maker Sparcraft, winch-maker Lewmar, leading hydraulic system manufacturer Navtec, and Brookes & Gatehouse, the top nautical instrumentation maker of the day. He would also go on to buy the British companies SP System and SP Technology, specialized in composite design and engineering, with which we had worked a few years earlier when building our one-tonner *Merope*.

Gardini's chosen designer for the various *Moro* was Germán Frers, who moved his whole office to Italy. Frers had no intention of going it alone: in their headquarters at Castellanza he assembled and coordinated a diverse team of designers. The times were changing, and Gardini and Frers were well aware of this fact. After the revolution of *Australia II*, designing for America's Cup had ceased to be the preserve of a single architect. It had become such a broad and complex enterprise that a team was now necessary, in which each member was in charge of a specific sector. From hull shape to appendages, structures to deck layout, and rigging, each sector had become so specialized as to have a life of its own.

The Este 39 was one of the first IMS designs by Studio Vallicelli in the early '90s. Compared to IOR designs, IMS yachts had a less extreme length to beam ratio and a higher prismatic coefficient. The Este 39 won the Copa del Rey and finished second in the IMS European Championship.

[39] www.morodivenezia.it. For more on what happened at the 1992 America's Cup, Beppe Donazan, *Il Moro di Venezia e il sogno di America's Cup*, Edizioni Mare Verticale, Grancona (Vicenza) 2016; Antonio Vettese, *America's Cup. San Diego '92 la leggenda cambia volto*, photos by Franco Pace, Calderini, Bologna 1992; Luca Bontempelli, *Da "Azzurra" al "Moro". Storia di dieci anni di partecipazione italiana alla America's Cup*, Longanesi & C., Milan 1992.
[40] During and after construction of the five *Moro di Venezia* yachts, around 130 specialized workers were employed at Tencara, divided into two departments: one conducting research into materials and resins, studying the chemical and physical side of materials, the other actually building the yachts.

Moro di Venezia, the Italian America's Cup contender, designed by Germán Frers and a large group of designers, was built at the Tencara yard. In 1992 she won the Louis Vuitton Cup but lost America's Cup in the final against the Americans in *America Cube*.

The designer-in-chief, in this case, Frers, was responsible for coordinating the various sectors. This specialization is the reason why, from the '90s onwards, we tend not to remember who designed a yacht, particularly in America's Cup, as if the team had prevailed over the author.[41]

The rest, as they say, is history. After winning the IACC World Championship in 1991, in 1992, *Moro* won the prestigious Louis Vuitton Cup, the challenger selection races, bettering Italy's previous record finished when *Azzurra* had reached the semi-final in 1983.

In America's Cup, *Moro* lost against the very fast *America Cube*, designed by the talented Doug Peterson and his team. The difference in speed between the two boats was considerable: as a wide boat with rounded sections, *Moro*'s design essentially revolved around form stability, whereas *America Cube* was, to the contrary, a narrower boat. Indeed, with more V-shaped hull sections, she was almost canoe-bodied, in some ways resembling the old 12 Meter designs. *America Cube*, with her high block coefficient, provided a template for the next generation of IACC designs, and whose influence would also give thinner hulls to the IMS.

Moro's epilogue was conditioned by the unfolding events in Gardini's private life. The year after his exceptional America's Cup exploits, Gardini was implicated in Italy's "Clean Hands" corruption inquiry and took his own life. Tencara carried on a few years, but before long, ownership of the many companies acquired by the Montedison Group returned to non-Italian hands.

Although Gardini's death seemed to spell the end of Italy's dream to become a sailing titan, this only partially turned out to be so. In the late '90s, *Moro* passed the baton to *Luna Rossa*. Interestingly, the *Luna Rossa* experience led to the foundation and development of what today is one of the world's most famous boatyards working in composites: Persico Marine.

Moro di Venezia coincided more or less with the end of the first stage in an IT revolution that had begun ten years earlier. Early experiences were, as is so often the case in technical and scientific revolutions, negative. After his second triumphant re-design of *Intrepid* for the 1970 America's Cup, Britton Chance Jr. (probably the most brilliant and unpredictable boat designer of the '60s and '70s) used the first ever IT tools for boat design to design a yacht for the following edition of America's Cup. The result, *Mariner*, looked completely different from all the other 12 Meters. *Mariner* had a canoe body with extreme lines to create discontinuities at the girth measurement points. Forward, below the waterline, the bow had a well defined knee with an embryonic bulb, a daring solution that would surface again in America's Cup in the early 2000s. Astern, the skeg, was at one with the abruptly cut-off bustle, with the rudder fully hung from this strange offshoot to the hull. The result of all these computer processing revolutions was a boat that was vastly slower than the classic 12 Meters that Olin Stephens had designed on intuition and experience alone. And yet despite her stunning slowness, *Mariner* offered a glimpse of an ever-more technological, ever-less intuitive future.

[41] The huge *Moro di Venezia* design team was organized as follows: Germán Frers, head of design; technologies and research: Robert Hopkins Jr. and Fernando Sena; head of operations, Laurent Esquier; design team: Giovanni Belgrano, Micky Costa, Nestor Fourcade, Debora Gallazzi, Claudio Maletto, Raffaele Marazzi, Leopoldo Murcho, Francesco Ricci and Oliver Yates; design technologies and research: Roberto Biscontini, Roberto Vergara, Walter Michelini, Mario Caponetto, Mark Downey, and Lars Larsson; mast designer, Chris Mitchell; sail designer, Guido Cavalazzi; and applied sail research, Michael Richelsen.

The Digital Revolution in Naval Architecture

In the early '70s, when I was visiting the offices of Jean Marie Finot, the eccentric designer of *Revolution* and the Comet 910, I saw him struggling with an early computer running one of the first modeling programs imported into the boating world from the car industry. This was the very first sight I had of a design office. Among the flexible battens, lead ducks, pencils and tracing paper used for centuries to draw the lines for boats, I got a glimpse of an alternative future – a future that was already being ushered in during the '70s.[42]

First, we need to take a step back. In the past, the classical method for designing a boat was based essentially on a procedure whereby, first of all, the hull was designed through a flat representation of the three families of curves that defined its overall volume.[43] Downstream from these two-dimensional drawings, known as the construction plans, measurements were made using special instruments to obtain the numerical data to calculate physical sizes, i.e., the volumes, centers of gravity, moments of inertia and so on.

CAD (Computer Aided Design) systems first appeared in mechanical engineering industries as far back as the '50s. When these systems first arrived in yacht design, they triggered a series of changes whose effects are still being felt to this day. In the early '70s, the main CAD programs were Unisurf and CATIA, developed by the Dassault aeronautics company together with IBM. Although these programs could quite easily create a three-dimensional drawing of a vessel, supplying the necessary numbers in real time, their diffusion was hindered by their very high costs – actually, prohibitive for a small-to-medium-sized practice like ours designing pleasure boats.

That all changed in the early '80s when personal computers became more powerful and affordable. Not only could they easily run CAD programs, they could also run CAM (Computer Aided Manufacturing) programs, whose software made hull modeling more straightforward and effective with CNC machines. Precisely because these programs were more accessible, they gradually began to simulate the characteristics and behavior of manual drawing instruments such as listels, plumb lines and splines.

At this point, we must draw a clear distinction between the various 3D programs for surface modeling and those for solid modeling. The former operates on the surface by configuring reticular frames in space; the latter construct an object starting from basic elements such as regular solids (in a process known as fairing). These programs are based on parametric functions (the most common perhaps being Bézier and B-splines), which generate the design of the vessel. This may be modified by the control points, or nodes, which taken together produce a grid that envelopes

[42] It is interesting to note that a computer program developed by National Aerospace of Amsterdam was used to fine-tune *Australia II*. This was one of the first programs to define and display the effects of a vessel's movement on the water, including the vortices caused by appendages and pressure along the hull surface. This final piece of data, obtained through sea-borne measurements of actual stress on the vessel, would enable structural design to take a huge leap forward.

[43] "The hull generally has a geometric form that is difficult to describe analytically. In traditional practice, not being able to rely on the tools offered by mathematics at all times, definition of this aspect required a laborious technique of graphical representation, along with methods of discretization offered by numerical analysis for calculating parameters regarding the bottom (volume, stability, etc.). So-called lines or body plans conventionally trace the vessel's forms, 'out of plating' if the construction is made of metal, or 'out of planking' if made of wood or composite material. As classically represented, body plans combine orthogonal projections from at least three families of curves, obtained by sectioning the solid into an equal number of parallel plane families lying along the corresponding principal plane, identified by reference to the orthonormal triad's axes... A construction drawing is considered to be technically well-executed when it is 'underway' and 'balanced', that is to say the entire surface is described by lines that have no unwanted discontinuities in their curvature (or 'singularities'), and when their representation is correct and unequivocal in the various orthogonal projections." Andrea Vallicelli, under "Yacht Design" in the *Enciclopedia Treccani*, p. 497.

[44] "Bézier surfaces and B-splines share the property of being contained within their own concavity: with the exception of marginal points, all of the control points are external to the surface." Andrea Vallicelli and

Andrea Vallicelli at the Apple station in his studio at Via Vittoria in 1982.

the hull.[44] A special function (blending) exists for every node, making spatial modification possible simply by moving the node, prompting the program to regenerate the drawing of the vessel as a whole.[45]

A second technique for generating a hull design is to start with pre-selected dimensional data such as length, width, displacement, etc. As the designer varies the position of the ends of the resulting grid in space, using these new parameters, the program automatically works out the surface best suited to the new configuration.

For a long time, many software packages used chromatic scales to offer an option for analyzing the hull's curvature progression and used shading to display the physics of the same as clearly as possible.

The impact of assisted design cannot be understood without putting it into relation with the more or less simultaneous development of hull performance computational analysis, those VPPs (Velocity Prediction Programs) we mentioned earlier when we were looking at the IMS rating rule and which, as we saw previously, are based on experimental data drawn from a variety of tests.

Usually, the hydro-dynamic component refers to the results of tests conducted in a towing tank, based on systematically varying the key parameters that govern resistance to motion. On the aerodynamic side, the data is based on results obtained from wind tunnel experiments.

Then there is what we may call the designer's library, that is to say, the pool of data each designer has assembled over the years from testing in tanks and on the water. Often jealously guarded, this data is constantly being reprocessed and improved. If we look back to *Azzurra II*, it was the lack of reliability of this data that prompted Peter Van Oossanen and ourselves to bark up the wrong tree – a mistake for which we paid dearly.

Another aspect that has undergone enormous development since the '90s is CFD (Computational Fluid Dynamics), which uses mathematical codes to predict the behavior of fluids around a solid: in this case, hulls and sails. This popular and accessible type of digital measurement has gradually been integrated with experimental tests on scale models in towing tanks and wind tunnels. In some environments, such as aeronautics, and now in sailboat design, CFD results have begun to supersede these other instruments.[46]

Overall, the structural design has advanced considerably as a result of the digital revolution, as it became easier to record reliable data from actual events (in particular, structural loads) and view that data via graphic software. Today, it takes less and less time to model and optimize a vessel's structure and materials.

It is also worth noting that advanced assisted design, VPPs, CFD and structure calculation programs (FEM, Finite Element Models) were already in use at some design houses in the early '80s, back in the days of *Azzurra*. With every year that goes by, these programs perform better, working to higher standards of operational precision and becoming more user-friendly and less costly.[47]

The whole raft of changes ushered in by IT has brought about a veritable revolution. The classic design route, based on tracing out a vessel's first lines by experience and intuition, and only then starting to make calculations, with the first check of the whole

Stefano Orlando, *Il disegno assistito da tecniche informatiche nella progettazione di uno hull*, "Disegno. Idee e immagini", no. 3, 1991.

[45] "In other words, shifting a node modifies the overall form, even though the effect diminishes in regions that are more distant. This typical characteristic of the Bézier curve is known as global control. Vice-versa, the blending function of B-splines, which have a more complex formulation, manages to circumscribe the effects of the modification to permit local control." Andrea Vallicelli and Stefano Orlando, *Il disegno assistito da tecniche informatiche nella progettazione di uno hull*, "Disegno. Idee e immagini", no. 3, 1991.

[46] "Thanks to progressively falling computational costs as computer technology evolved, in recent years CFD has undergone significant advancement in the field of marine hydrodynamics, becoming an important complementary and even replacement tool for statistical and experimental methods. During design research a number of different methods are chosen depending on their ability to approximate physical phenomenon, and on their ease of use. One of the most commonly used is BEM (Boundary Element Method), which is more simply known as the 'panels method'. The solution here is based on potential non-viscous, irrotational flow, in which speed may be expressed as a gradient of a scalar function known as 'potential'. Its main advantage is rapidity of calculation: computational dominance is simple to discretize and renders preprocessing (preparing for numerical calculations) very easy. Its limitation is in modeling fluxes in which viscous effects have an impact, such as the stern area of the hull. BEM may successfully also be used for analyzing seakeeping-related phenomena, which requires significant calculation. One of the most advanced sectors of naval architecture, seakeeping is the study of swell and how hulls behave subject to it. To understand this requires starting from probabilistic and not just deterministic criteria, and then building

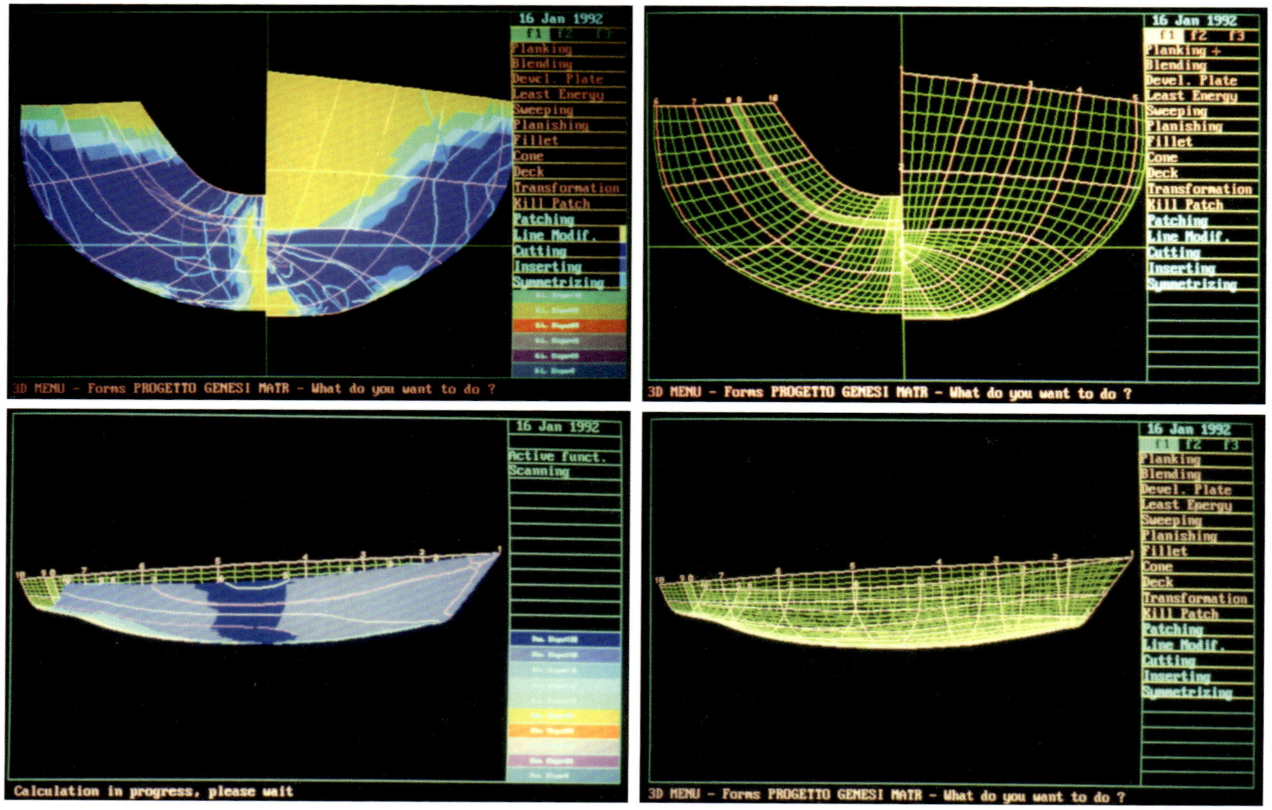

The Genesis project for Comar was developed in the late '80s by Studio Vallicelli with the extensive use of digital technologies.

thing at sea or, in more fortunate cases, with tank-testing before construction, has fallen by the wayside. Today, what we may define as the digital design has limited intuition even further, and relies more and more on data and the immediate display of data effects on hulls. We have moved from a design path based on verification to one that we may describe as a spiral: data obtained analytically (from tank tests and on the water from similar vessels), is compared and updated, and then compared and updated again, in a continual process that, given its fragmentation into specializations, can no longer be managed by a sole designer, but requires an integrated team.

IT has not only enabled the exponential increase of available data, it has done so alongside an equally exponential increase in data processing speed. Once project data has been entered into the computer, it is almost instantly possible to access a more or less infinite number of versions. The scope of this possibility makes it harder and harder to settle on a final configuration. The ability to intuitively synthesize experience in designing a hull – something that was once considered a designer's finest quality – has been replaced by the ability to manage data to choose a final configuration.

This does not mean that designing fast and efficient hulls is easier today than it was yesterday.

highly complex oscillatory models. To resolve a viscous flux, RANS (Reynolds Averaged Navier-Stokes) codes are used. These are based on a solution of Navier-Stokes equations developed by Reynolds, which are capable of describing a flux, whether it be laminar or turbulent. These methods have only recently been applied to hydrodynamic hull analysis. In the past, their use has been rather limited as a result of difficulties in generating the mesh that, because it refers to complex geometries, required long calculation times... Another area in which RANS methods are often applied is studying sails, in particular downwind sails which generate forces that to a large extent may be ascribed to flux separation, creating a pressure that differs from the upwind portion. Given that these phenomena are dominated by viscous effects, these methods are useful for studying the point of separation and analyzing the consequences of the turbulence generated." Andrea Vallicelli, under "Yacht Design" in the *Enciclopedia Treccani*, pp. 499-500.

[47] At the Studio Vallicelli, in the '80s we never used CATIA, just CIRCE, derived from CATIA to run on less powerful mainframes. It should be noted that CATIA had been developed for the aeronautics sector. Conceived by Pommelet, CIRCE was a program that built surfaces using transverse and longitudinal frames. The control points were placed at their intersections, which remained inside the generated 3D surface. Max Surf, another program, had control points external to the surface. Although it was more user-friendly, it wasn't as accurate as CIRCE.

Open Classes and Fast Cruisers

Riviera di Rimini may be considered one of the most successful designs by Studio Vallicelli. She was originally an Open 55 which, following modifications, was transformed into an Open 60. She won the Middle Sea Race, the Barcolana, the Giraglia, the Rimini-Corfu-Rimini, the Roma per Tutti and the Roma per Due.

We have a long history with the Open Classes at Studio Vallicelli. We have always believed they represent a pure state of evolution in naval architecture, without the specific regulation-related restrictions imposed by measurement rules. We are interested in them because they are at the vanguard in yacht design.

Open Class development truly took off in Italy in the '80s, when Bruce Farr designed two captivating boats, *Grifo* and *Farrneticante*, for the Centomiglia on Lake Garda, drawing inspiration from the Australian 18 Footers with which he began his career. Although these designs brought nothing completely new to the fore, they did leverage the artisan approach that until then had characterized Centomiglia boats, transforming it into leading-edge naval architecture.

Our first Open design was *Stradivaria*, a 1992 yacht in the least extreme class (Class 0, without racks or trapezes), which Soleri built for us from composite materials. She did very well, winning three Centomiglia races in class (1992, 1993 and 1994). *Stradivaria* was followed by *Gaia*[3], *Gaia Legend*, *Amor*, and finally, *Riviera di Rimini*, one of our most successful designs.

Here too, our practice's design philosophy eschewed the extreme: for us, it was about bringing into balance all of the various different factors. Our Opens were not exaggeratedly wide and flat, like the *Flying Dutchman*; instead, they had a low metacentric height, which meant that they did not exceed in form stability and retained a reasonable block coefficient. To ensure sufficient stability, we did not rely on beam or large immersed volumes. Instead, we chose to shape the topsides to harmoniously unite the racks with the hull: as well as the greater stability generated by the weight of the crew, the flared sections offered the option of housing water ballast tanks, further increasing righting moment.

The Open Class world underwent a period of change in the early '90s; its success meant that races were no longer confined to inland lakes, but were increasingly held in open waters. In addition to the historic Middle Sea Race and the Barcolana, now there were the prestigious Rimini-Corfu-Rimini, the 500 per Due, the 500 per Tutti and others.

Our non-extreme design approach attracted new demand for seaworthy Open Classes; a sector in which, given our past experience, we had plenty to commend us. *Riviera di Rimini* was the most impressive result we achieved during this period. Conceived by a group of businessmen to promote the Romagna Riviera, the boat belonged to the Open 55 class. Its architecture revolved around a handful of parameters: a high block coefficient (compared to boats from the same category at that time), a very narrow bow entry angle, reduced wetted surface, and a rather limited displacement (5.8

tonnes). *Riviera di Rimini* also had flared cross sections to increase the righting moment generated by crew weight and water ballast. Aft, the hull became wider and the bottom flatter to encourage planing when reaching and running. The sail plan was flexible: the 85% fractional rig offered the option of hoisting a large genoa to the masthead for light winds, and a bowsprit for the gennaker. The appendages consisted of a deep, narrow keel fin of steel with a lead bulb, a high aspect ratio rudder, and a small canard (a small centerboard) forward of the mast.

In 1999, to allow the boat to compete in the Fastnet and the Tour de l'Europe, *Riviera di Rimini* went back to the boatyard. Lengthened by a meter and a half, she became an Open 60, a class conceived for solo ocean racing with plenty of sail and great form stability. The boat's lateral stability was improved by increasing the weight of the keel bulb; the underwater plane was made more effective by replacing the small canard from the first version with a pair of asymmetric-section canards, further astern and with a greater surface area.

The story of *Riviera di Rimini*, whose magnificent management was comparable to Pasquale Landolfi's management of *Brava*, was a litany of distinguished achievements. In 1998, the yacht won the Middle Sea Race, the Barcolana, the Giraglia and the Rimini-Corfu-Rimini; in 1999 a second in the Fastnet Race and the Tour de l'Europe, once again winning the Rimini-Corfu-Rimini; in 2000, she won the Roma per Tutti, and then the following year the Roma per Due. The yacht's winning times in many of these races set new records that wouldn't be bettered for years, when a new generation of longer Open class boats appeared, twenty meter craft such as the stellar *Alfa Romeo*, a 90-footer from Reichel & Pugh that was the class benchmark for several years following her 2002 launch.

Riviera di Rimini offers an example of how sailing changed during the '90s. The changes were driven by demand, which continued to expand at a breakneck pace. So as demand and supply both changed, they

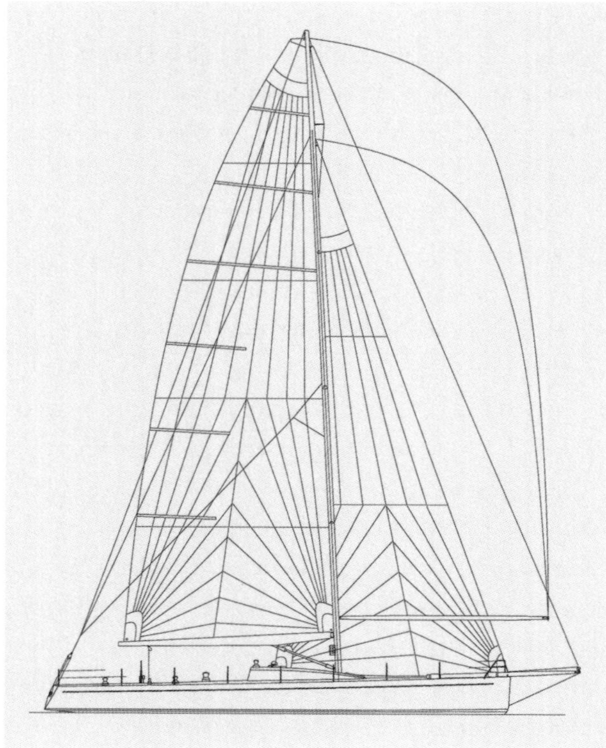

The design characteristics of *Riviera di Rimini* were a high block coefficient, a very narrow bow entry, reduced wetted surface, and wineglass shaped cross sections, all within a limited displacement.

Walligator II, the forerunner of today's large, fast cruisers, was a project by Luca Brenta for Wally. Innovative for her streamlined lines and high-tech equipment, she was fitted with electro-hydraulic control systems that tremendously facilitated sailing the boat.

met in the middle with an ever more varied set of solutions no longer confined to just one category, as had been the case in the past with the IOR.

The winds of change were also felt in the large yacht sector. We had an inkling of this in 1994, with the launch of a ground-breaking vessel: *Wallygator II*, a 105' ketch designed by Luca Brenta, who in the late '80s had designed the successful 43' one-tonner *Marisa*.

Wallygator II stood out for her extremely clean and streamlined lines, achieved by not worrying too much about the reasons underlying the rating rule. Although not expressly conceived for racing, *Wallygator II* imported a number of solutions from racing yachts for both her architecture and her hull form; moreover, like more extreme designs, she was built completely out of carbon, including her masts. However, the boat's most exciting new development was that sail handling was made as automated as possible, using electro-hydraulic servo mechanisms on winches and hydraulics. The idea of making sail handling as user-friendly as possible was

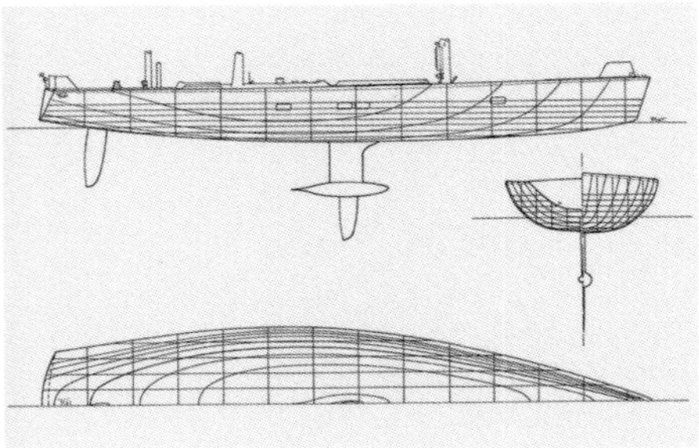

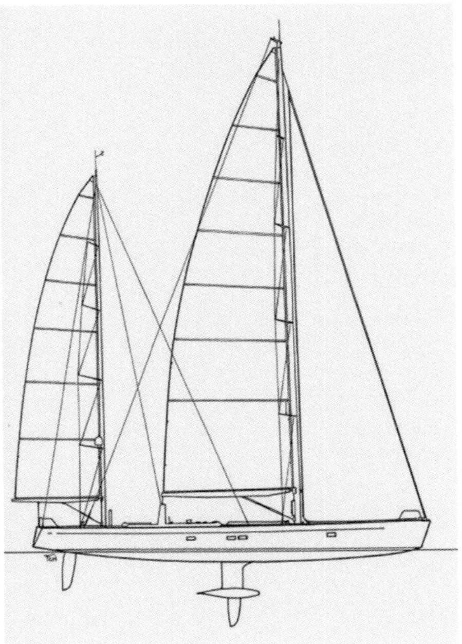

evident in the choice of a self-tacking jib and swept back spreaders. The resulting rig could be sailed with a bare-bones crew, and a yacht that had no need for a big crew was a boon not just to owners' bank balance, but to life on board.

As one of the first yachts of its kind, the technology on *Wallygator II* – indeed, on all of the later Wallys – simplified operation enormously. Building on *Wallygator II*'s success, Wallys started to focus more aggressively on style, espousing a self-declared minimalism that remains in vogue today. The yacht looks like a hugely upscaled dinghy; a strange effect, yet undeniably seductive. The boat's minimalism is emphasized by the deck plan, where all the lines are hidden under protective carters, as little in evidence as possible. The approach is reiterated through interiors that echo this dry, essential style, characterized by large undecorated *partitura*, with a few details elevated to pure elements of design, typical of the London approach.

This minimalism contrasts with the vast size of these boats, unthinkable just a few years earlier, harking back to the glories of the immense J-Class. One comparison is significant: in the early '80s, the Swan 65 (yes, a 65-footer) was the queen of big yachts. By the late '90s, yachts had grown to 100' or more. This impressive change in scale took just twenty years, in times of burgeoning earnings for the financial sector in an ever-more globalized world. It's no coincidence that the beneficiaries of these earnings were the new clients for these yachts. It is also interesting to note that with Wally's design based on the unification of style for the boat and all of its components found its full expression with a sailing yacht. The effects of this would, in the future, trickle down to other decidedly smaller boats.[48]

In the late '90s, we were asked to design a Wally-type boat, a category that would soon earn the evocative name of "fast cruisers" by Carlo Perrone. (He had given us the commission for *Dioscura* in the late '80s: a lovely IOR yacht, fast but not specifically designed for racing, we had her built in cold-molded wood at the highly skilled Bani boatyard at Monte Argentario.) It was 1997, the same year as our *Riviera di Rimini* when we launched Perrone's new yacht. *Virtuelle*, 24 meters long, was built by Tencara, the *Moro di Venezia* boatyard, which unfortunately closed down soon after her launch.

When we began work on the design of *Virtuelle*, we proposed something similar to *Orsa Maggiore*, where for the first time, we had inserted panoramic windows around the coachroof to maximize light per square meter. At our meetings with the owner, the choices we had to make ran the entire gamut from naval architecture to superstructures. He wanted the design to be more and more radical, to the point – at Perrone's suggestion – that we picked well-known designer Philippe Starck for the interiors.

Starck wasn't new to yachting, having collaborated with Beneteau on cruising interiors. We began working with him creating innovative solutions and noteworthy aesthetics. For instance, the cockpit's sides were rounded, a solution of great elastic impact. This effect was heightened by curved benches whose size could be adjusted, highlighting the fluidity inspired by the rounded forms. The result was an object in which every component blended together, creating a single shell seemingly cast in a foundry. Starck was also able to infuse *Virtuelle* with his hallmark trait of joyful and subtle irony. For example, he placed an abiding symbol of yachting, a circular porthole, on each side of the hull: using a vintage design offsets the yacht's avant-garde feel and mitigates an extreme purism of design that often makes yachts look over-polished or aseptic. Starck was also keen to play with semantic ambiguities. For example, the spokes of the twin helms are twisted, evoking both bicycles and airplane propellers. He specifically created an interesting fabric of cotton impregnated with a special reddish pigment to clad the interior, a color that reminds us of the first plastic of all, Bakelite, or to be more precise, Bakelite's simulation of mahogany.

[48] Designers have always worked with naval architects, many of whom have also been excellent designers. In the late '70s, Gae Aulenti created the graphic design for a fast two tonner, Carriero's *Viola*. At the Studio Vallicelli, in the same period we worked with Pierluigi Spadolini on two yachts he owned. In more recent years, new generation fast cruisers include the Wally 80's *Nahita*, *Aori* and *Wally B*, with input from Lazzarini and Pickering; Luca Brenta and Piero Lissoni's *Ghost*, *Esense* with interior design by Odile Decq, *Nina J*, on which Ivana Porfiri and Tommaso Spadolini worked, as well as other yachts on which well-known architects of the calibre of Norman Foster and David Chipperfield, or designers like Philippe Starck and Pininfarina, participated.

Moving on to *Virtuelle*'s naval architecture, although the hull was wider than a Wally's, it still maintained a respectable block coefficient. It also did not bow to the all-pervasive fashion of the IMS days for exaggeratedly straight topsides; on the contrary, her topsides were slightly rounded, to avoid the boxy look of many contemporary hulls.

Virtuelle also had two special features rarely found on fast cruisers, even today: the first was a twin rudder and helm system we imported from the BOCs which ensured excellent efficiency upwind (the leeward rudder is always fully immersed when the boat heels) and easier steering with a following breeze – this reduced strain on the wheels, which could consequently be geared down. Another new development that was just as important (if less showy) was the masthead rig, which remains unusual for a yacht of this type. The trend then – and indeed now – is for a fractional rig, giving a competitive advantage, though not necessarily enhanced efficiency. Among other things, fractional rigs on very large yachts have the disadvantage of huge mainsails, which are not only harder to handle but put the rig and fittings under great strain.

For *Virtuelle*'s keel, we opted for an almost butterfly-shaped profile with a central belly-like bulge. We changed this configuration for the next keel, which was lighter and with a higher aspect ratio for improved light wind performance.

Virtuelle was not only the yacht with which we went head-to-head with integrated design (for which *Wally* was the benchmark), it was also the project on which we fine-tuned the naval architecture choices we would be offering over the next ten years. Without a doubt, this yacht was the forerunner of a line that we may sum up as follows: volumes harmoniously distributed along the longitudinal axis, a limited wetted area, a high metacentric radius, and more rounded aft sections to ensure stability and effectiveness when reaching and running. This was the approach we followed when we designed the successful Comet series for the 2000s, and then

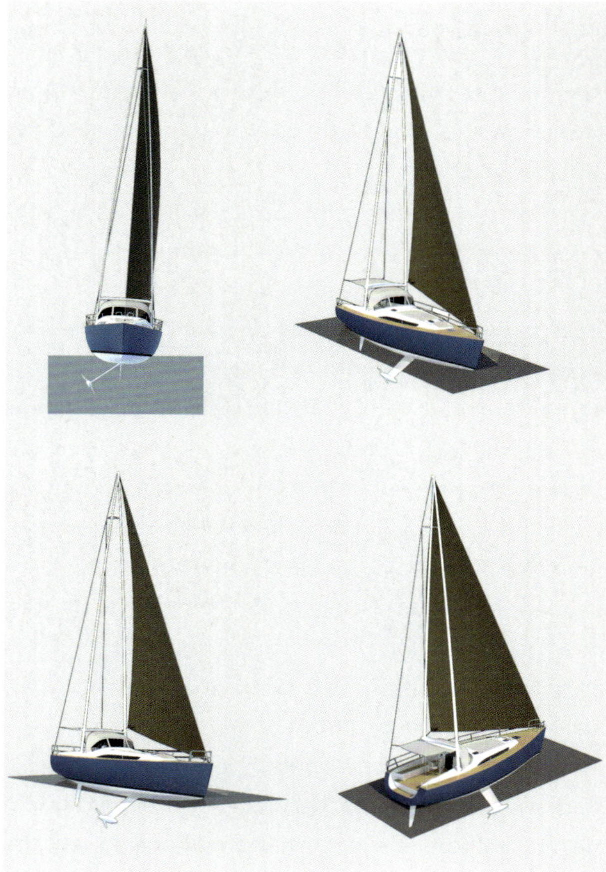

In the 2005 and 2006, together with the Polytechnic of Milan and the Faculty of Architecture of Pescara, Studio Vallicelli produced a prototype for power-assisted sail featuring an A-type mast placed well aft, with a large roller-furling genoa. Opposite in a design solution by Massimo Di Nicolantonio.

In the late 1990s, *Virtuelle* was the first fast cruiser designed by Studio Vallicelli. She introduced innovative solutions such as twin rudders and a masthead rig. For the design of the superstructures and interiors, the Studio worked with the well-known designer Philippe Starck.

Along with *Virtuelle*, *H_2O* marked a turning point for our studio: compared to previous boats, she had more harmonious volumes distributed along the longitudinal axis, less wetted surface, and rounder stern sections: solutions that would be adopted in subsequent designs.

again with our fast cruising yachts and other fast cruisers like *Wizard* and *H_2O*.

All of these yachts evoked *Virtuelle*'s sleek lines. They also featured an important new development with the deck saloon. This was a raised indoor area near the companionway, at an intermediate level between the cockpit and the interiors with the dinette and other amenities. This solution perfectly matched the fast cruiser spirit, combining the need to be as much in contact with the sea and boat as possible, yet with the greatest possible shelter and comfort. Comfortably sitting in *Wizard* or *H_2O*'s deck saloon, you have access to it all from a sheltered yet spectacular semi-enclosed space. These two yachts reaffirm our desire to simplify sail handling and make it increasingly accessible. On both of them, for example, all sail handling can be carried out via a battery of electric winches placed astern of the helm: they both adopt every possible device on the market to rationalize and simplify sailing.

Our interest in what has been defined as power-assisted sailing – something that, lest we forget, is only possible thanks to greater availability of materials like carbon, Kevlar, Spectra and others, as well as new electronic and hydraulic technologies – led Alessandro Nazareth and me to base our yacht designs on these radical new criteria.[49]

One example is a model we developed and subsequently patented in the 2000s, in partnership with the Mechanics Department at the Politecnico di Milano and the IDEA Department at the Pescara University's Faculty of Architecture. The roots of this boat may be traced all the way back to the vessels that sailed the Nile in ancient Egypt using an A-type mast, a rig in which the mast consists of not one but two poles stepped to the sides of the hull, with their tops joined together.[50] We sited the A-type mast well astern to hoist a single sail, a large genoa (or at most two) suited to sailing close to the wind. This so-called "primitive" rig was developed using contemporary materials and techniques to maximize its advantages: eliminating the mast's aerodynamic interference on the sails, disposing of the boom and mainsheet traveler, simplifying deck hardware and the structures anchoring the mast to the hull (reinforced frames, tie rods, counter keels, chainplates, etc.) and, lastly, increasing usable deck space.

[49] Experiments with which our practice has been involved: in the early '80s, we designed a windsurfer, and in the early 2010s *Gala*, a cabin-less boat around seven meters long conceived for mass production for match racing and sailing schools.

[50] "The historical origins of 'bipode' or A-shaped masts as we prefer to call them, date back to at least 2500-2000 BC. Documentary sources and portrayals of Egyptian river craft show the adoption of this type of mast, which puts less strain on hull robustness. Twin masts spread the compression loads onto the vessel's sides. With flat bottoms and no central keel, these boats were made from lightweight materials such as woven papyrus. In more recent times, in the pleasure boating sphere there have been a number of more or less original offerings taken to different degrees of experimental testing. One of the most interesting of these was *Amoco Procyon* (1971, USA), a vessel designed by O. Harken and B. Chance, built with an 'A' mast (patented as the Bi-POD mast and designed by E. Hall) and a central stay onto which the main sail was fastened like a jib. The Bi-POD mast reduced interference with windage of the spar on the leading edge of the sail, albeit without completely doing away with it. It should be noted that the idea for this mast type has cropped up every now and then over the years, for example in the '30s in designs by M. Curry, and in the design (N. and G. Siron) that won the 1971 Philips competition. Other experiments in the '70s, for example D. Carter's *Luna* or *Vendredi 13* projects, were built to a schooner-type sail plan with just jibs and no main sail. The perhaps more appealing idea of resolving aerodynamic issues by adopting an A-shaped mast on the boat's aft body, is

We tested our A-type rig by comparing it to traditional sloop rigs at the Milan Politecnico Wind Tunnel facility, with the help of the late Prof. Fabio Fossati. The resulting data was comforting, confirming that our model, with its twin genoa configuration, performed better than sloop rigs. This advantage increased considerably if a canting keel was added.[51] This was achieved despite critical issues regarding mast compression, as its position towards the stern tended to overload the backstay.[52]

Our patent is one of many examples of how today's pleasure boating is going through a very interesting period, as new requirements and new technologies push research to reformulate the principles on which boats of the past were designed and made. Does this mean that we stand on the verge of a season of radical design and a general rethinking of everything? Well, it's not exactly like that: as I mentioned, the forerunners of our A-rig date back to craft that sailed the Nile in 2500 BC, something that should give us food for thought not just about sailing, but about the future of design in general.

rather old. A number of pioneering experiments date back to the '50s, including C. O. Walker's *Marara* catamaran, along with a number of very similar patents." Maria Carola Morozzo della Rocca and Francesco Tiboni (eds.), *Atti del secondo convegno nazionale. Cultura navale e marittima transire mare*, September 22-23, 2016, Editore goWare, Florence 2017.

[51] Experimental research undertaken in the Politecnico di Milano's wind tunnel compared the performance of a traditional sloop-type sail plan and a simplified sail plan of the same dimensions, both in single and twin sail configurations. Two 1:10 scale models were built to carry out this comparative analysis, one for the sloop-rigged boat, the other for the "A" mast rigged boat (located astern). The sail plan of a traditional cruiser-racer was employed for the first model: the Comet 51s (designed by Vallicelli Yacht Design & Co., winner of the IMS 2006 Italian Championship). For the second model, a double configuration was designed for the rig: a single and a double jib (with and without overlap). The models were completed with spars, rigging and sails, and the hull cut above the waterline and mounted on an adjustable plate in the wind tunnel. Tests were conducted with the vessel in a vertical position. Furthermore, to obtain the fullest possible range upwind, measurements were also taken at apparent wind angles of 22, 27, 32 and 42 degrees. The correct size of the sails and optimal adjustment was calculated to achieve maximum efficiency for each of the four different solutions investigated (fractioned sloop, single jib, overlapping double jib, and non-overlapping double jib). Data from the tests clearly show that the simplified sail plan double jib (overlapping and non-overlapping) yields better performance than a traditional sloop sail plan. These results are presented in graphics from the VPPs drawn up on the data generated during tunnel testing.

[52] It is worth noting that the position of the mast, on the vessel's aft body, considerably reduces the backstay's leverage compared with a sloop. This prompts greater compression force for the A-shaped mast. This issue may, however, be attenuated by the fact that the poles under strain are not one but two, and may be sized with longitudinally lengthened sections which, if correctly designed, would also offer lower aerodynamic resistance. Furthermore, engineering-side experimentation could very quickly advance this type of mast, eliminating the disadvantage of greater compression loads. As we have had occasion to note, in the past there have been many attempts to reduce the problem of the mast's aerodynamic interference on the sails, both for sloops and multi-mast sail plans.

Boat Design Evolves: From Production to Design-Oriented Boats

The advertising page from the '70s for a high-class boat designed by Olin Stephens and built in Italy: the Alpa 11.50.

The Canados 33 and Canados 37 were production boats that perfectly represented the typical trend of the '70s, where cruising hulls were derived from racing designs: a trend made possible by the fact that racing boats were not as extreme as they would soon become in the following decade.

Over the last forty years, coinciding more or less with the life of our practice, cruising yachts have undergone a quantity of substantive changes comparable to racing yachts. The very term "cruisers" has been subject to a significant semantic shift over this period, and almost completely disappearing from the lexicon.

It is instructive to take a brief look at how these changes unfolded. In the late '70s, cruisers were a product of what we might term a double emulation: the fiberglass construction now used to make the majority of cruisers emulated the stylistic approach to wooden construction – not only did the earliest fiberglass boats mimic the shapes of wooden boats, the influence of the previous technique was also evident when, often without rhyme or reason, wood would be used to ennoble what were considered the boat's most valuable parts; in general, all of the superfluous elements were in some way affected by this legacy.

This kind of inertia is typical of design: for instance, for decades, cars mimicked horse-drawn carriages, often unconsciously. Throughout the history of design, the general rule has been that such inertia is overcome when, almost by magic, the public finally metabolizes new technologies and accepts their implications.

In the early '60s, the French were the first to understand the expressive potential of fiberglass construction and bring it to market. With extensive use of counter-molds, Michel Dufour, Jean Marie Finot and Philippe Harlé were the first to understand that a fiberglass boat could have it's own aesthetic, without any sterile emulation. With the boats they designed and built, they aimed to show that the typical polished, enveloping shell-like look of fiberglass construction had its beauty, to be exalted rather than masked. They were punctually proven right by the market, and their success was reinforced by the lower price of a fiberglass boat, and it's ease of maintenance.

As we saw at the beginning of this book, boats like the Arpege, Dufour 35 and Comet 910 were fully-fledged industrial design objects: not only did they fully express the techniques used for their construction, these techniques also took full advantage of the potentials offered by mass production.

After this great French leap forward, for the next decade, things carried on much in the same vein. By the late '70s, the feeling was that production boat standards had stabilized. And yet the second stage of emulation was coming for cruising boats, this time derived from racing yachts. This emulation was widespread in Italy, yet only marginally found in countries like France or England, and was almost wholly unheard of in the United States. This may be attributed to the

Italian love of anything that has the idea of Gran Turismo attached to it.[53]

It is worth noting that in the '70s, the IOR was well-suited to cruisers: wide beams and generous volumes made for relatively spacious interiors. The area where the emulation of racing yachts triggered critical issues was with overhangs: the IOR required them to be highly accentuated, at odds with comfortable cruising.

Throughout the '70s and beyond, then, cruising boat design could be divided into two schools. The first, the French school, had Jean Marie Finot as its undisputed leader. His particularly voluminous boats, with great form stability and generous beam – for example, *Ecume di mer*, the afore-mentioned Comet 910 and the Grand Soleil 34 – weren't just comfortable, they sailed beautifully too, and even acquitted themselves well in club racing. The other was the American school with Sparkman & Stephens, then Dick Carter and Doug Peterson. Their designs were narrower and with higher displacement, and therefore higher block coefficients.

During the '70s, many of the boats built in Italy were penned by U.S. designers: the Alpa 9.50, 11.50, 12.70, and the Impala 35 were by Stephens; the Impala 36, derived from *Ganbare*, and the Show 34 were by Doug Peterson. French and American designers alike arranged their interiors in a classic manner, with V-shaped bunks in the bow cabin, a walk-through bathroom, amidships the dinette with convertible bunks, and then near the companionway the galley to one side and the other a chart table, with quarter berths aft, under the cockpit.

At the Studio Vallicelli, we followed these criteria and referred to the American school. From the beginning, we ensured that our racing hulls could adapt to the needs of cruisers. The first two boats we launched in 1976, *Gattone* and *Ziggurat*, went into production in Gran Tourism versions. The V-CAT 38 (*Gattone*'s production name) and production Ziggurats won race after race in the following years.

In the late 1970s, the Marchi 39 was the first boat to be designed by Studio Vallicelli specifically for cruising, although the design was directly derived from racing hulls.

Between the late '70s and the mid-'80s, we modified a whole series of boats for pleasure boating: *Ziggurat*, which stretched became the *Ziggurat 916*; *Argento vivo* became the Canados 33; *Blu Show* became the Show 42, and so on. In 1979, for the first time, we began work on a boat purpose-built for fast cruising. That boat was the Marchi 39. Our underlying principle was to emulate a racing yacht, but without extremes: in consequence, the boat's overhangs were less pronounced, and the volumes easily catered to interior needs.

The differences between a Marchi 39 and a Canados 33 (whose shape was derived from a racing hull), may not be many at first glance but are nevertheless recognizable. Racing and cruising hulls diverged during the '80s. The radicalization of racing yachts into ever more extreme boats, with ever decreasing internal volumes, coupled with an increasing demand for comfort in cruising boats, pushed Italian cruisers down a differ-

[53] The attractiveness of Gran Turismo yachting explains the success of X-Yachts in Italy since the '80s, designed and built since 1979 by Dane Niels Jeppesen. These craft continue to perform remarkably well, while at the same time offering spacious and commodious interiors. The X-Yachts line may be considered the only one to have survived the radicalization of racing yachts that began in the late '80s, leading to the final sundering of the connection between racing and cruising fleets.

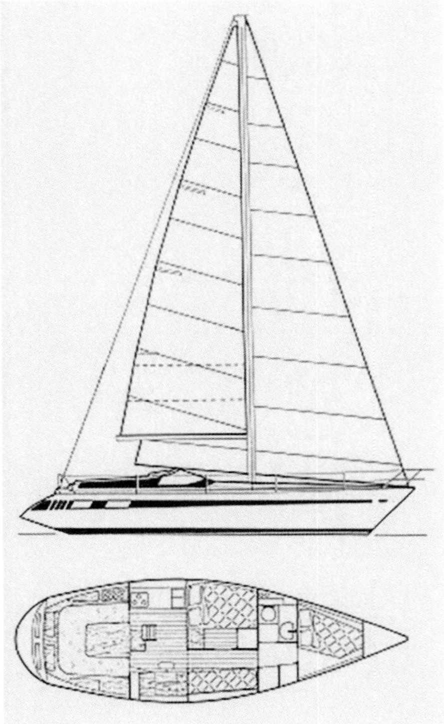

The Comet 11 was a highly innovative boat designed by Jean Marie Finot in 1977, one of the first production boats built in Italy specifically for cruising, with a high freeboard and an unusual dinette placed beneath the cockpit. The boat was also one of the first to feature a deckhouse with a panoramic window.

ent path from the Gran Tourism models, and they began to accept what they really were. Boats began to be purpose-built for cruising in the early '80s. Their appearance was not that attractive: with high topsides and *ad hoc* bulges to maximize internal volume, they sometimes lacked grace. This was also because back then, boats were much smaller than they are today: to install all of the amenities people wanted, designers had to be ready to sacrifice the purity and grace of a boat's lines. For the average '80s hedonist, comfort was non-negotiable. People no longer wanted to camp on the water as in the untamed '70s: they wanted to go sailing in a comfortable country house.

A number of Italian boats (though not designed by Italians) set the pace. Jean Marie Finot's Comet 11 from 1977 was one of the first boats to be designed specifically for cruising, with deliberately higher freeboard and an unusual dinette beneath the cockpit; among other things, the Finot design featured a wraparound panoramic window that heralded the popular solution much favored today. Finot also designed the Grand Soleil 41, one of the first boats with twin stern cabins, and was notable for the smooth transition between the coachroof and the coaming. Lastly, the Grand Soleil 39 was a successful 1983 design by another Frenchman, Alain Jezequel, which managed to combine comfort and speed elegantly; this yacht's hull was one of the first to eschew emulating a racing yacht completely.

The first open transoms began to appear around this time. Offering easier access to the sea, they were a clear sign of changing tastes and expectations. These were soon followed by systems for power-assisted sail handling, notably jib and mainsail furlers and electric winches. Lastly, the arrival of GPS would make navigation much easier for those less skilled in it.[54]

There was one major difference between French and German products and Italian products. Since the very beginning of fiberglass construction in Dufour's day, French and German boat design was oriented towards large-scale production to exploit economies of scale, while in Italy, production always took place within an artisan or semi-industrial environment. This was (and still is) partly because investments were decidedly lower, and partly because the heritage of crafts-based and family-run companies was the norm in large part of Italy's industrial and social structure. This ensured that Italian boatbuilding was in its element making one-off cruisers or limited production runs, with yachts that were part of the bespoke design that characterized the top end of the market over the previous twenty years.

Italy's boatbuilding industry perfectly reflects the vices and virtues of the Italian economy as a whole, never quite striking a balance between the excellence of tailor-made design and high-end craftsmanship on a semi-industrial scale, and the limitations of what is known in Italy as "dwarfism," a disparaging reference to the difficulties of achieving production runs large enough to generate economy of scale and solid production chains.

The first yacht we designed specifically for fast cruising was the Comet 12, in 1983. This was the boat with which we began our relationship with Comar, a relationship that has continued for over three decades. Incidentally, the Comet 12 was one of the first yachts to be designed in Italy and conceived for mass production. This boat, which did very well for itself indeed, was a balanced and stable all-round yacht, with a generous but not overly wide beam, smoothly faired to her extremities. Nor did she have excessive overhangs, something that, on the contrary, was common to racing yachts of the day. We drew the freeboard to limit its height as much as possible.

The Comet 12 was made in a number of versions, including one with an open transom. The interior had two stern cabins with a bathroom, a wide horseshoe-shaped dinette with the galley and chart table near the companionway and, at the bow, the classic cabin with V-shaped berths plus another head.

For our design of the boat's superstructures, we adopted a classic deck plan, importing it from our rac-

[54] In the early '80s, Baltic yachts were the first with internal bulkheads no longer perpendicular to the axis of the keel, but rather to the hull, generating a significant increase in internal space. In the late '80s, the first one piece fiberglass blocks began to appear for bathrooms and galleys; Swan made the first boats with large panoramic windows. During the same period, Beneteau started making the first yachts with a kitchen on runners, opposite a horseshoe-shaped dinette.

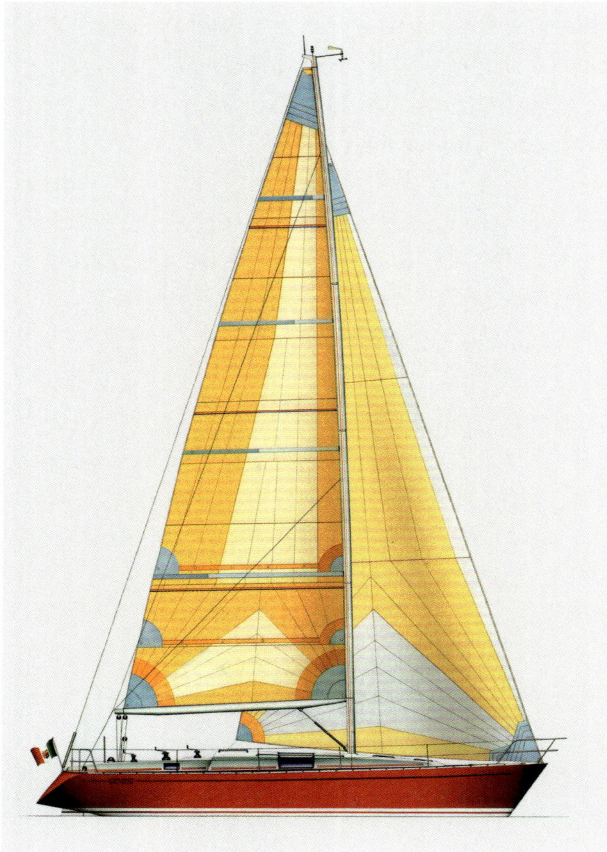

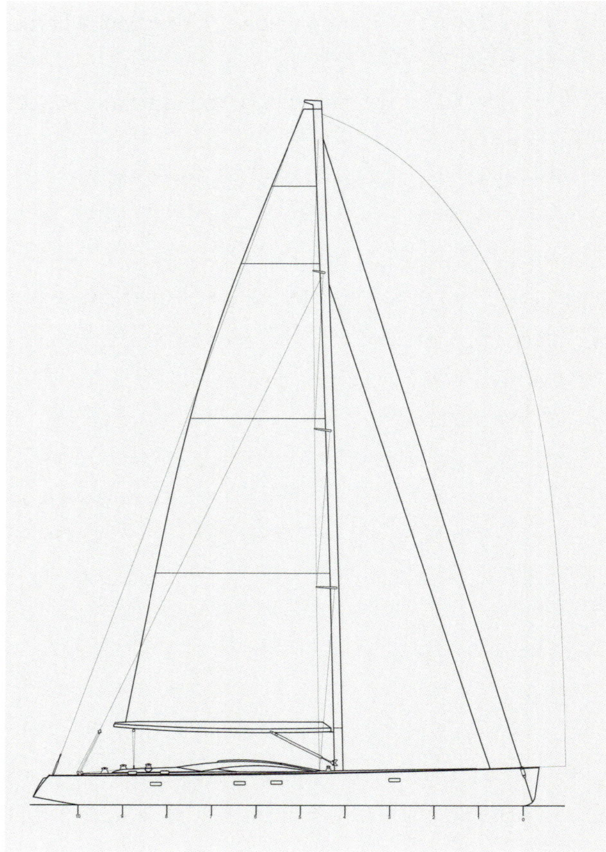

Apart from her classic lines, Genesis 43 offered substantial innovations, unusual for cruising boats of the time: substantially lighter components, a self-tacking jib, a rig with raked spreaders, and, below, twin cabins aft.

Comet 12 is a 1983 design by Studio Vallicelli where the choices that would characterize the subsequent boats designed for Comar in the 1980s were developed.

The Comet 100 was born in 2011 as an evolution of the successful Comet 52 RS and 62 RS models for fast cruising. It is the largest production boat designed for the COMAR in Fiumicino.

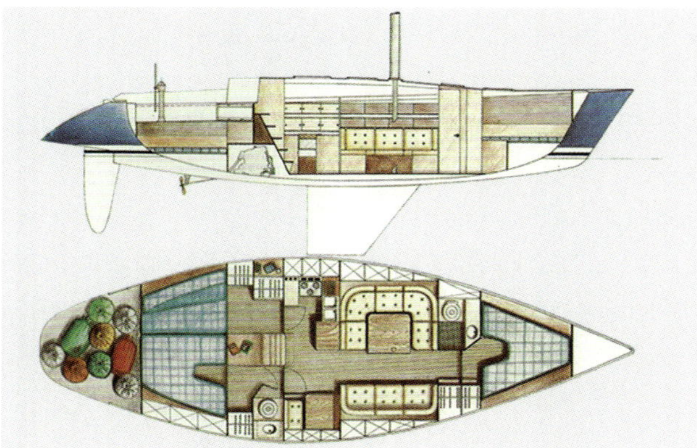

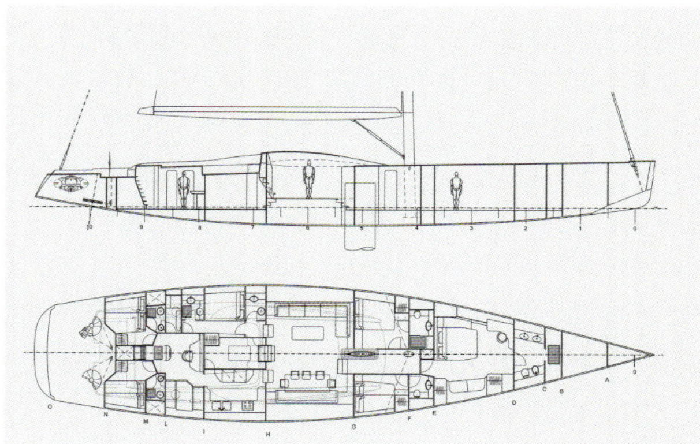

ing yachts *Brava* and *Filo da torcere*, along with a flared wedge-shaped coachroof descending slowly towards the bow; aft, the cockpit coaming was used to support the winches.

The commercial success of Comet 12 prompted Comar to continue with the line. In short order, the boatyard asked us to design a whole series of boats: the Comet 333, 303 and 285. The Comet 12 was the reference model for all of these boats in terms of naval architecture, superstructure design, and interiors. For this series of designs we envisaged extensive use of counter-molds, but unfortunately, production costs could not be reduced enough for smaller boats. We also made a mistake when we chose to scale down the designs from the Comet 12. We were going against the trend in a market that was increasingly interested in upsizing, and that over the coming years would lead to today's gigantism.

In the late '80s, we designed a yacht for Comar (the company had since been taken over by new owners and hired Paolo Villani as commercial director) that would be one of the first experiments in Italy for a design-oriented product. She would have many of the attributes of a one-off design while offering an array of top-tier yacht features in mass production form. The story of this design, the Genesi 43, began with the yacht that preceded it, the *Phoenix*, which Comar had commissioned from Britton Chance. The boatyard managers were not overly enamored of the result delivered by the well-known U.S. designer; we were commissioned to redesign the superstructures and interior.

We applied our experience with *Phoenix* to the Genesi design: sleek lines, a guardedly classic look, and free from the impositions required under the IOR rating rule. Along with moderate maximum beam amidships, gently faired to the yacht's extremities, we put a great deal of work into drawing the sheer, something that was highly unusual for a production boat of the day. To ensure that the coachroof and coamings merged as harmoniously as possible with the hull, we designed a variable sheer-line, similar to the 12 Meters; this made it possible to limit freeboard significantly. The yacht also had a wrap-around window that completely enveloped the coachroof over the dinette.

Genesi boasted a number of innovations: the substantive reduction in weight of components and hardware, a self-tacking jib, and swept-back spreaders to do away with running backstays – a prelude to today's power-assisted sailing. She also had a rather unique internal layout, with twin cabins astern and the bathroom beneath the companionway, shifting the engine amidships. The dinette was also organized in a new way, with the galley to one side and the dining area, with a U-shaped seating arrangement, on the other.[55] Working with Paolo Villani, we also organized the involvement of Nemo, a Comar subsidiary with experience in nautical components, so that it could reorganize its output to produce components for the yacht. Genesi was not so much series-produced as high-end, semi-series design, conveying something of the Gran Turismo approach that somehow had been lost in the comfort-seeking but hardly exciting cruisers of the '80s.

[55] Interestingly, jib roller furling, a true revolution for cruisers, had been developed for single-handed round-the-world yacht races.

New Projects for a New Scenario

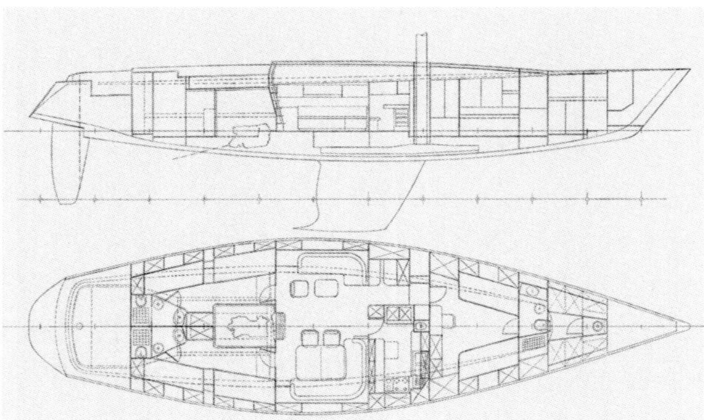

Designed at the end of the 1980s, *La Cardinala* was a fast cruising sloop whose design appears to be influenced by the IOR, but whose fluidity of lines and harmonious proportions would characterize the designs of the following decade.

A new season opened for our practice with the Genesi design. In the early '90s, the new IMS measurement rule was created, pushing back against the liberalization of offshore sailing that had followed the IOR's twenty-year monopoly. In a much-changed industry, our Studio adapted, catering to new demand. In those years, we had been making three different types of boats: Open Classes, fast cruisers and production boats. Subsequently, we branched out into large cruisers as well.

During the '90s, our design approach used *Orsa Maggiore* as a reference framework: smooth, unstrained lines, a moderate maximum beam, and reduced wetted surface, all combined to ensure speed and excellent seakeeping qualities. Prior to *Orsa Maggiore*, we had designed *Koala VI*, our second yacht for Pierluigi Spadolini, and *La Cardinala*, for the Astaldi family. Both of these yachts were very different from *Orsa Maggiore*, their conception still influenced by the IOR rating rule style. The change of direction after *Orsa Maggiore* was plain to see: *Mia* was an aluminum sloop built by Cassanelli-CNB; the Cutter 57 and Oceano 54 were yachts for fast cruising, freed from a rating rule approach, endowed with what at the time was a major new development for cruisers: a deck saloon.

We responded to emerging demand for ULDBs (Ultra Light Displacement Boats) with medium-light displacement designs like *Nastro Azzurro;* compared to American models, she had a wider maximum beam and more rounded stern sections. After winning the 1992 Columbus Week, and coming second overall in the Genoa to Porto Rico race, the VR Nautica yard put *Nastro Azzurro* into series production. Characterized by the long, wedge-shaped coachroof we had created in the early days of our career, the series version coachroof also included a sequence of portholes. This vintage detail gave the boat the allure of a high class yacht, further bolstered by a teak deck.

The Este 35 and Este 39, built by the D'Este boatyard in Fiumicino, were our first designs specifically conceived for the brand new IMS rating rule. Even though they hadn't been designed exclusively for racing, both won the Copa del Rey; the Este 39 also came in second in the IMS European Championship held at Punta Ala in 1995. Looking back at the drawings for the Este 39', especially the interiors, it is amazing how such a comfortable, if basic, yacht – and a production boat to boot – managed to come second in a European Championship race. This was not an isolated case: the racing success of production boats, with results that should have been impossible for anything but a one-off, had begun towards the end of the '80s with the X-Yachts line. Designed and built by Niels Jeppsen of Denmark, they won several Ton Cups, even if they were wholly mass produced. In Jeppen's designs, all

Nitissima was an ultra light displacement boat designed in the early '90s. Compared to US designs of the period, she had a wider maximum beam and more rounded stern sections. Note the long deckhouse, with a row of portholes giving the boat a vintage touch.

Studio Vallicelli's long term relationship with Comar continued in 1999 with the Comet 51. Designed taking the IMS into account, but without going to extremes, the Comet 51 won the 2006 Italian IMS championships.

The design line inaugurated with the fast cruisers *Virtuelle* and H_2O, and continuing with the Comet 51 in the early 2000s, was adopted on a smaller scale with the Comet 38.

the components are reduced to essentials, from hull lines to superstructure design and interiors. It is precisely this essentiality coupled with the quality of their hull design, that is the secret behind their success, heralding the convergence of racing yachts and cruising yachts that finally became possible through the introduction of new regulations.

Although the IMS rating rule indubitably contained some critical issues and was complex to implement, at least for a time – and it was not a short time either – the rule succeeded in closing the gap between racing and cruising yachts that, towards the end of the IOR, had become an unbridgeable chasm.[56] The X-Yacht, Beneteau, Jeanneau and Grand Soleil designs, along with our Este and Comet designs, show this with great clarity.

However, there was a high price to pay for this confluence. In the past, the IOR represented the leading edge of offshore sailing, the ultimate in experimental and hi-tech design: the latest developments, previews of the future of sailing were to be found on IOR yachts, especially those that raced in the One Ton and Admiral's Cups. Yet as we have seen throughout this book, new developments appeared every year, often so radical that they caused a level of obsolescence for older yachts that was worrying (if not to say discouraging) for many owners. From the '90s on, the annual array of new developments was no longer just to be found on the quayside at IMS races, they were also showcased in the Open classes, BOC and, of course, America's Cup.

So, looking back and comparing the various different measurement and rating rules, without fear of being contradicted, we may say that the IOR rating rule was highly creative and allowed designers to express themselves with great freedom. The differences between a Peterson or a Farr yacht were vast. With the advent of the IMS and subsequent measurement and rating rules, we have arrived at the present day's frankly embarrassing homogeneity of design: although of high-quality, differences in design are fewer and farther between.

This trend towards similarity is also visible in today's motor vehicles. In the past, the difference between an Alfa Romeo Giulietta and a Citroen DS was so huge it was like comparing two contrasting approaches of living. Today, where homogeneity has the upper hand (different car manufacturers often share the same chassis), it makes no difference whether somebody drives a Toyota or a Renault.

After the Este's we returned the next season to Comar, which in the meantime had changed ownership and moved to Fiumicino. The new Comets we began designing in the late '90s reflected choices deriving from *Virtuelle*: medium-light displacement, sleek lines

[56] The IMS rating rule also managed to narrow the gap between old and new boats. Indeed, at the 1994 IMS Italian Championship, two of our 'old' designs, the 1980 half-tonner *Hipazia* and the 1985 two-tonner *Dioscura* acquitted themselves well, finishing in the upper echelons, which was usually the preserve of newly-launched yachts.

capable of maintaining a respectable block coefficient (to the relative disadvantage of form stability), reduced wetted surface, and rounded sections continuing all the way to the transom. We researched these lines using computer simulation programs to ensure that they did not deform too much at different angles of heel, while at the same time ensuring that the boat did not yield too much to the inevitable loopholes that gradually appeared in the IMS rating rule. While *Filo da torcere* and the original *Brava* were designed to exploit the IOR rating rule and its loopholes fully, our IMS yachts were freed from the tyranny of the measurement rule. This was not just our approach, either; it was shared by other designers like Germán Frers. However, it was opposed by radical IMS designers like Botin of Spain.

In any case, these designs were successful. The Comet 51, a yacht designed in 1999 with very pure, balanced, and by no means extreme lines, won the Italian IMS championship in 2006, a full seven years after it was conceived. A special case is another Comet, the 45, a design from 2003. Here we were seeking the greatest possible comfort, even if it meant foregoing certain architectural choices and sacrificing at least some of the purity of line that had characterized the Comet 51. The Comet 45's freeboard was high, and the articulated coachroof design made it possible to carve out plenty of interior space, allowing people to stand fully upright even in the head at the bow – albeit to the detriment of its external appearance. And yet the Comet 45 was highly competitive. Comar and we decided therefore to improve some aspects of the yacht, upgrading the appendages. These changes were sufficient: in 2008, a Comet 45 won the World Championship in Athens. Then we took the design philosophy behind the 45-footer down a category with the Comet 38S. That design won the Italian Championship and came second in the European ORC World Championship.

It is interesting to note how our practice's design approach evolved through these boats, especially their superstructures and deck plans. Twenty years after the Comet 420, we had moved from a classic wedge-shaped coachroof to the more organic design in the Comet 38S. The wedge-shaped roof design ended up morphing into a shell without being too radical because the memory of the wedge was still visible. This is a hallmark feature of our practice's design approach: a style characterized by a constant mediation between innovation and tradition, between new developments and the familiar.

In my opinion, yacht design cannot but confront itself with its traditions. If that wasn't the case, yachts would inevitably lose the aura they have acquired over the centuries in the collective imagination, and that would be a terrible loss. It is clear that in some cases, in what is known as extreme sailing (for example, with foils), this mediation has no sense, but these are special cases that should not be taken as a more general rule.

Over our long career, through the ups and downs that have characterized our experience, our framework of reference has always been clear, though perhaps a bit cross-eyed, based as it is on looking both backwards and forwards. Yes, this does have an element of the cross-eyed about it, but we have always been convinced that this "optical challenge' is well worth it. Indeed, over the years, it has become a source of pride.

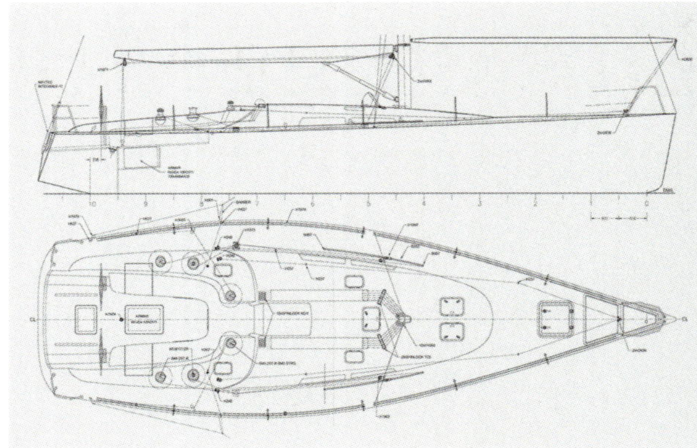

Comet 45 was designed by Studio Vallicelli in 2003 with the aim of assuring a very high degree of interior comfort, but even so this the boat won the IMS World Championship in Athens in 2008.

Nastro Azzurro winner in 1992 of the San Pellegrino Trophy in the waters of the Hudson River in front of Manhattan on the occasion of the Columbiadians.

Designing Motoryachts

In the early '90s, we began designing a very special category of powerboats: small military patrol vessels and pilot craft. We worked with Intermarine and then collaborated with their technical office on a number of calls for tender. Intermarine was the Montedison department specialized in composites that worked for the Italian Navy. We won one of these tenders for the design of a hydrographic research catamaran, and this took us indirectly into the world of motoryachts, at a rather interesting time as the sector was going through a major upheaval.

To understand the extent of these changes, we must take a step backward to the '60s and the beginning of Italy's industrial development boom, with the country already active in the design and construction of this type of boat – a position that over time consolidated into a true position of leadership. Today Italy generates the highest revenue of any nation in this sector and is the absolute uncontested leader in megayacht construction.

In the 1960s, the top shipyards were located along the coasts of the northern Tyrrhenian Sea, in Tuscany and Liguria: Baglietto, Benetti, Picchiotti and the Cantieri di Pisa, whose high standards of craftsmanship put them on par with Sangermani and Carlini. Designs were developed in-house by these yards, integrating creation and construction. The only exceptions were for finishings and trim, often sub-contracted to individual craftsmen working nearby the boatyards. By the '80s, these highly skilled artisan boatyards had evolved into industries, particularly boatyards like Ferretti and Azimut, which ramped up their production significantly during that decade.

The industrialization of sailboats in France in the '60s came about twenty years later for motoryachts in Italy, where, despite the impressive numbers, there was no comparison with other industrial sectors because of the difference in scale. This had a number of consequences: for example, motor vehicle company research departments not only performed R&D, for the most part they also oversaw product integrity; if they worked with external designers, they would never dream of delegating the final say on the product. However, in the world of motoryachts, services undertaken by research centers were outsourced, often to multiple contractors, in fragmentation of skills that inevitably had significant effects. For instance, hull design was often contracted out to naval engineers, architects were hired for the superstructures and interior layouts, and owners often commissioned a trusted interior decorator. It comes as no surprise that this product conception lacked overall stylistic unity.

Things began to change in the '90s when the design was booming across all sectors. The leap in scale we had seen for sailboats was also common to motoryachts, as they grew in size and became increasingly customized. Here we began working in the motoryacht industry when the ISA boatyard of Ancona called us in to modify and re-engineer a Philippe Briand designed 45-meter sailboat. Though never built, this marked the start of a long and intense working relationship with the boatyard, similar in many ways to our relationship with Comar. ISA was owned by the Rodriguez group, a French company working to interpret the changes taking place in the sector optimally.[57]

Company management next asked us to develop a design for a 120′ mass-produced fiberglass motoryacht, based on a hull conceived by their technical department. We carried out research into its superstructures, interior layout, and a number of hallmark features. The ISA 120, as the design came to be known, boasted a number of stylistic features uncommon at the time. To absorb the volume of the superstructure, towards the bow, we created a solid bulwark around the lateral walkways that gave the boat a dynamic appearance. This dynamism was also highlighted by the design of the cornices connecting the sides of the transom with the bridge windows. This design element encompassed the stairway connecting the afterdeck to the sundeck. Previously developed by Franchini, this solution had characterized ISA designs for some time now. We decided to extend this look to other components of the yacht: the communications mast, side arches, and bulwarks. The result was a synthesis of its various elements, creating a stylistic unity unusual for boats at that time.[58]

We applied a similar approach to designing the ISA 140, an even more stylized yacht than its predecessor, as we were also involved in designing the topsides, which we harmonized optimally with the superstructure. That the ISA 140 was a ground-breaking product was borne out by the fact that, though the boat was launched a full eight years after its conception, it was still a commercial success.

Over the years, we also designed a series of steel hulled motoryachts with aluminum superstructures, between 40 and 70 meters in length, for the ISA boatyard. The most significant was *Okto*. As is often the case with motoryachts, these yachts were assembled from separate elements, each of which corresponded to compartments of the vessel, with ducting for the primary systems inserted prior to assembly. This construction system makes it possible to modify overall length by varying the individual blocks. Designing these blocks was a complex operation, a form of tailor-made design that meshed with industrial production, requiring designers to be in continuous contact with the boatyard. We developed two product lines for ISA, Gran Turismo and Classic. The Classic line was characterized by flowing lines, with a longitudinal profile vaguely resembling a sailboat. On *Kolaha*, for example, a customized 63 meter ISA yacht, we designed an unusual closed transom, slender with no visible stairways, perhaps a homage to the '70s IOR look. With the more experimental Gran Turismo line, we sought to achieve an acceptable synthesis between the organic soft curves and the angularity of the boat's multifaceted geometries.[59]

As is true of sailboat design, designing motoryachts is also a team game. From conception onwards, the designer, with representatives from the boatyard and owner, meet around a table. Generally, choices are taken together, under the direction of the designer whose main task is to reconcile the different positions.

Project development also follows a different path. We start with a concept, a general outline that establishes the boat's character, before moving on through various stages to the final configuration. For mega-yachts (between 60 and 80 meters in length) and giga yachts (100 meters plus), one major technical and commercial criteria must always be borne in mind: this type of vessel has to be unique, luxury and one-off, which means that production runs are out of the question. This raises a commercial issue that heavily impacts both design and production engineering: on average,

The ISA 120 was the first motoryacht designed for the ISA yard in Ancona: the intent was to give the boat a style that could balance dynamism with a certain classic look.

[57] ISA Yachts/Rodriguez drew on the services of employees with acknowledged technical and commercial prowess, to name but a few: Richard Dixon, Marcello Maggi, Daniele Socci, Gianluca Fenucci and Antonio Longobardi.

[58] The ISA 120 immediately attracted customers, as may be seen from the fact that the boatyard managed to sell seven yachts though the craft existed only on paper at the time.

[59] At the time of writing, our practice is working on two apx.70 meter mega yachts in Turkey (Antalya and Istanbul), built on a hull platform with a swept-back look and three decks above the main deck, with the novel feature of side racks to help extend the area dedicated to the owner's cabin. We are also working with Fincantieri on a number of mega yachts. One of these is hydrogen-powered and has a SWATH (Small Waterplane Area Twin Hull) architecture – a special type of catamaran equipped with a very narrow emerged area that rests on two large submerged bulbs to reduce pitching.

Okto is a steel motoryacht with aluminum superstructures, assembled from separate elements. Note the open transom with a swimming pool projecting over it.

the start-up to prototype phase takes between a minimum of two years to a maximum of four before a project is ready to launch. Clients are simply not prepared to wait that long. To get around this problem, boatyards and designers develop what are known as platforms: basic hull designs that, although only outlines, are already equipped with essential engineering, architecture and systems solutions. The end-design for the product is subsequently conceived on the basis of these platforms.

The designers' job is to be smart enough not to limit the freedom of expression that customers demand. It's as if a top tailor draws outlines for clothes and later personalizes them, giving the impression that the end product was dreamt up from scratch for that particular client. Lead times are reduced with pre-prepared platforms, and today the competition in this sector is in making delivery times as rapid as possible.

Returning to the evolution of motoryachts as a whole, we should remember that the increase in demand, and therefore supply, really took off for sail and motor boats in the '90s, and has continued to rise ever since. Today the offer for motoryachts includes series production designs, generally made from fiberglass, along with custom designs in steel, with compartmental construction and aluminum superstructures, with the latter offering more scope for significant design innovation.

New categories are also appearing: Explorer and Expedition yachts are displacement hulls whose appearance, often through direct citation, mimics vessels designed for scientific and geographical expeditions, or workboats like tugs and supply vessels. These yachts offer a great deal of external and internal space, and are very seaworthy, although not particularly fast.

Another new category is in military-inspired yachts, which mime the aggressive stance of military naval design. Top designers such as Norman Foster, Philippe Starck and Espen Oeino, who designed what is considered one of the most successful experiments in this genre, the *Skat*, have all tried their hand at this.

These categories, as well as the successful "small ship" design type that, albeit on a smaller scale, echoes the ocean liners of yore, base their appeal in evoking worlds distant from the traditional idea of yacht cruising. This crops up in much of contemporary design, which increasingly expresses itself through analogy, seeking to intercept the unconscious desires of a clientele no longer willing to accept a designer's taste and choice passively.

Last but not least, we have the fast motoryacht category, in which aeronautic and sometimes even space technologies are applied and taken to their most extreme consequences.

The motoryacht sector continues to expand, becoming ever more eclectic as design strives to create the perfect seductive object, often expressed through extreme and spectacular taste. An outstanding example of this is *Guilty*, a vessel designed by Ivana Porfiri with the artist Jeff Koons, a man considered the wizard of extreme (if not kitsch) taste.

RACING YACHTS

In 1976, Studio Vallicelli launched its first boats, *Ziggurat* and *Gattone*. *Ziggurat* finished third at the Half Ton Cup held in Trieste, while *Gattone* won the Giraglia and the Mediterranean Championship. The following year, *Argento vivo* came second at the Three Quarter Ton Cup in La Rochelle, and several of these yachts achieved excellent results in various other regattas.
The Studio made a rapid advance with *Filo da torcere* and *Brava*, establishing itself internationally: *Filo da torcere* won the One Ton Cup in 1980 in Naples; *Brava* participated twice in the Admiral's Cup, and won the prestigious Fastnet Race in 1983. In 1982, the first Italian America's Cup challenger, *Azzurra*, was launched. In the waters off Newport the following year, she became a household name in Italy, the first time that sailing became a mass phenomenon in the country. In the years that followed, Vallicelli Studio's design research evolved to match changes in the IOR, and we began producing racing boats other than offshore yachts, like the Open Class boats. These included *Stradivaria*, *Brancaleone*, *Gaia Legend* and especially the very successful *Riviera di Rimini*, whose list of victories include the Rimini-Corfu-Rimini, the Giraglia, the Barcolana, the Roma per Tutti, the Roma per Due and the Fastnet. In the '90s, our practice designed a number of boats conceived for the new IMS rule and later for the ORC and IMC rules. These included *Libertine*, which in 2008 won the ORC World Championship held in Athens and *Verve*, which may be considered a showcase boat for our more recent design choices.

Ziggurat

ZIGGURAT

Year 1975
Builder Alaver Verona – Tecnomar Roma
Construction GRP
LOA 9.15 m
B MAX 2.95 m
D 1.72 m
DSPL 3.1 t
Mainsail 16 mq
Headsail 30 mq
Spi 65 mq

This was the first boat I designed while still an architecture student: a half tonner born with the aim of participating in the Half Ton Cup of 1976, held that year in Trieste. *Ziggurat* had a medium displacement and was equipped with a generous masthead sail plan. Compared to the canons of the American school, Carter and Peterson, in particular, offered some peculiarities: less displacement, slightly wider stern sections and a tapered trapezoidal keel with a rounded lower entry. Furthermore, the rudder was attached to a fixed skeg, an integral part of the hull, and had a higher percentage of compensation than normal. The longitudinal distance between the center of effort and the center of lateral resistance was also greater than usual. This made the boat very light on the helm, with less of a weather-helm than the American boats. This was a characteristic that we also kept in our subsequent designs. The hull was made in Verona by Giovanni Sigovich in fiberglass with a balsa sandwich core, laminated on female wooden mold. This was an innovative choice for a prototype and offered the advantage of reducing the amount of filler required to fair the hull, reducing weight. *Ziggurat* was completed at the Tecnomar yard in Fiumara by myself together with Patrizia Ferri (who was formally the owner), Nicola Sironi and other friends (some of whom later became part of the studio) using almost exclusively Italian materials and technologies.
The available budget was very small, but we managed to complete the work thanks to the support of some suppliers who believed in this project (Carozzo & Peer sailmakers, Canclini for the mast, Gary Redfern for the Barlow winches and deck equipment, etc.). After getting through the selections, we arrived at the Half Ton Cup in Trieste: the crescendo of races saw us win the last race, an offshore race, and so conquer a third place on the podium. *Ziggurat* won the Best Italian Design Prize, established by the FIV (Italian Sailing Federation) in 1976. The boat was sold at the end of the event, and the design was put into production by the CPR yard in Fiumicino. Many races were won by the Ziggurat 9.10 production boats, including two Italian Championships in 1978 and 1979, and the Giraglia in 1977.

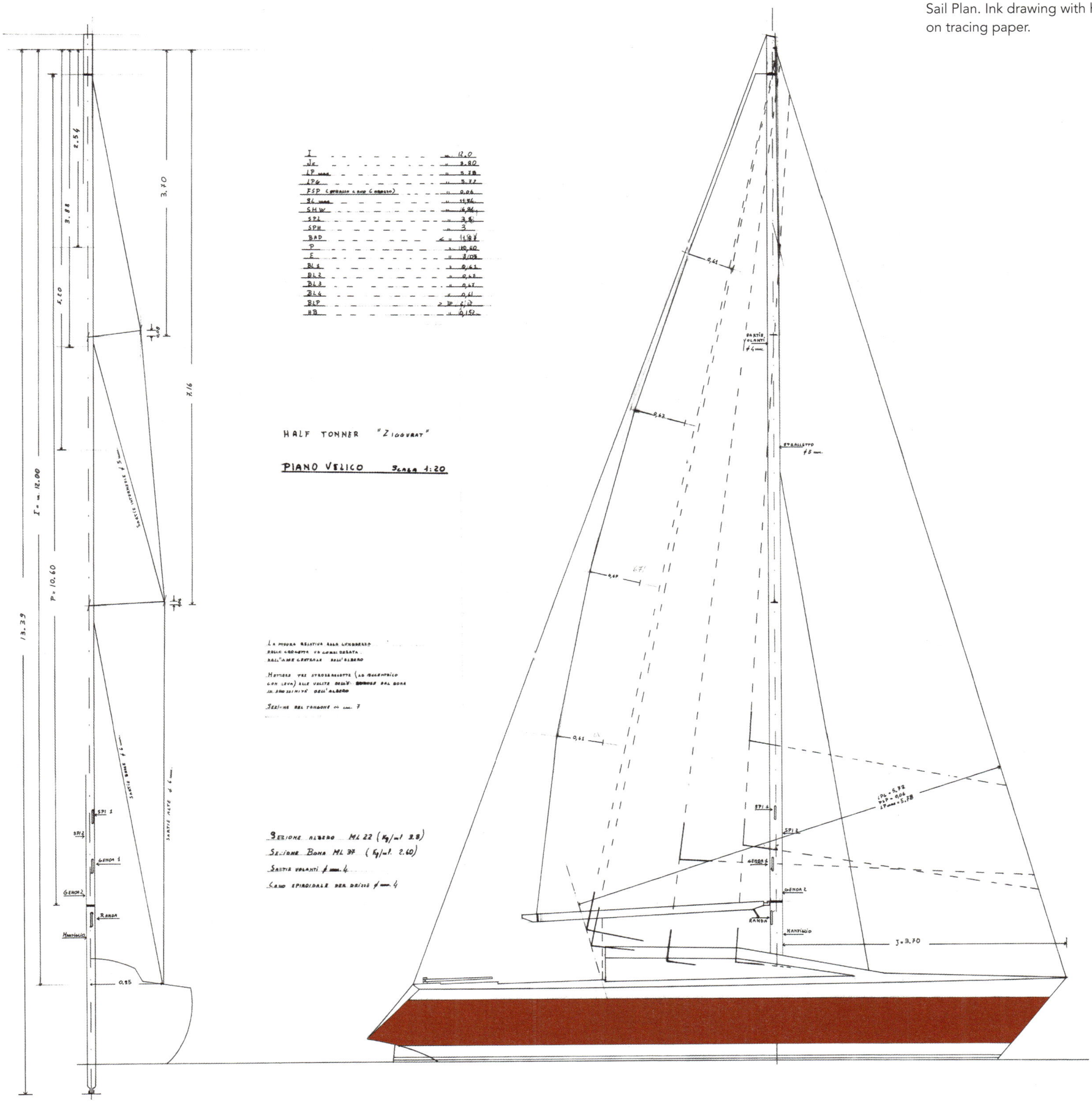

Sail Plan. Ink drawing with halftone screen on tracing paper.

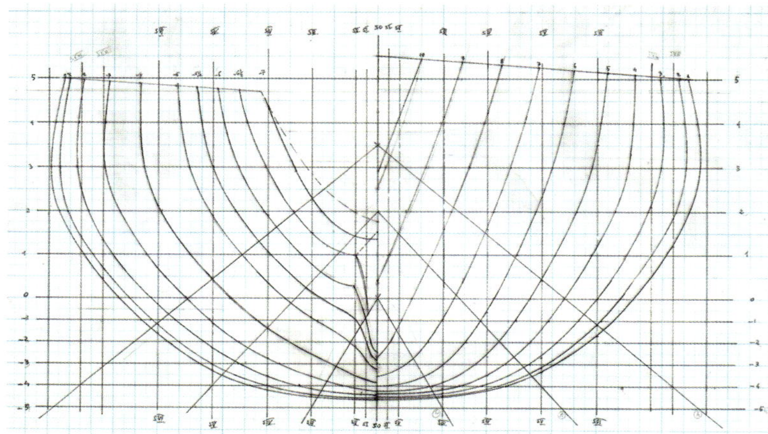

Ziggurat at the 1976 Half Ton Cup in Trieste.

Vertical sections. Pencil drawing on graph paper.

Lines plan (waterlines, buttocks, diagonal sections). Ink drawing on tracing paper.

Andrea Vallicelli and Paolo Bitetti are checking the frames for the hull's wooden mold at the Alaver shipyard in Verona.

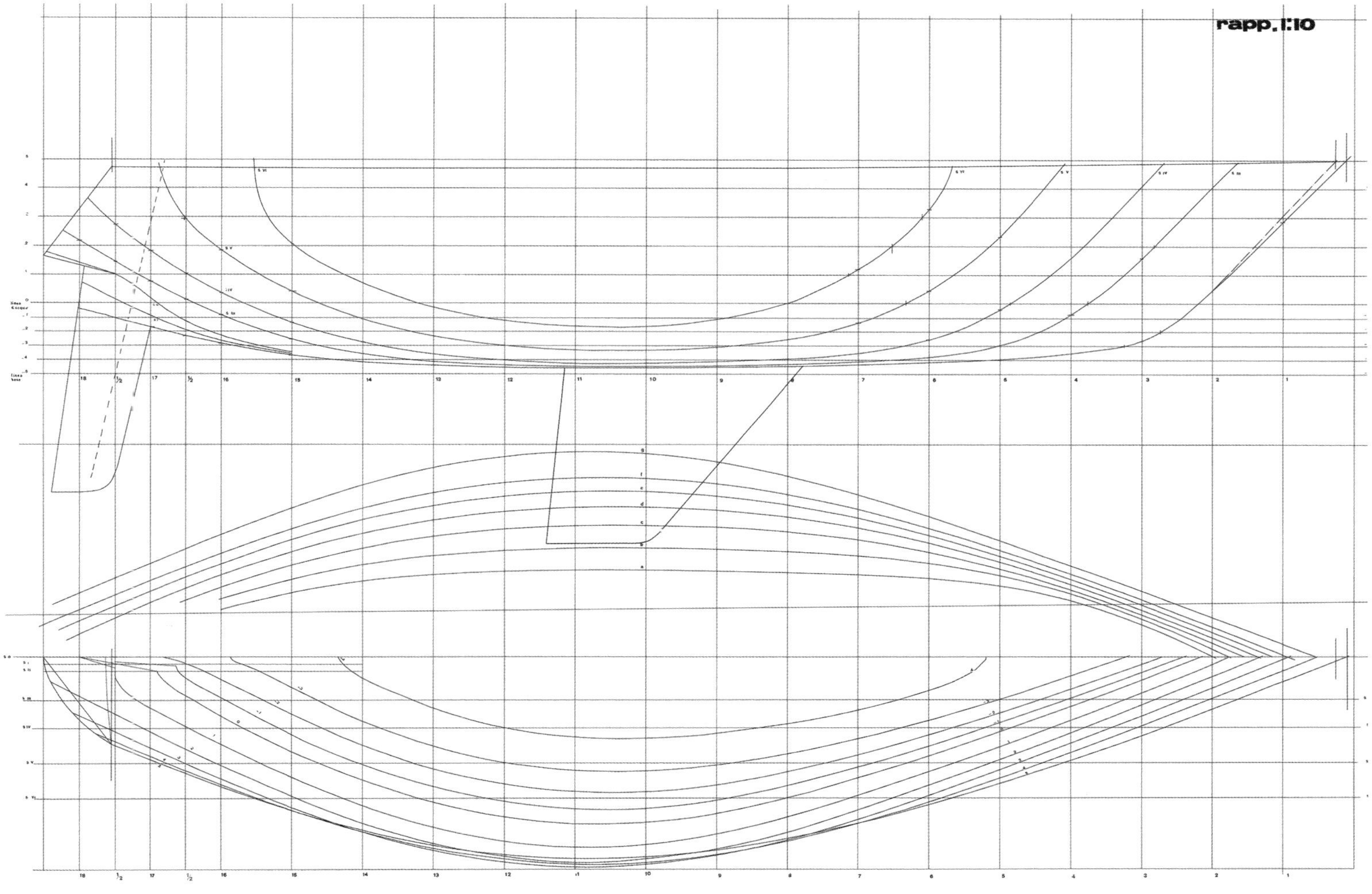
rapp.1:10

171
00

Gattone

gattone

Designed together with Carlo Alberto Tiberio, a physics professor and co-owner in the Artmare boatyard, *Gattone* was a medium displacement One Tonner, born in parallel with *Ziggurat* at the start of my career. Compared to the latter, *Gattone* had more extreme lines: her maximum beam was set at 65% of the waterline length (LWL), while in *Ziggurat* this was at 60%, and she had a narrower, almost V-shaped transom. She also had conspicuous tumblehome, with beam at deck level narrower than halfway up the topsides. The deck layout featured a stern cockpit, separate from the central cockpit, where the helm was located. The hull of *Gattone* was constructed using a method optimized by Artmare: the corrugated sandwich. This is a type of sandwich lamination with a core made up of polyurethane omega-shaped strips placed side by side. These strips support a thin fiberglass skin, connecting the external and internal skins. Even though this construction technique requires some skill, it makes it possible to produce a light and robust hull and at a fairly low cost. After winning the Mediterranean Championship, the Giraglia, and the Coppa del Rey, *Gattone* participated in the 1976 One Ton Cup in Marseille. The series version, which was produced from 1976 to 1983 as the V CAT 38, had two deck layout options: a single cockpit and the double cockpit like the prototype.

Year 1975
Builder Artmare Roma
Construction GRP
LOA 11.26 m
B MAX 3.5 m
D 1.605 m
DSPL 5.6 t
Mainsail 24 mq
Headsail 56 mq
Spi 120 mq

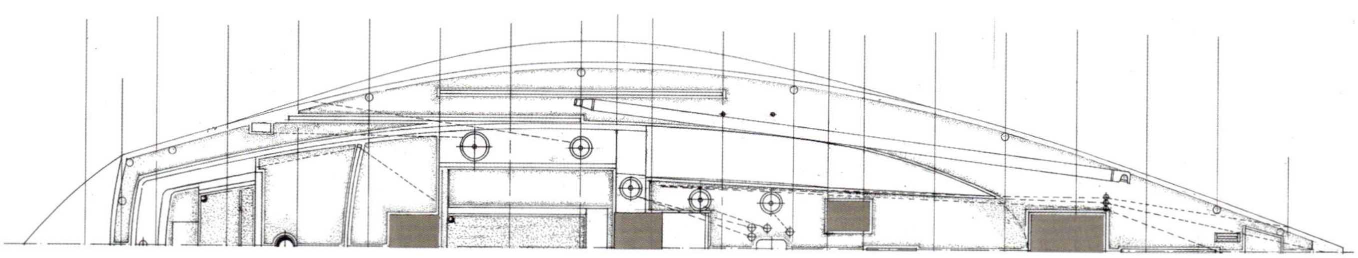

Deck plan. Ink drawing with halftone screen on tracing paper.

Lines plan (stations, waterlines, buttocks, diagonal sections). Ink and pencil drawing on polyester paper.

Building the wood model for the female fiberglass mold of the CAT 38 at the Artmare shipyard: fitting the strips of marine plywood.

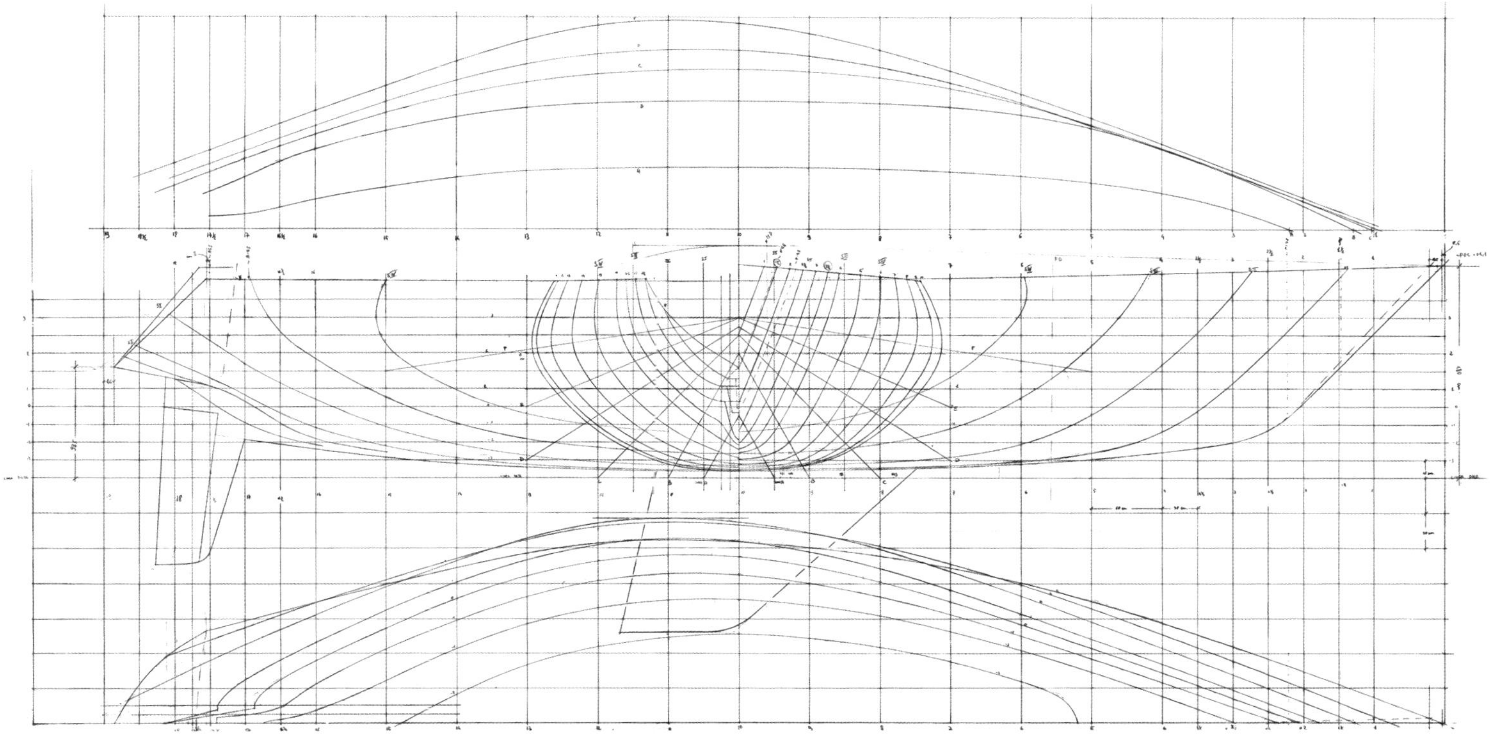

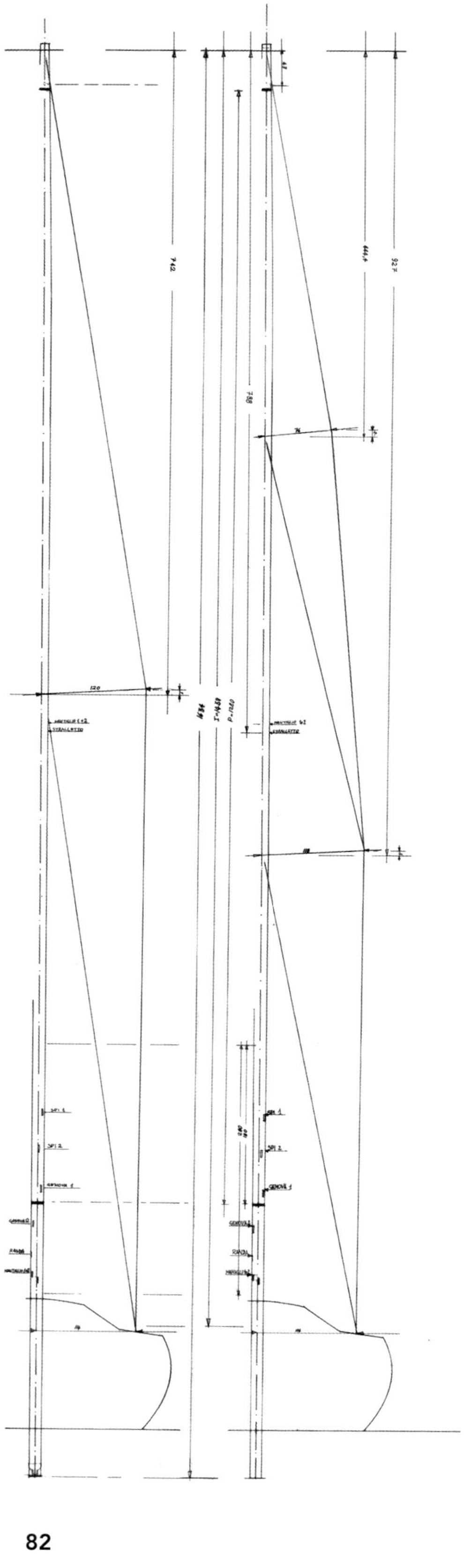

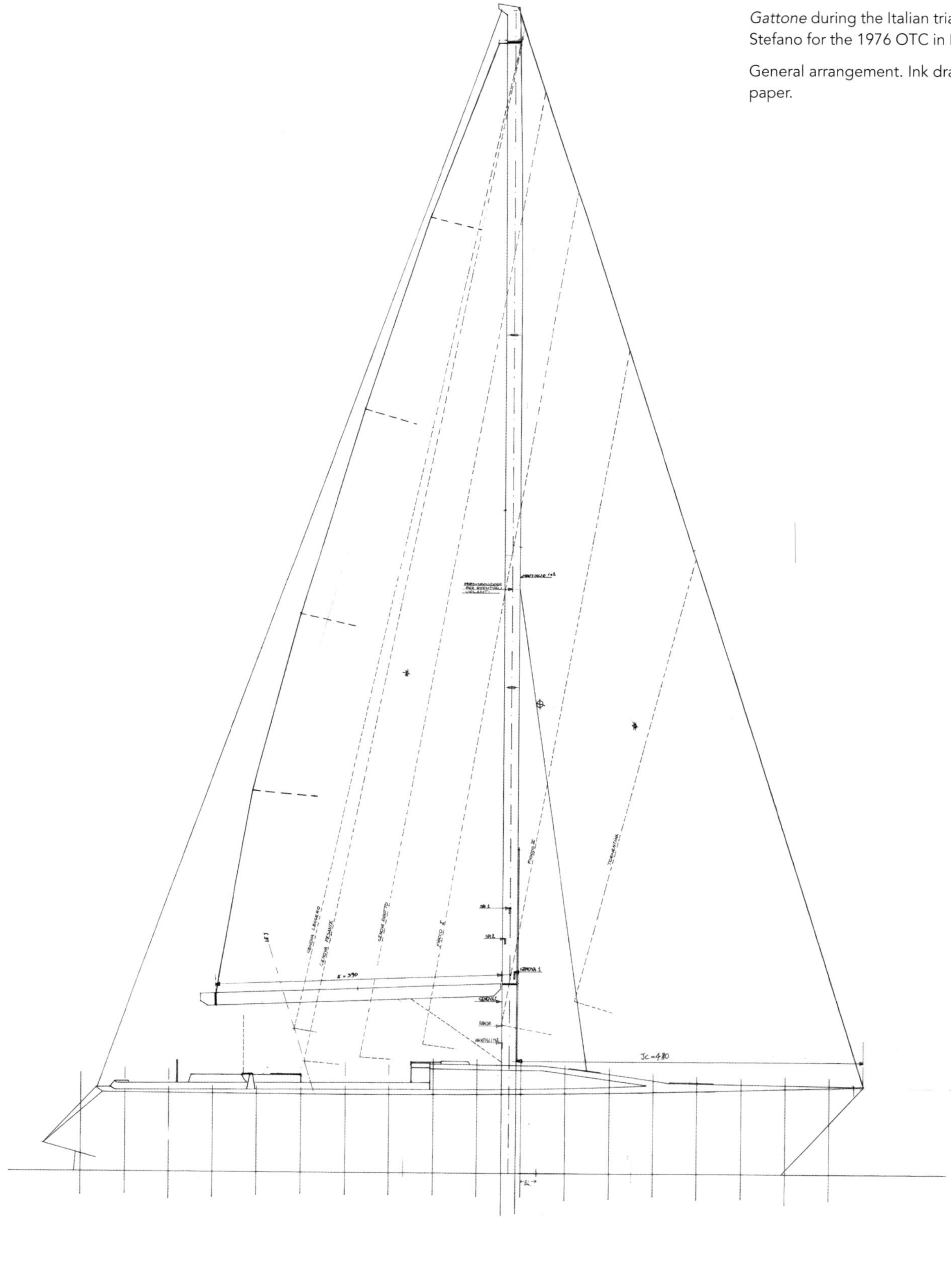

Sail plan. Ink drawing on tracing paper.

Gattone during the Italian trial races at Santo Stefano for the 1976 OTC in Marseille.

General arrangement. Ink drawing on tracing paper.

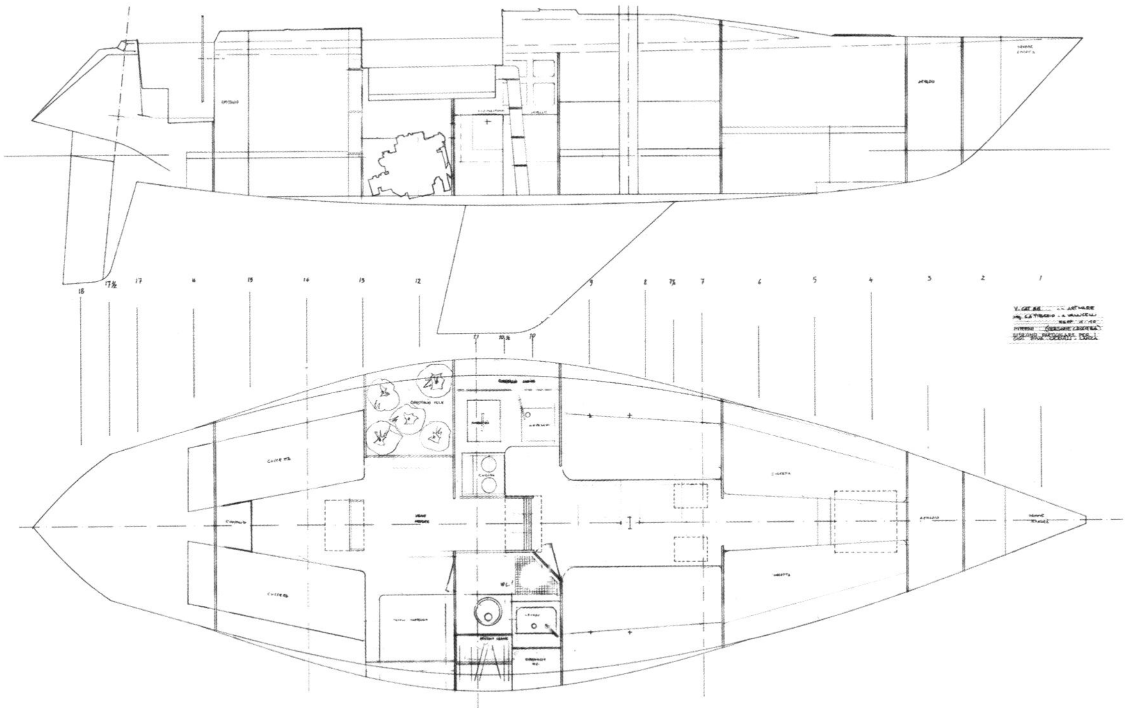

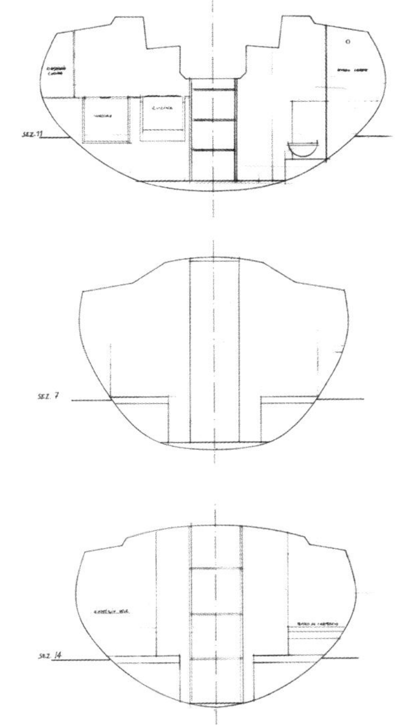

Argento vivo

With *Argento vivo*, just a year after our debut, our studio made a qualitative leap thanks to the serious approach and commitment of her two owners: Pigi Vigliani and Riccardo Trionfi. They commissioned us to produce a design for the 1977 Three Quarter Ton Cup at La Rochelle. Compared to *Ziggurat*, *Argento vivo* had a finer bow and a fuller stern. She was an all-round boat of medium displacement, without excesses, able to combine – as we always tried to do in those years – a good block coefficient with good form stability. For the structure of the boat, built of cold molded wood by De Cesari, we worked with a naval engineer of our generation, with whom we later often collaborated: Carlo Bertorello. The boat was built by gluing four layers of 4 mm thick okumé strips, set at different angles. The innermost layer consisted of very thin plywood used for aircraft (Adriano De Cesari, the yard's owner, had a long experience in the construction of hulls for seaplanes). The deck was made of marine plywood, on which thin strips of teak were overlaid to beautify the boat without adding too much weight. Shortly after launching in 1977, *Argento vivo* won the Italian Championship for her class. She returned immediately to the yard to lighten her keel bulb and improve performance in medium to light winds and subsequently finished second in the Three Quarter Ton Cup in La Rochelle. The following year she won the Italian Championships a second time.

Year 1976
Builder De Cesari Cervia (RA)
Construction Cold molded wood
LOA 10.155 m
B MAX 3.218 m
D 1.8 m
DSPL 3.9 t
Mainsail 20 mq
Headsail 42 mq
Spi 85 mq

Argento vivo in navigation.

Argento vivo after her launch in the Milano Marittima canal.

Construction plan. Deck sections and keel plan. Ink drawing on tracing paper.

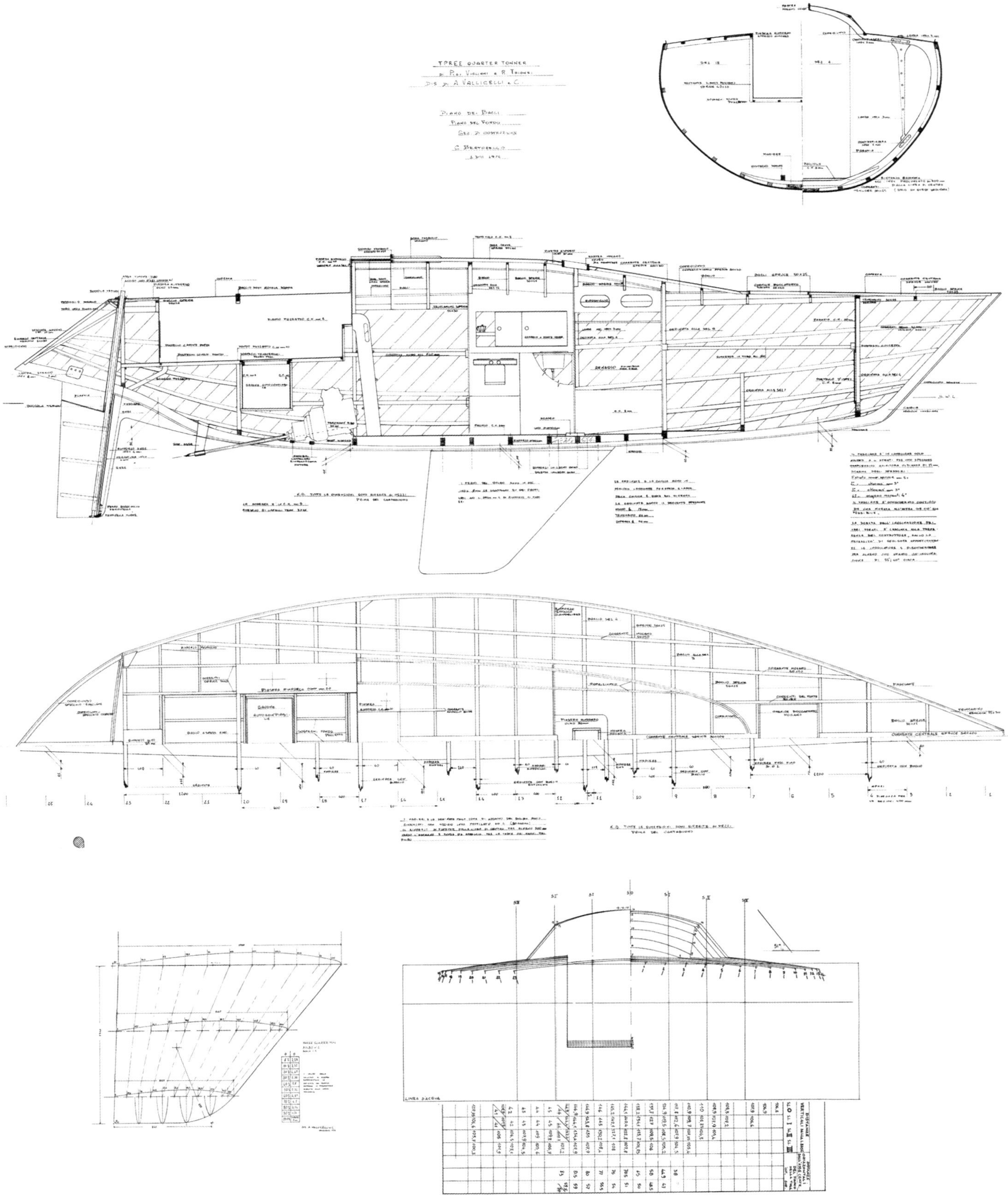
TPREE QUARTER TONNER
DIS. DI A. VALLICELLI e C.
PIANO DEI BAGLI
PIANO DEL FONDO

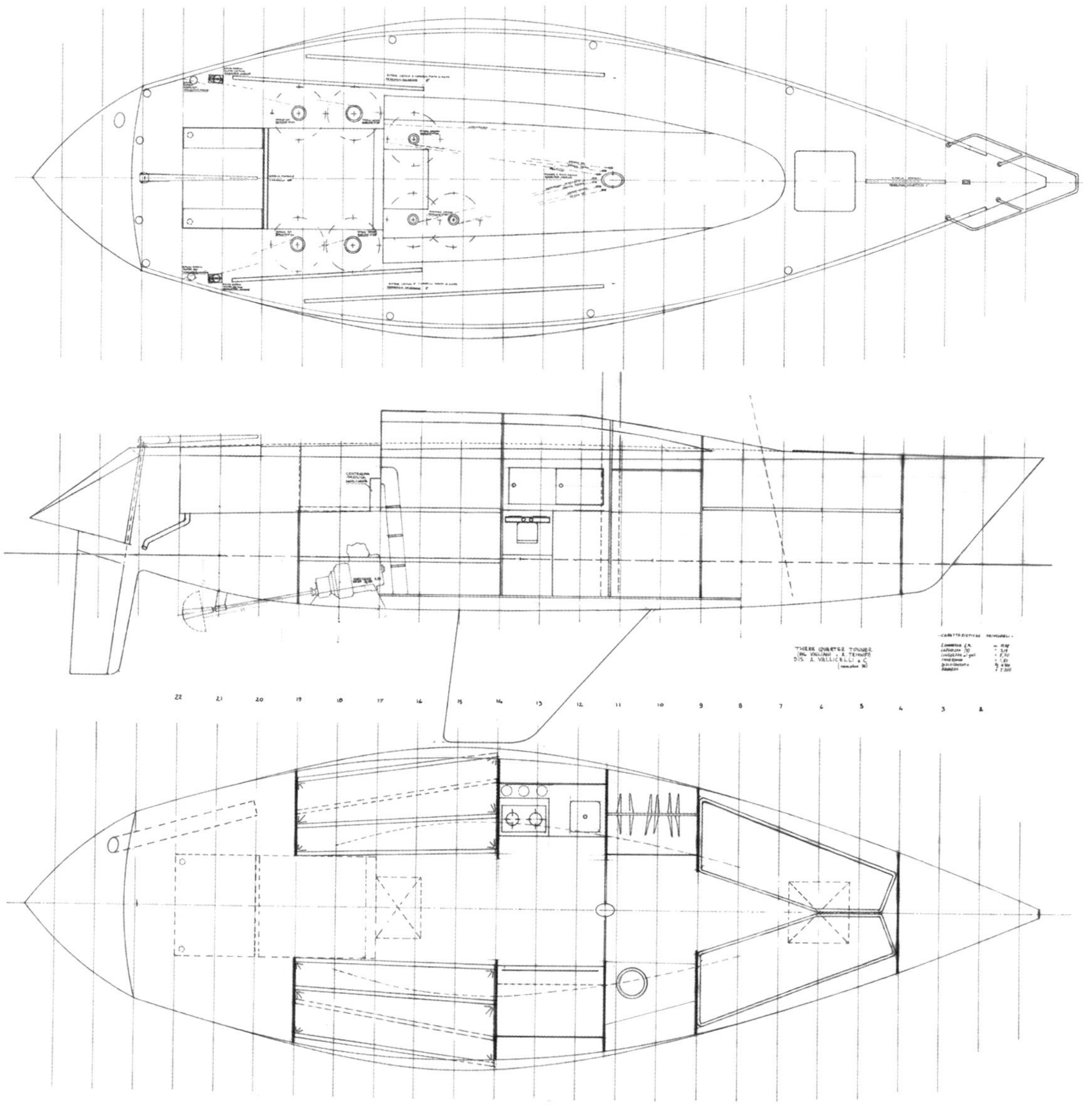

During construction and at the launch of Argento vivo.

Deck plan and general arrangement. Ink drawing on tracing paper.

Sail plan. Ink drawing with halftone screen on tracing paper.

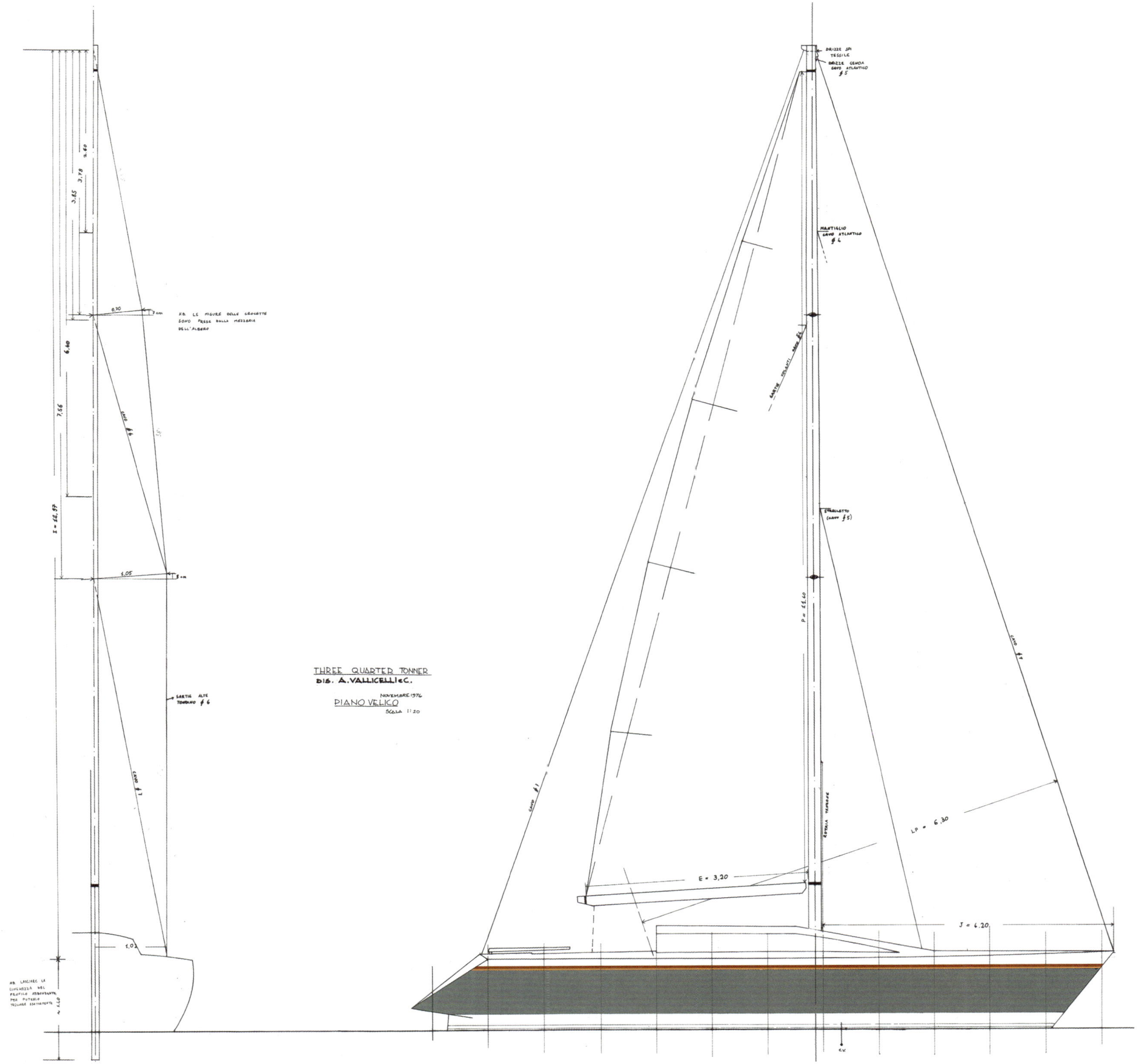
THREE QUARTER TONNER
DIS. A. VALLICELLI e C.
NOVEMBRE 1976
PIANO VELICO
SCALA 1:20
E = 3,20
LP = 6,30
J = 6,20

El Cid

The success of *Ziggurat* in the 1976 Half Ton Cup drew the attention of Trieste's yacht owners to our studio; Franco Zago, a passionate sailor and racer, commissioned a one tonner from us, *El Cid*. Constructed in cold-molded wood at the Sergio&Nico boatyard, *El Cid* borrowed and developed some concepts from *Gattone*, but compared to our first one tonner. She had a slightly fuller stern and a more classic deck layout with a single cockpit. Carlo Bertorello helped with the structural plans of this design too. For this yacht, we used a steel structure for the frames near the mast step and the keel to distribute the load transmitted by the chainplates. This solution, which gave the hull a mixed structural core of wood and steel, proved very effective.
El Cid was a very well managed boat, and for years, she performed well in many regattas, especially in the upper Adriatic, even winning the Barcolana in real time several times, despite her small size.

Year 1976
Builder Crisman & Giraldi Trieste
Construction Cold molded wood
LOA 11.295 m
B MAX 3.46 m
D 1.95 m
DSPL 6.3 t
Mainsail 28 mq
Headsail 48 mq
Spi 110 mq

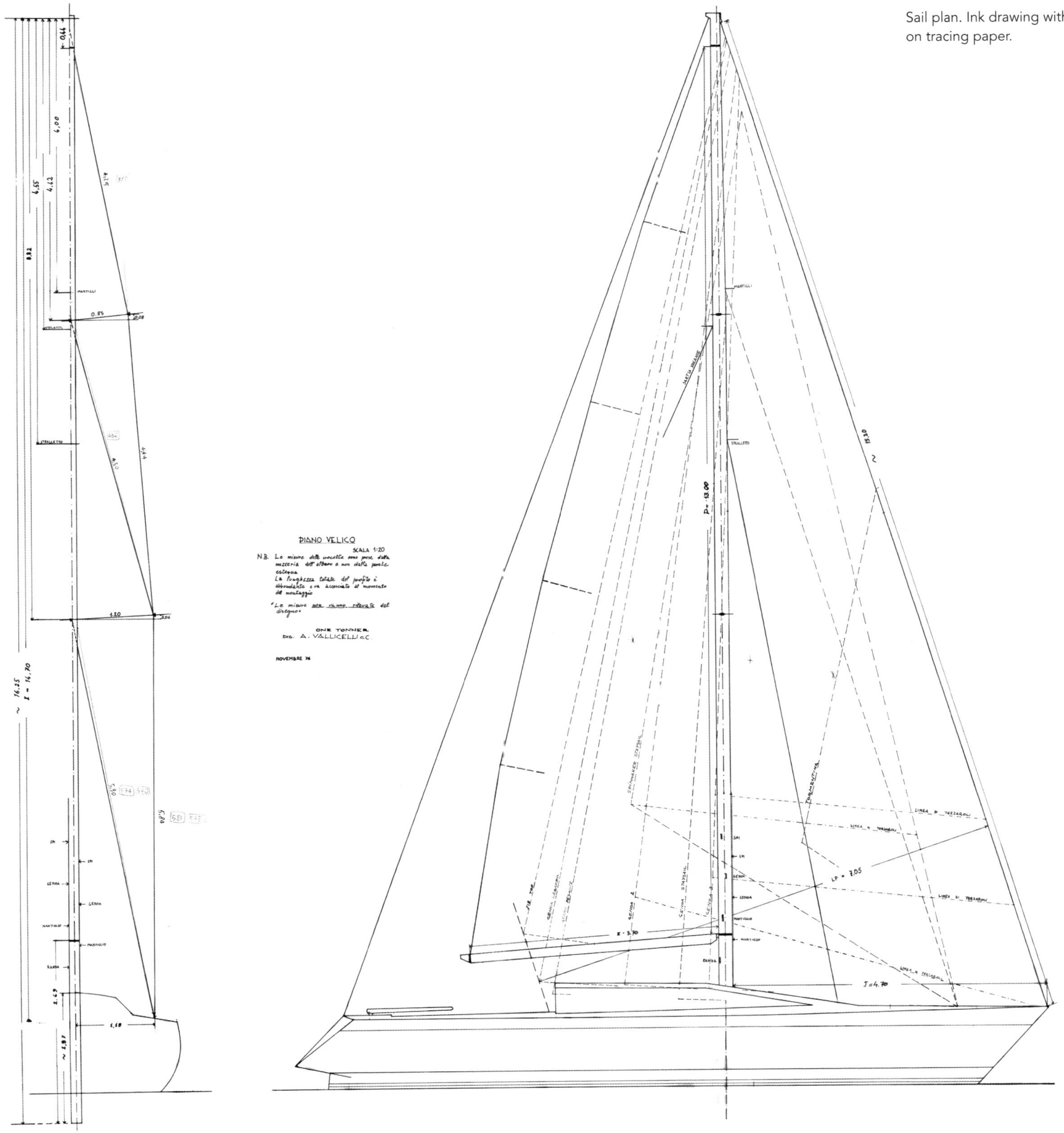

Sail plan. Ink drawing with halftone screen on tracing paper.

Caccia alla volpe

Caccia alla volpe is the first light displacement boat we designed. The 13.4 meter boat was built using marine plywood by the Bani yard at Monte Argentario. She had a chined hull, so was cheap to build, and a fractional rig, a novelty then but one we often used later on. She was commissioned to participate in an ocean regatta that was very popular at the time, the Cape Town to Rio, but she was destined never to participate in it. However, she has always done well in regattas in the Mediterranean, proving to be very solid and safe. From 1992 onwards, fourteen years after her launch, *Caccia alla volpe* won various editions of the Antigua Race Week.

With this project, and with *Asabranca* the following year, we began experimenting with light displacements, but without overdoing the concept as was common back then. *Caccia alla volpe* is, in fact, a little wider and more stable than the very slender American boats: she does not have an ultra-light but rather a medium-light displacement: so she was an LDB, not a real ULDB (Ultra Light Displacement Boat). Coupled with a generous sail area, she did well in the light wind conditions of the Mediterranean regattas. The design experience with *Caccia alla volpe* would prove to be very useful for our future designs for the Open classes.

Year 1977
Builder Bani Monte Argentario (GR)
Construction Marine plywood
LOA 13.4 m
B MAX 4.1 m
D 2.29 m
DSPL 6.72 t
Mainsail 48 mq
Headsail 58 mq
Spi 120 mq

Remo Bani, Andrea Vallicelli, Patrizia Ferri during a survey in the yard.

Sail plan. Ink drawing with halftone screen on tracing paper.

Deck sections and line plan offsets.

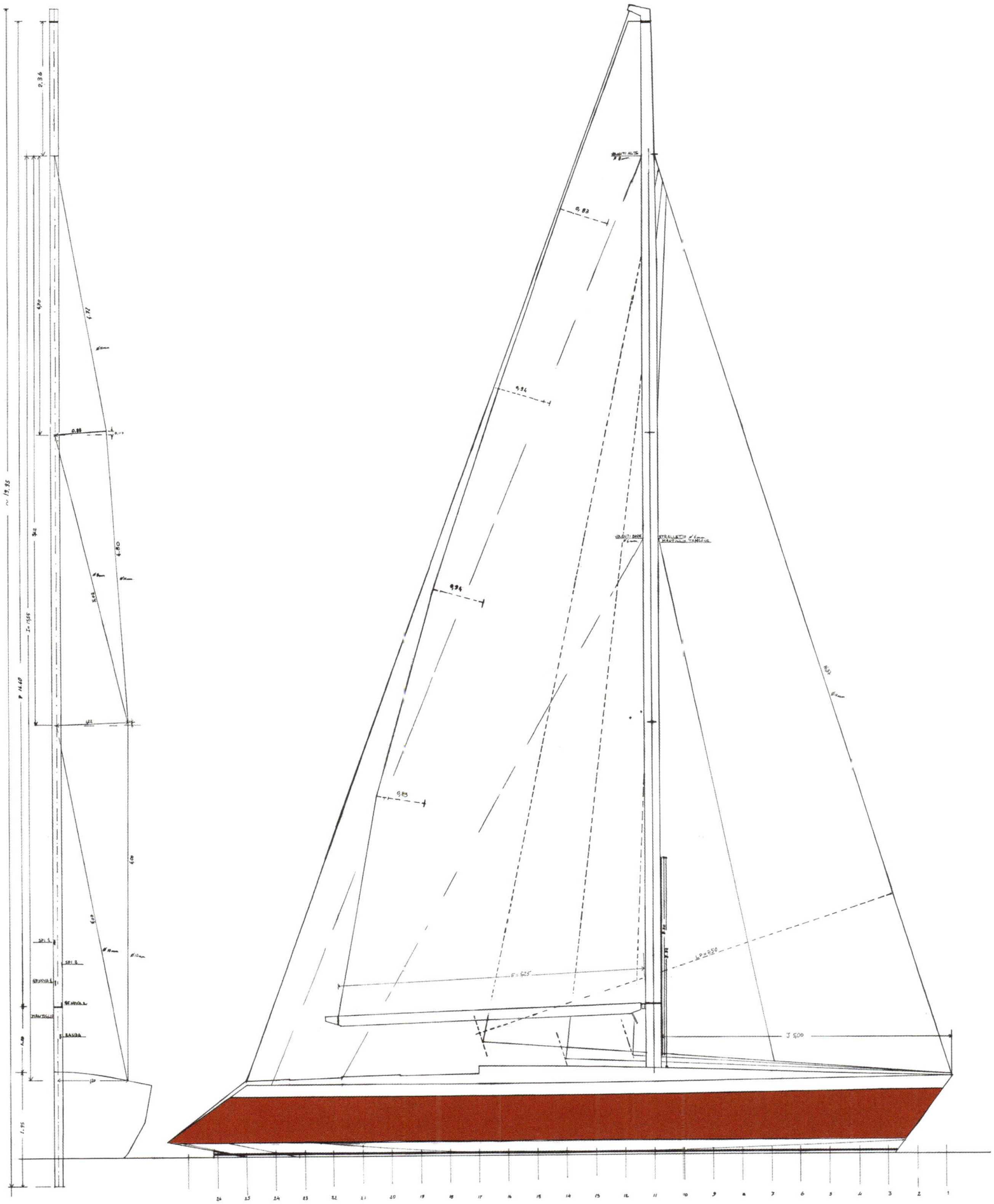

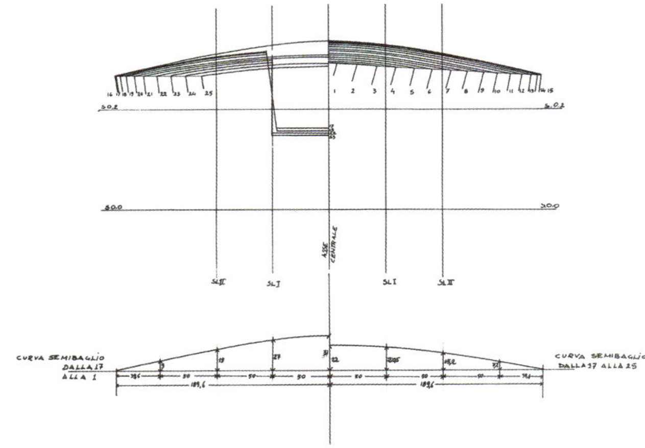

ORDINATE	QUOTE ORIZZONTALI dell'ASSE CENTRALE								
	COP	s.o. 2	s.o. 1	s.o. 0	s.o.-1	Sp. A	Sp. B	Sp. C	Sp. D
2	24,4	13							5,8
3	42	31	18,4	3,4			2,4	9,6	23,2
4	59	48,4	36	18,4		6	15,8	25,8	40,4
5	75,6	65,6	53	32,4	14	14	27,6	40,4	57,6
6	91,8	81,6	68,4	47	24,4	20,4	39	55,6	73,4
7	107	98	85,6	61	36	26,2	49,2	70	90
3.	122	113	100	74,4	46	31	58,8	84	105
3	136,2	128,2	116	89	57	34,6	68	97,8	120,6
0	149	142	130	102	68	38	76	109,6	134,4
1	161	154	143,2	113	78	40	83	120	147,8
2	171	164,6	155	124	89	44,6	89,8	130	158,8
3	179	174	165	134	96,6	43	95	138,2	168,6
4	184,6	179,8	171,6	139,6	101	44,2	98	143	175,2
5	188,2	184	175,2	142	104	44,4	100	146	179
6	189,6	184,6	176,2	143,8	103	45	100	148,4	180,2
7	188,4	183,8	175,4	142,4	102	44,4	99,4	147,8	179,6
8	185,4	180,4	171,2	138,8	98	44	98	146	176
9	179,8	175,4	165,8	134,4	87	43	95,8	142	171,2
0	172,4	168	157,8	122	73	41,6	92	136,4	164
1	163,4	159,8	148	109,6	57	39,8	88	130	156
2	152,4	148	136	94,2	37	36,6	82,2	121	144,6
3	140	135,8	123	78,4		34	76	111,6	132,4
4	126,4	122,4	108,4	52		30	68,4	101	119,8
5	111,8	108	93,2	26		26	60	89,6	106
6			77			22	51,8	77,8	92
7			54			17,6	43,2		

ORDINATE	QUOTE VERTICALI dalla LINEA BASE							
	COP.	s.l. 2	s.l. 1	CHIGL.	Sp. A	Sp. B	Sp. C	Sp. D
2	189,2			107,6	107,6			125,8
3	188			48,8	48,8	57,2	80,4	124,4
4	187	156		35,4	35,4	53,2	78,4	123,8
5	186	103		30	30	49,6	76,8	122,6
6	185,4	66		25,8	25,8	46	75	122
7	184,4	44,4	157,6	22	22	42,6	73	121
8	183,8	33,4	110	18,8	18,8	40	71,6	120
9	183	25	75,6	16	16	36,8	69,8	119,4
10	182,4	20	58,4	14	14	34,4	68	118,4
11	182	17	48	12,2	12,2	32,2	66,8	118
12	181,6	14,6	39,6	10,6	10,6	30	65,8	117,2
13	181,4	12,4	33	9,8	9,8	28,8	64,4	116,4
14	181	11	29,6	8,6	8,6	27,8	64	116,2
15	180,6	10	28,4	8	8	27	63,8	116
16	180,2	10	28	8	8	28,2	64	116,6
17	180	10,4	28,8	9	9	29	64,2	117
18	180	12,6	32	10,6	10,6	30,4	65,6	118
19	180	16	37	13,8	13,8	34	68	119
20	180	22	43,8	17,8	17,8	38	71,6	121
21	180	28	52,4	23,4	23,4	43,6	76	123
22	180	36,8	64,6	30,2	30,2	50,4	82	126
23	180	46,4	78	39	39	58	88	128,8
24	180	59	94	49	49	68	95	132
25	180	72,8	123	60	60	78,4	102,8	135,8
26		88		69,8	72	89,6	110	139
27		106		79,6	84,2	100,2		

Lines plan (stations, waterlines, buttocks, diagonal sections). Ink drawing on tracing paper.

General arrangement.

Deck plan.

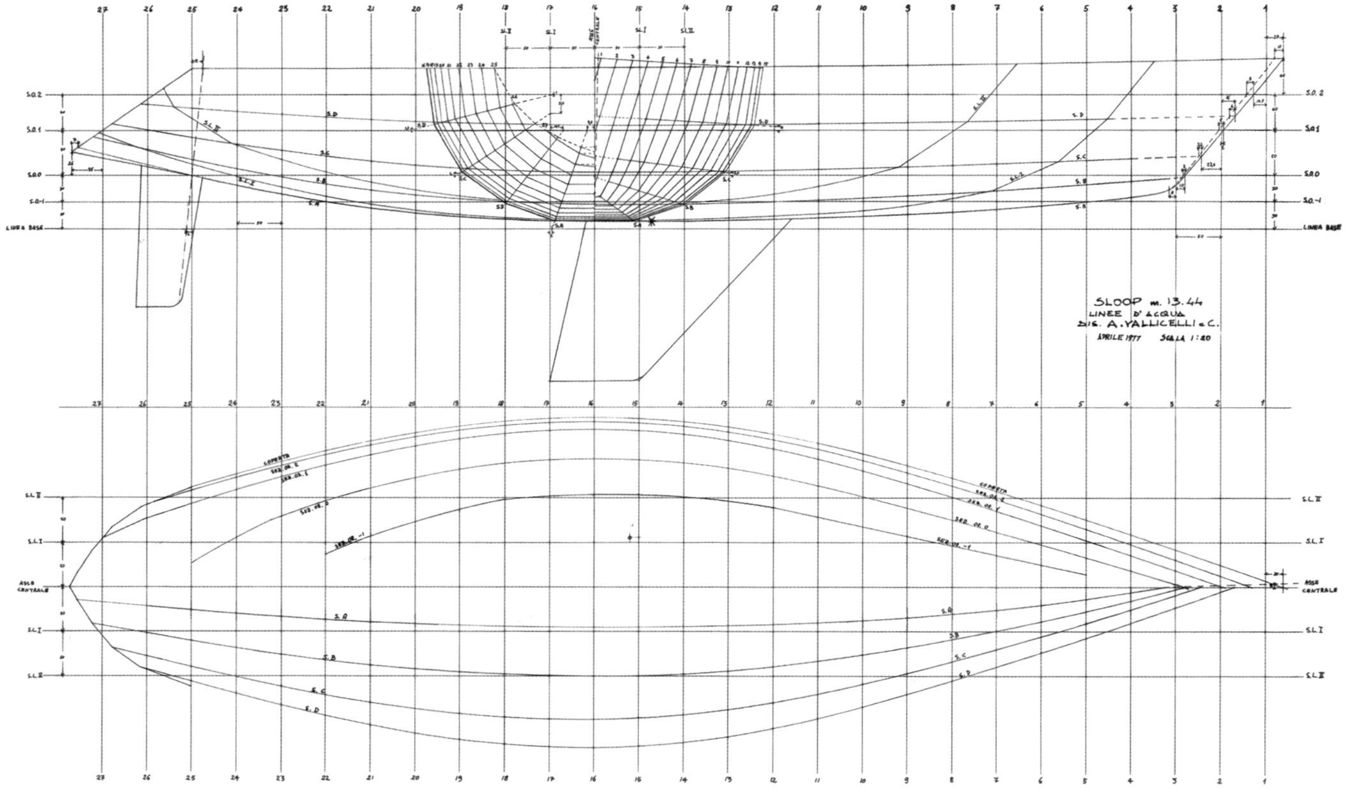

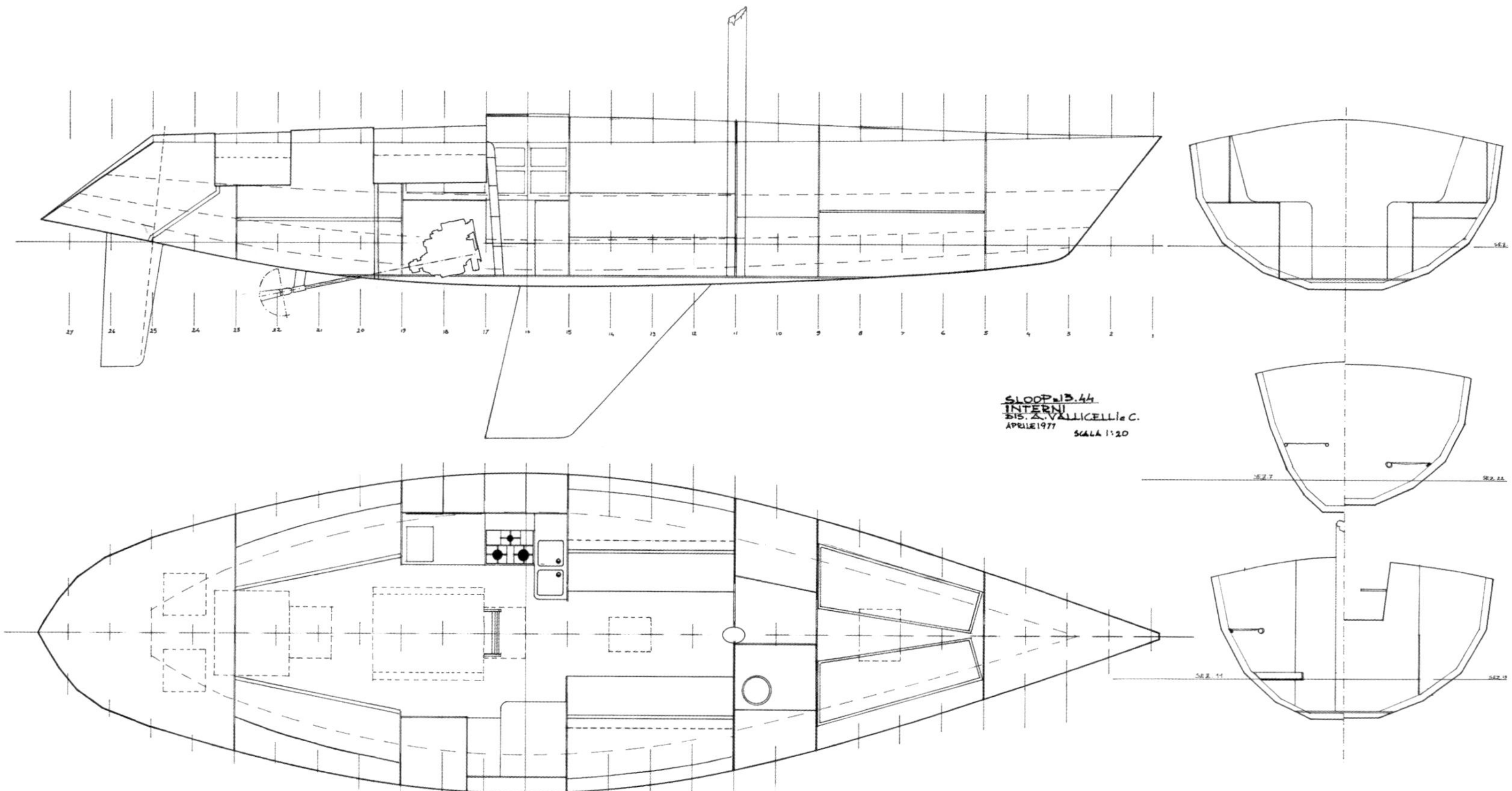

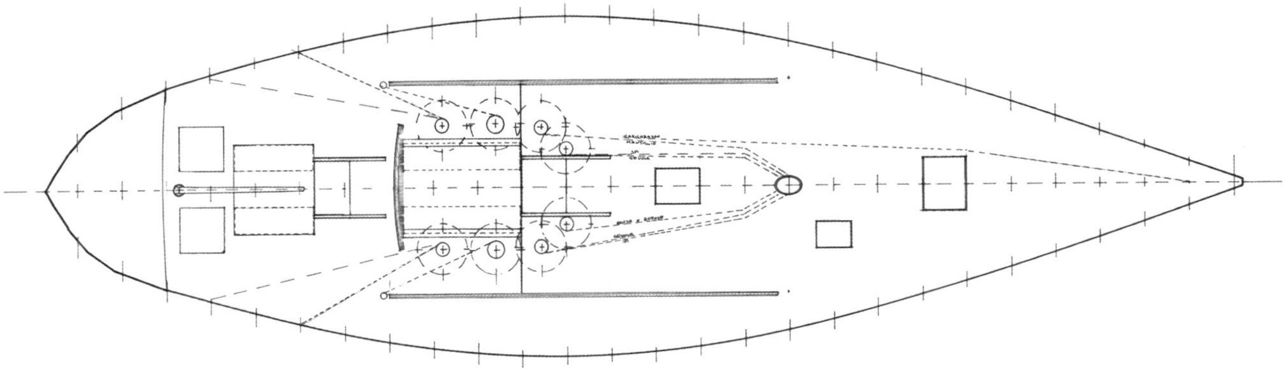

Asabranca

Right after the excellent second place of *Argento vivo* at the Three Quarter Ton Cup in 1977, we decided to explore new design options. The first boat born of these changes was *Asabranca*, a light displacement half tonner, equipped with a centerboard, wide stern and good form stability. With *Asabranca*, we took on the New Zealand school, led by Bruce Farr, whose boats had won four Ton Cups in 1977. Compared to the canons of New Zealand design, our half tonner presented some marked differences: the most significant consisted in less extreme stern sections, more rounded at the knees between the bottom and the topsides. The idea was to reduce wetted surface when heeled. The typical boats of the New Zealand school had squarer sections with flatter counters that on the one hand, facilitated planing in a strong wind when off the wind, but on the other hand compromised performance in light airs. Our experimental half tonner proved to be a fast boat, despite the fact that our inexperience with light displacements had caused us to make a few mistakes. The most serious was to build a centerboard case that was not only over-engineered but also difficult to use. *Asabranca* did not participate in many regattas, but at the 1978 Half Ton Cup in Poole, she demonstrated her good qualities, often sailing at the front of the fleet.

Year 1978
Builder CPR Fiumicino (RM)
Construction GRP
LOA 9.4 m
B MAX 3.2 m
D 1.8 m
DSPL 3.1 t
Mainsail 24 mq
Headsail 22 mq
Spi 48 mq

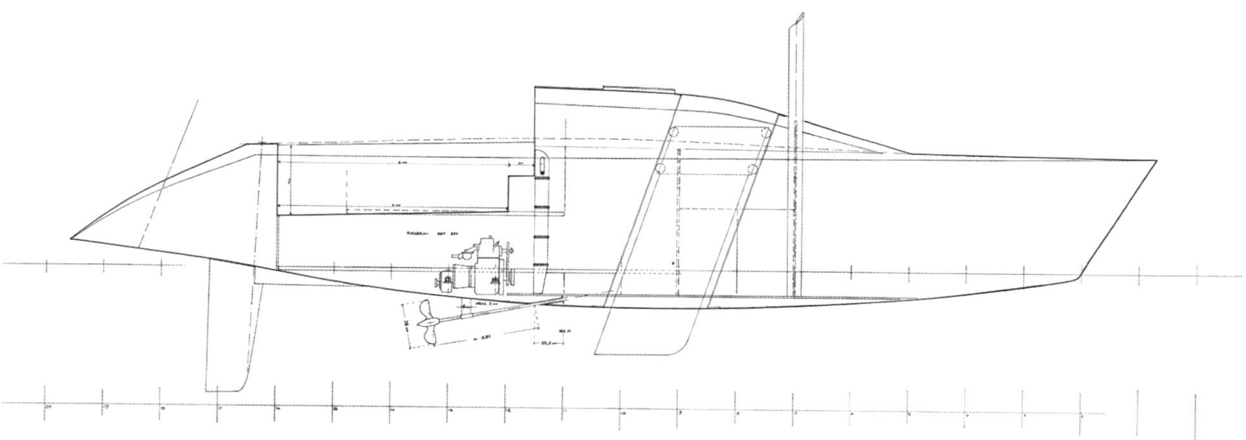

Longitudinal section.

Asabranca at the 1978 Half Ton Cup in Poole.

Sail plan.

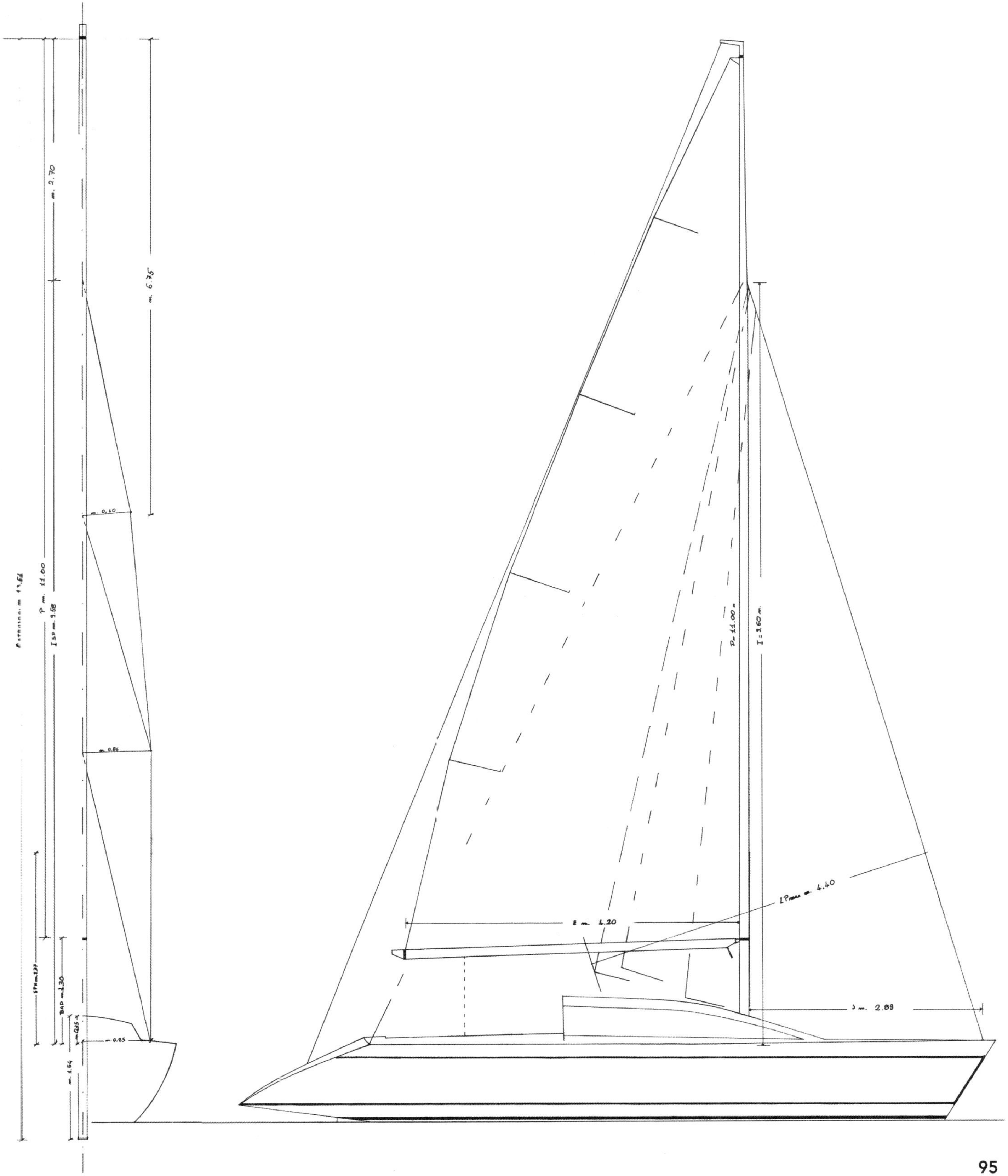

P= 11.00 m
I= 3.60 m
E m. 4.20
J m. 2.88
LPmax m. 4.40

Oro fino

Oro fino was perhaps one of the most elegant boats we ever designed. Our goal was to create a real all-round boat with very sleek lines, without extremes of any kind, with a perfectly balanced relationship between overall beam and length. When designing her, even if she was destined to compete in IOR races, we did not take advantage of the forced hull geometry that was fashionable at the time with the so-called bumps, protrusions that deformed the shape of the hull at the measurement points. The curious shapes that had characterized the designs of *Brava* and *Filo da torcere* disappeared in *Oro fino*, which can be considered the synthesis of the experience gained in the first four years of our studio's work. The boat was given a fractional rig and the deck fitted with the best hardware available. The construction was of particularly high quality, made in cold-molded wood in Rimini by Luigi Morri and Dino Para, owners of the boatyard bearing their name, where they built boats as if they were musical instruments. The light and rigid construction was one of the strong features of *Oro fino* as, unlike many boats of the time, she never twisted or lost her shape in heavy weather. We worked on other projects with the same boatyard a few years later, including the one tonner *Brava*. *Oro fino* won the Fourth Class Italian championship in 1979 and 1980 and is still well maintained and regularly races.

Year 1979
Builder Morri & Para Rimini
Construction Cold molded wood
LOA 10.15 m
B MAX 3.34 m
D 1.93 m
DSPL 3.9 t
Mainsail 34 mq
Headsail 34 mq
Spi 68 mq

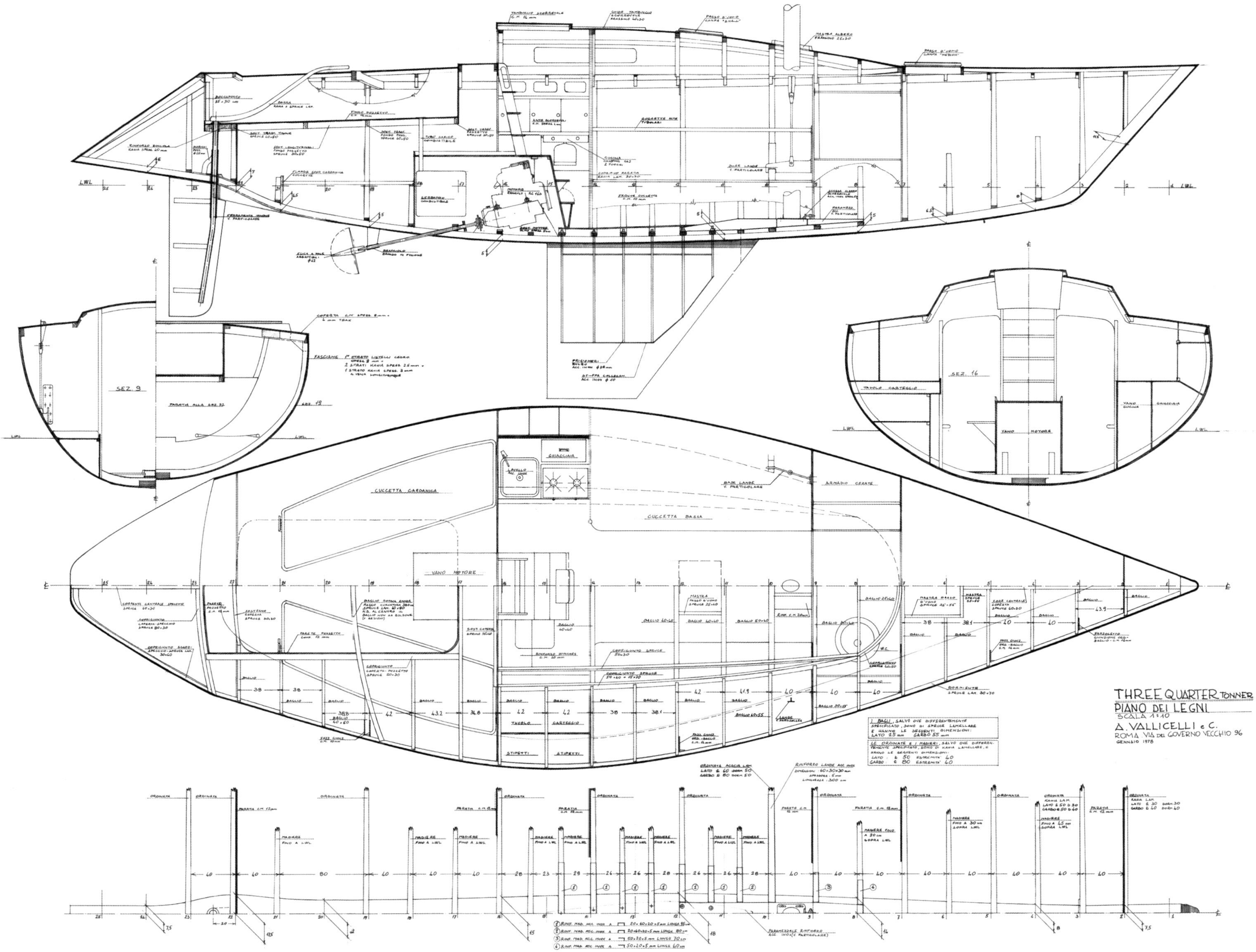

Oro Fino in the yard during construction.

Oro fino at the Italian Championship fourth class in 1979 and 1980.

Construction plan.

Nonno Gigi

The Florentine boat owner Giorgio Tantussi commissioned us to design a boat for cruising, capable of participating in club races at the most. Instead, we convinced him to create a racing boat, in which we further developed the design trend started with *Asabranca* the year before. The result was a small Second Class yacht of medium-light displacement, little wetted surface, the maximum beam set well aft, and finishing in a very wide transom, with a fractional rig. *Nonno Gigi* was built in cold-molded wood by the Bani yard in Monte Argentario and was the expression of our desire towards the end of the 1970s to find a compromise between the medium-heavy displacements of the American school and the light displacements of the New Zealand school. Our hull lines were not characterized by radical solutions as in Farr's boats but were designed to reduce wetted surface and maintain a good block coefficient, even to the detriment of form stability. We participated in the selections for the 1979 Admiral's Cup with Paul Elvstrom at the helm. After four regattas with medium strength winds, *Nonno Gigi* was at the top of the standings but missed out joining the team by just a few points due to a poor finish in the long-distance race. In Cowes that year, the edition of the Admiral's Cup that went down in history for the Fastnet tragedy, the boat that drew the most attention was a design by Ed Dubois, *Police car*, very similar to *Nonno Gigi:* both boats demonstrated that a compromise between the two schools – American and New Zealand – was not only possible but also a winning solution.

Year 1979
Builder Bani Monte Argentario (GR)
Construction Cold molded wood
LOA 12.065 m
B MAX 3.85 m
D 2.17 m
DSPL 6.5 t
Mainsail 49 mq
Headsail 51 mq
Spi 99 mq

Nonno Gigi during the 1979 Punta Ala races. Left: Paul Elvstrøm at the helm. Right (L to R): Giorgio Tantussi, Gary Redfern, Andrea Vallicelli, Paul Elvstrøm, Vittorio Mariani, Riccardo Trionfi.

Construction plan.

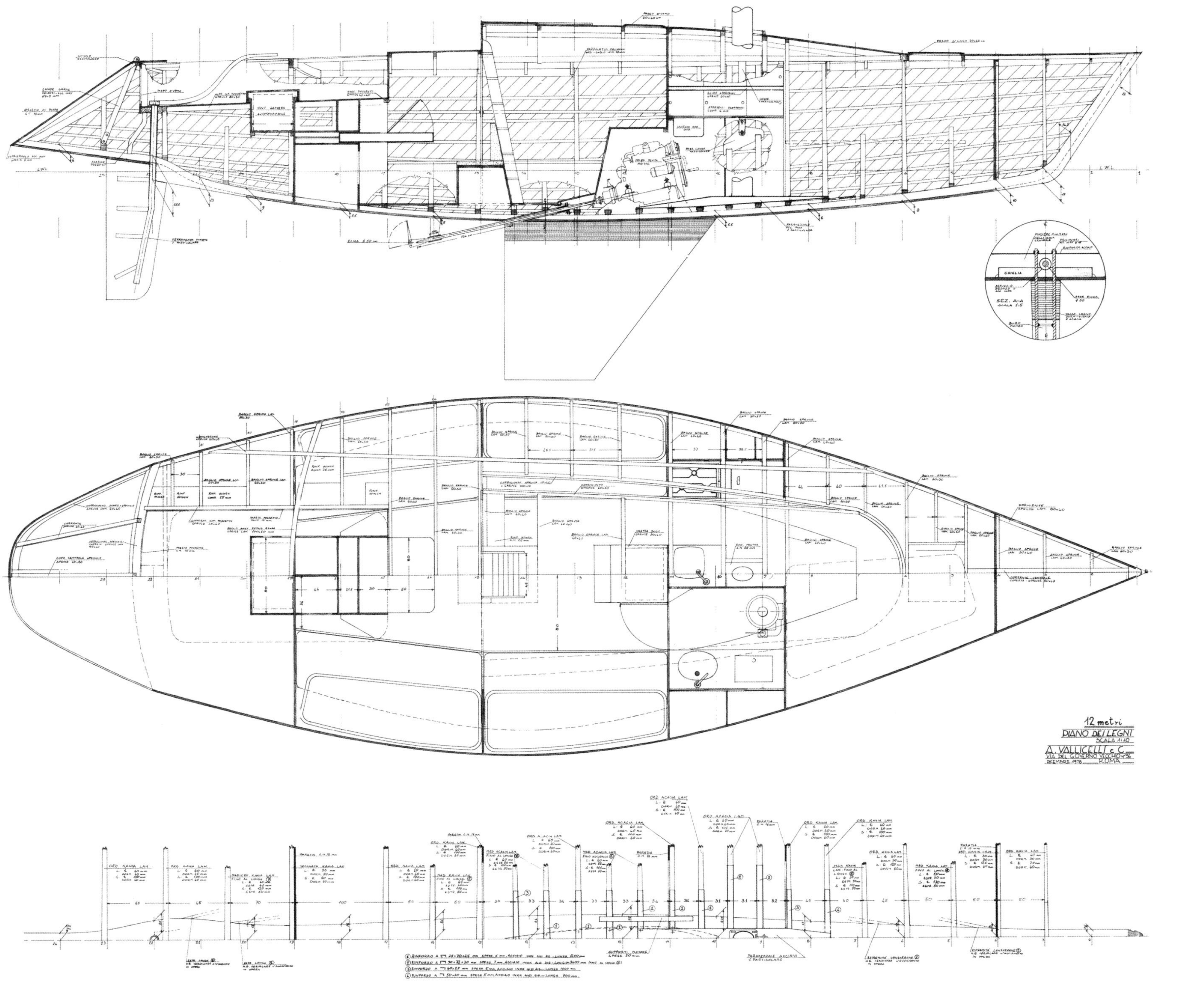
12 metri
PIANO DEI LEGNI
SCALA 1:10
A. VALLICELLI e C.
VIA DEL GOVERNO VECCHIO n°56
DICEMBRE 1978
ROMA

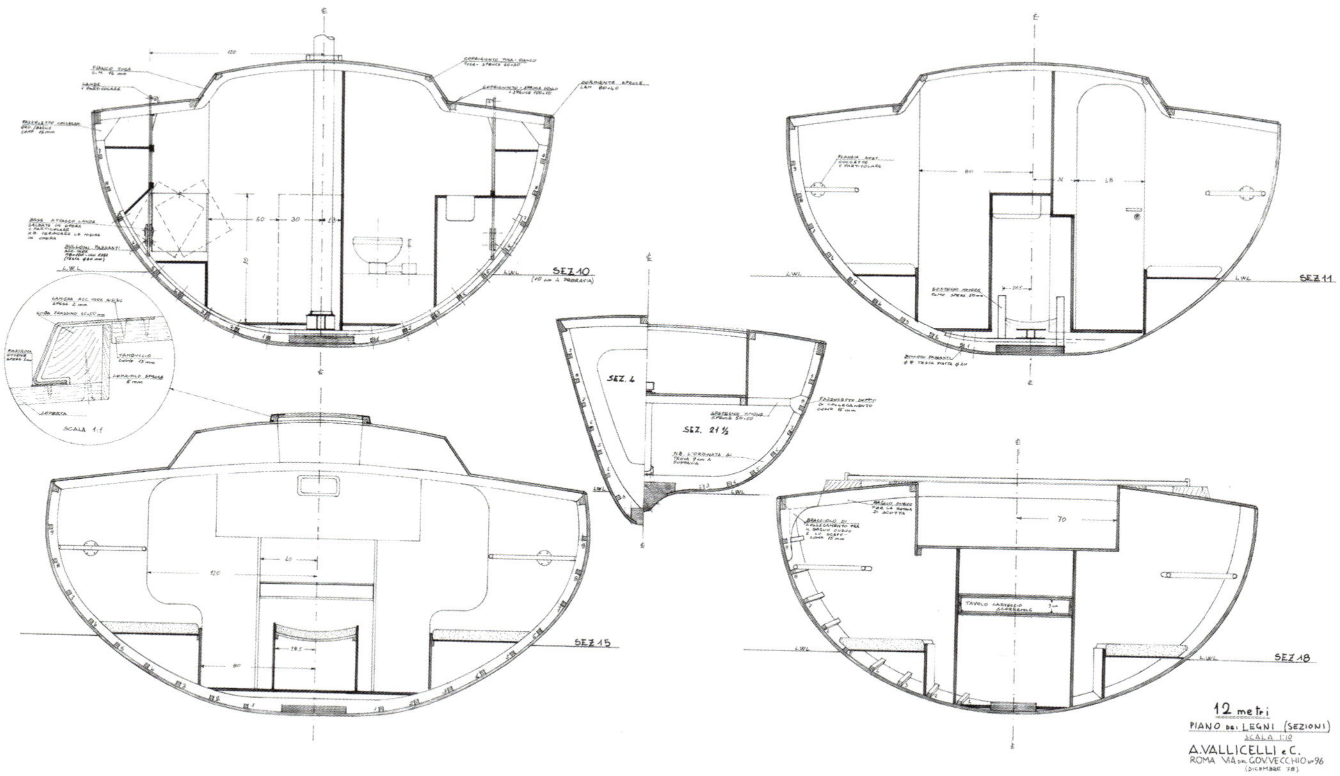

Extract from Rating certificate.

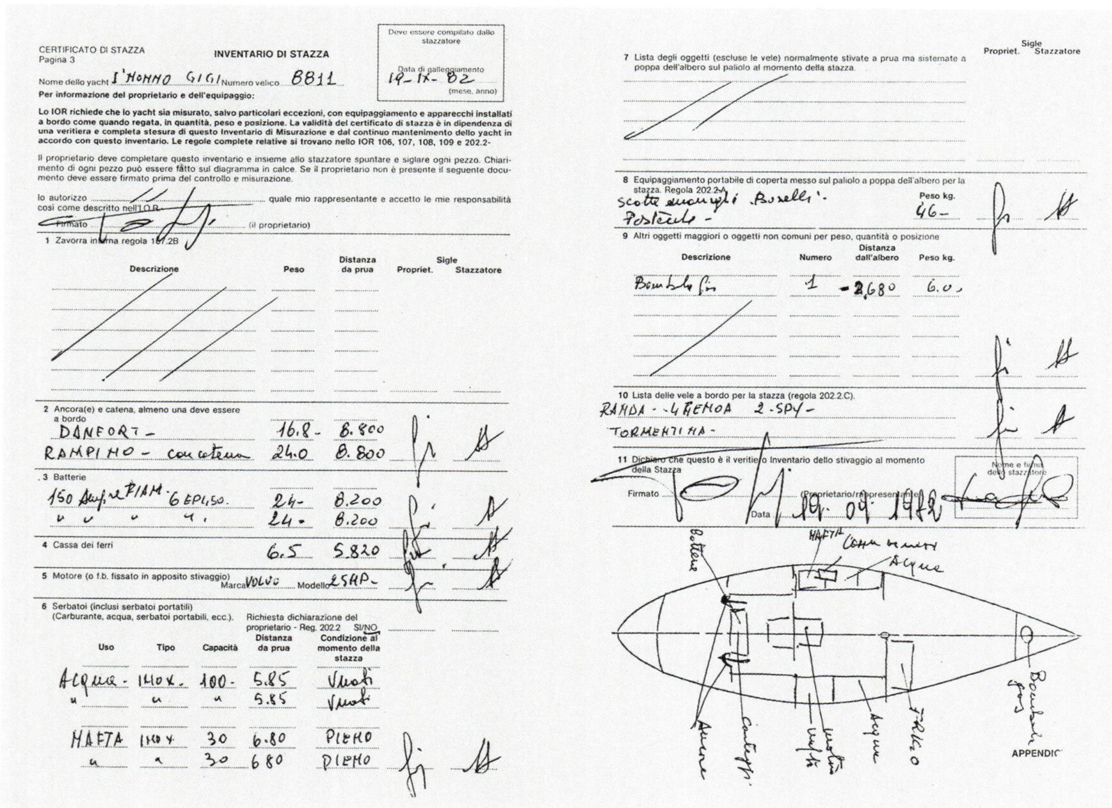

CERTIFICATO DI STAZZA
Pagina 3

INVENTARIO DI STAZZA

Deve essere compilato dallo stazzatore

Data di galleggiamento 19-IX-82 (mese, anno)

Nome dello yacht I'MOMMO GIGI Numero velico 8811

Per informazione del proprietario e dell'equipaggio:

Lo IOR richiede che lo yacht sia misurato, salvo particolari eccezioni, con equipaggiamento e apparecchi installati a bordo come quando regata, in quantità, peso e posizione. La validità del certificato di stazza è in dipendenza di una veritiera e completa stesura di questo Inventario di Misurazione e dal continuo mantenimento dello yacht in accordo con questo inventario. Le regole complete relative si trovano nello IOR 106, 107, 108, 109 e 202.2-

Il proprietario deve completare questo inventario e insieme allo stazzatore spuntare e siglare ogni pezzo. Chiarimento di ogni pezzo può essere fatto sul diagramma in calce. Se il proprietario non è presente il seguente documento deve essere firmato prima del controllo e misurazione.

Io autorizzo .... quale mio rappresentante e accetto le mie responsabilità così come descritto nell'I.O.R.

Firmato .... (il proprietario)

1 Zavorra interna regola 107.2B

Descrizione	Peso	Distanza da prua	Sigle Propriet.	Sigle Stazzatore

2 Ancora(e) e catena, almeno una deve essere a bordo

DANFORT-	16.8-	8.800		
RAMPINO- con catena	24.0	8.800		

3 Batterie

150 Amp. FIAM. 6 EPL 150.	24-	8.200		
" " " ".	24-	8.200		

4 Cassa dei ferri — 6.5 — 5.820

5 Motore (o f.b. fissato in apposito stivaggio) Marca VOLVO Modello 25HP-

6 Serbatoi (inclusi serbatoi portatili) (Carburante, acqua, serbatoi portabili, ecc.). Richiesta dichiarazione del proprietario - Reg. 202.2 SI/NO

Uso	Tipo	Capacità	Distanza da prua	Condizione al momento della stazza
Acqua-	INOX-	100-	5.85	Vuoti
"	"	"	5.85	Vuoti
NAFTA	INOX	30	6.80	PIENO
"	"	30	6 80	PIENO

7 Lista degli oggetti (escluse le vele) normalmente stivate a prua ma sistemate a poppa dell'albero sul paliolo al momento della stazza.

8 Equipaggiamento portabile di coperta messo sul paliolo a poppa dell'albero per la stazza. Regola 202.2A

	Peso kg.
scotte ... Boselli. Postecalo-	46-

9 Altri oggetti maggiori o oggetti non comuni per peso, quantità o posizione

Descrizione	Numero	Distanza dall'albero	Peso kg.
Bombole gas	1	-2,680	6.0.

10 Lista delle vele a bordo per la stazza (regola 202.2.C).
RANDA - 4 GENOA 2 - SPY -
TORMENTINA -

11 Dichiaro che questo è il veritiero Inventario dello stivaggio al momento della Stazza

Firmato .... (Proprietario/rappresentante)

Data 19. 09. 1982

Nome e firma dello stazzatore

APPENDICI

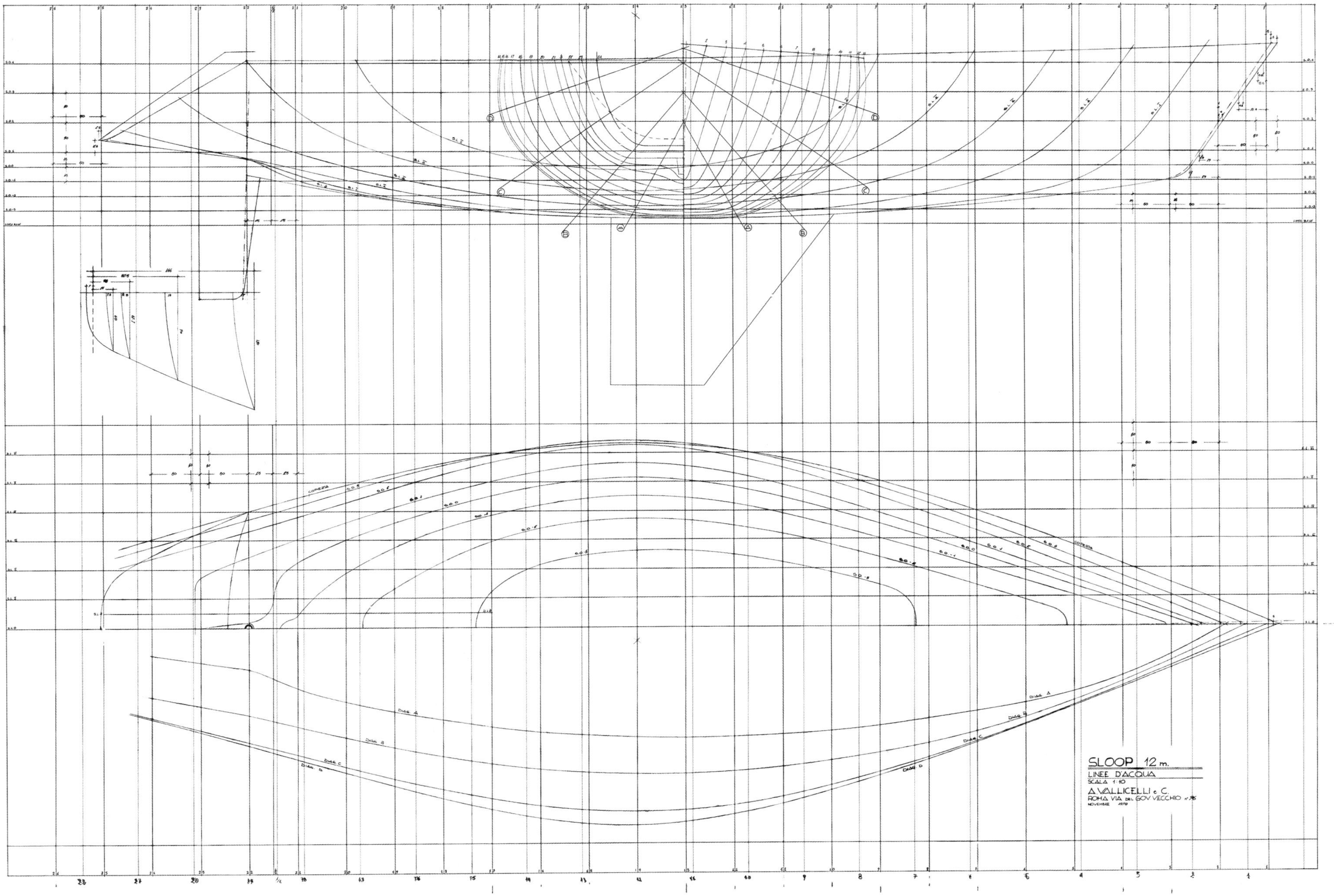

Construction plan sections.

Lines plan (stations, waterlines, buttocks, diagonal sections). Ink drawing on tracing paper.

Filo da torcere

We designed several boats for the 1980 One Ton Cup, which was to be held in Naples: among these was *Excelsior*, a production Canados 37 with Paul Elvstrom at the helm and *LSD*. We took first place with *Filo da torcere*, the first Italian designed and crewed one tonner to win the prestigious trophy. The owners were once again Pigi Vigliani and Riccardo Trionfi, the same as *Argento vivo*, the boat that three years earlier had come second in the Three Quarter Ton Cup in La Rochelle. Like her contemporary, *Brava*, *Filo da torcere* was designed with an acutely angled stem, and a maximum beam similar to other boats of the same generation, but further aft. The hull had rather wide and flat stern sections and was fitted with a large rudder and keel with NACA profiles, less tapered than in our previous boats. The design was optimized for the medium-light wind conditions in the Gulf of Naples and, therefore, with a large sail plan, able to give its best when running free. With Stefano Roberti at the helm, *Filo da torcere* won the One Ton Cup in the last stretch of the long offshore race, when the wind had dropped and backed to give a light breeze, the exact conditions for which she was designed. *Filo da torcere* was built by De Cesari in cold-molded wood glued with the West System and, like *Argento vivo*, had an elegant deck covered with a thin layer of teak.

Year 1979
Builder De Cesari Cervia (RA)
Construction Cold molded wood
LOA 11.292 m
B MAX 3.7 m
D 2.043 m
DSPL 5 t
Mainsail 43 mq
Headsail 43 mq
Spi 87 mq

Filo da Torcere during a race.

Construction plan.

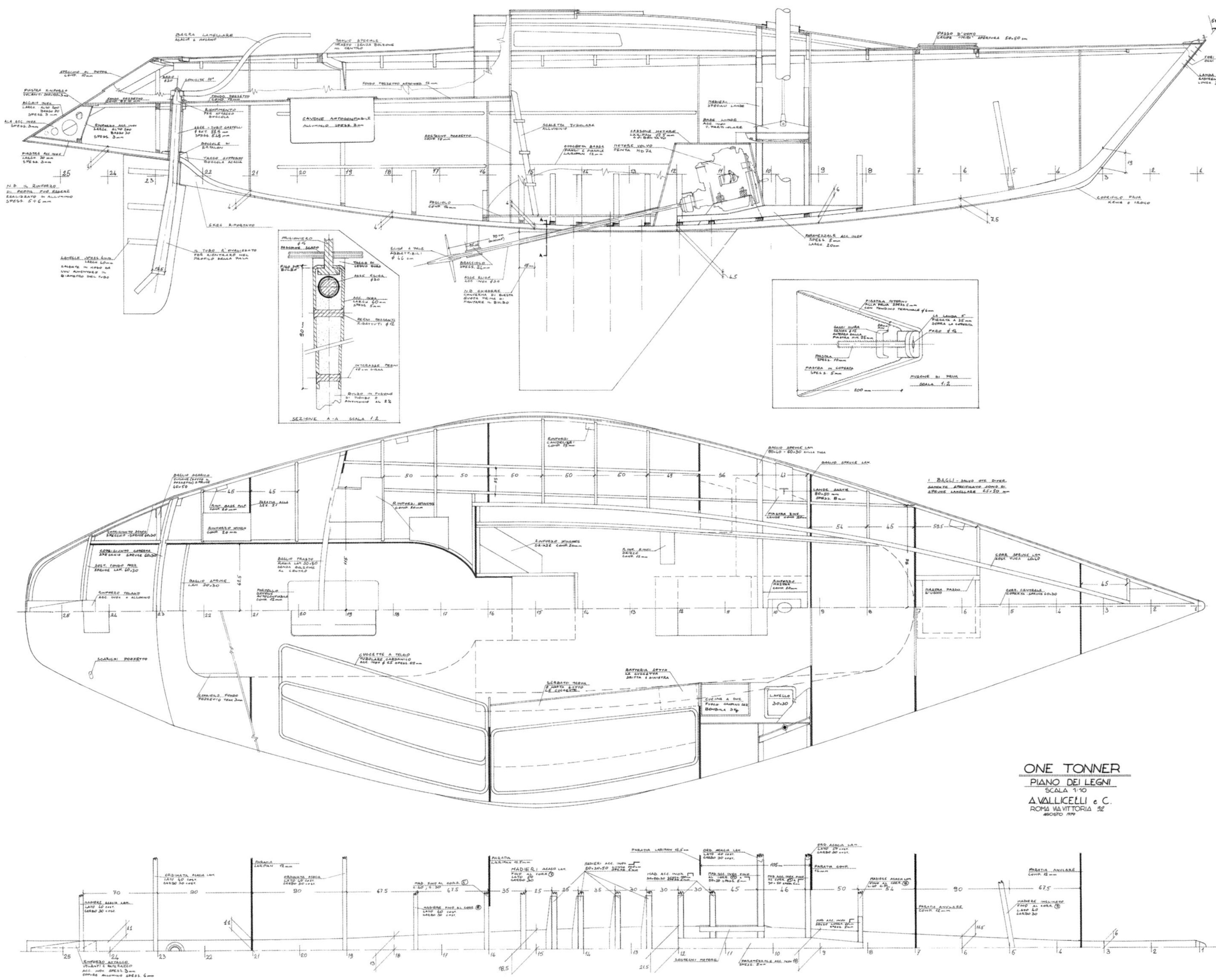
ONE TONNER
PIANO DEI LEGNI
SCALA 1:10
A.VALLICELLI e C.
ROMA VIA VITTORIA 32
AGOSTO 1979

Construction sections.

Filo da Torcere during a race.

Interiors: chart table, hull structure in the forepeak.

Lines plan (stations, waterlines, buttocks, diagonal sections). Ink and pencil drawing on polyester paper.

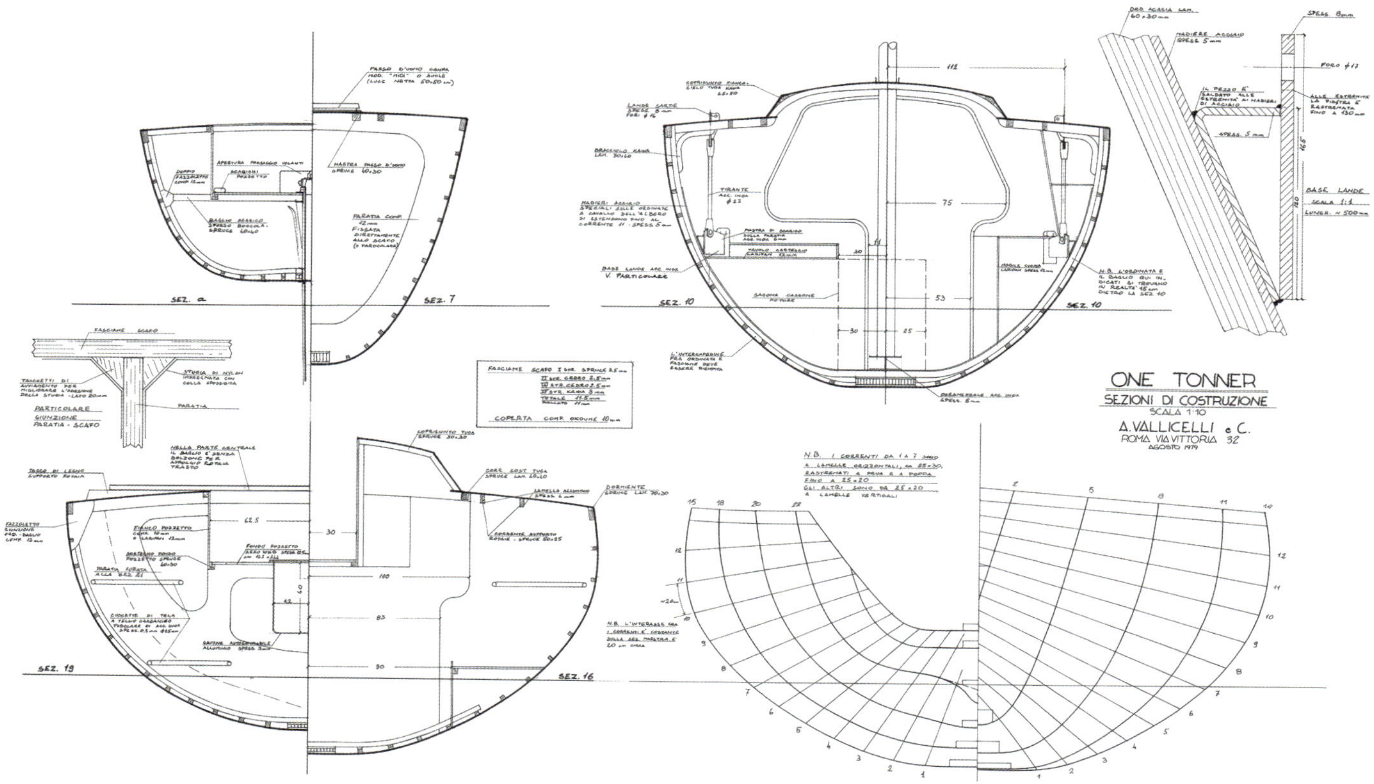

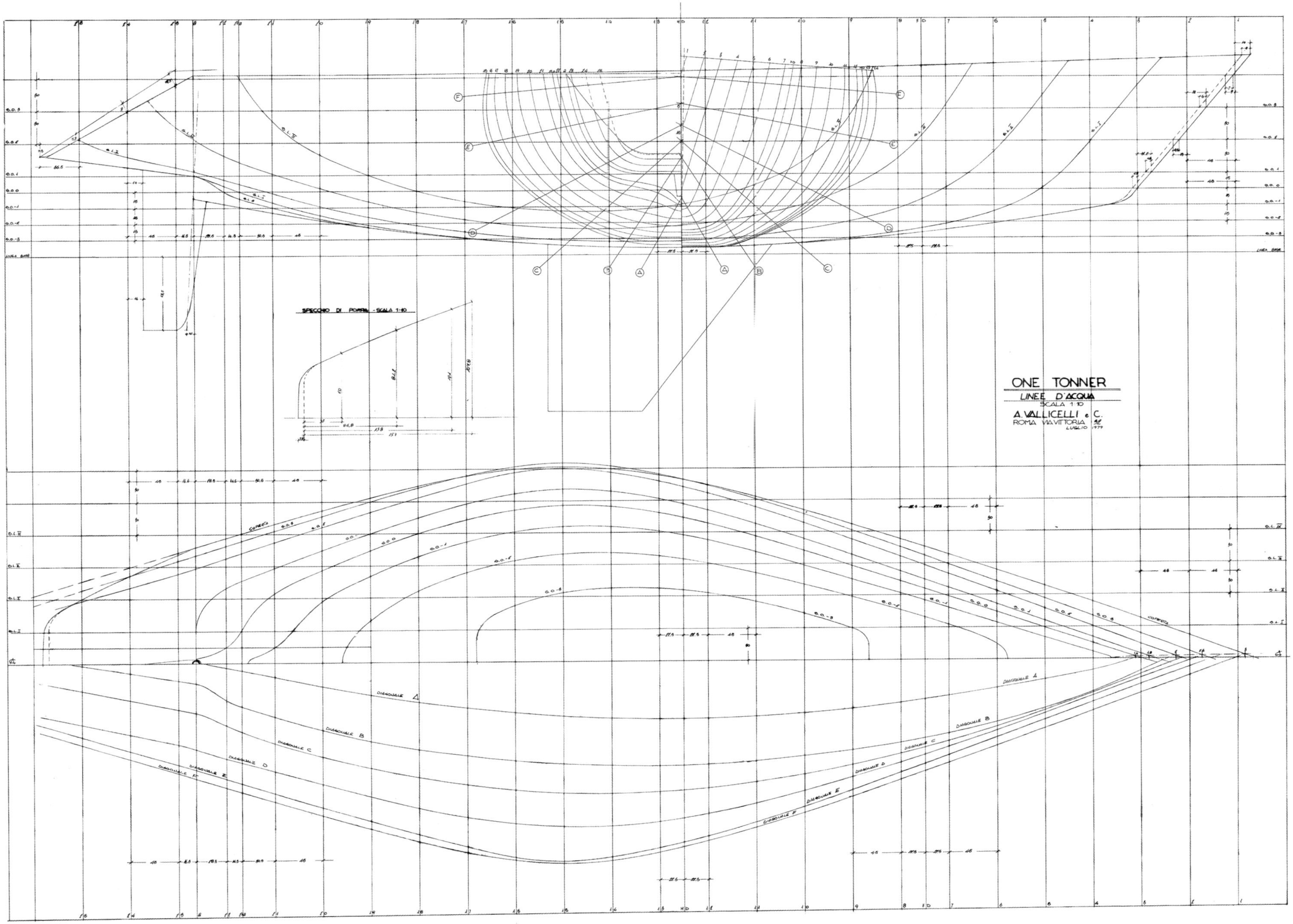
SPECCHIO DI POPPA - SCALA 1:10
ONE TONNER
LINEE D'ACQUA
SCALA 1:10
A. VALLICELLI e C.
ROMA VIA VITTORIA
LUGLIO 1979
DIAGONALE A
DIAGONALE B
DIAGONALE C
DIAGONALE D
DIAGONALE E
DIAGONALE F

Brava 13.40

With *Brava*, designed in 1979, our studio took a leap forward. Pasquale Landolfi, until then owner of one of our cruising boats (*Stranezza*, a production Canados 33), commissioned us to produce a First Class and allocated a budget sufficient for an ambitious program. *Brava* was a 44′ boat whose design was, in some respects, an evolution of *Manua*, which had won the Mediterranean Championship in 1978. Brava's design was rather unusual, and almost a gamble as a detailed interpretation of the IOR Rule. The result was a medium-heavy displacement, with a larger maximum beam than most of the First Class yachts of the time. She has a diamond-shaped hull, with a very acute stem angle, similar to what we had adopted for *Filo da torcere*. The aluminum construction was undertaken by the famous Minneford yard under the direction of Chuck Sadler. Right from her first outings, *Brava* proved to be a fast boat, winning the Alassio regattas and finishing fourth in the 1980 Sardinia Cup. She was selected in 1981 to participate in the Admiral's Cup, but tactical errors kept her away from the podium. On the other hand, she won the Italian Championship, a success repeated in 1982, the year in which she contributed to the Italian victory at the Sardinia Cup. At the Admiral's Cup of 1983, helmed by Vittorio Mariani and with the tactical contribution of Gary Weisman, she not only came in second with the Italian team but also won the prestigious Fastnet Race overall and even the Grand Prix d'Elegance.

Year 1979
Builder Minneford New York
Construction Aluminum
LOA 13.45 m
B MAX 4.158 m
D 2.57 m
DSPL 10 t
Mainsail 40 mq
Headsail 76 mq
Spi 153 mq

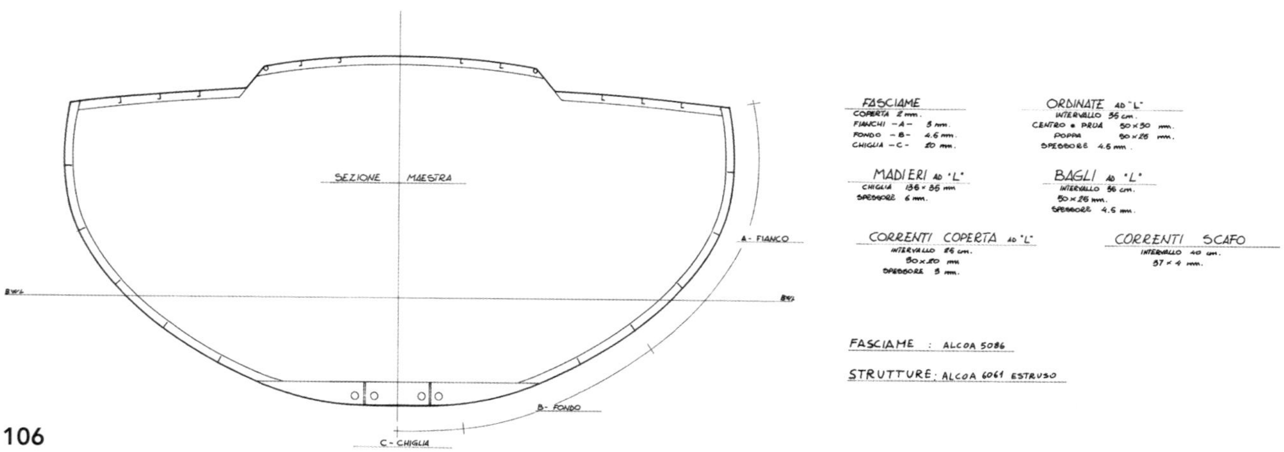

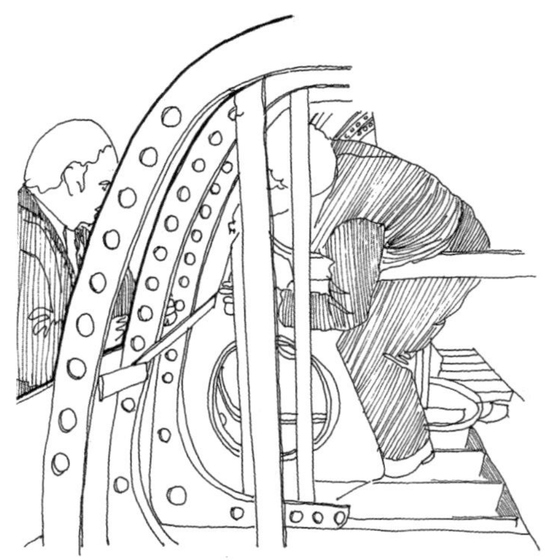

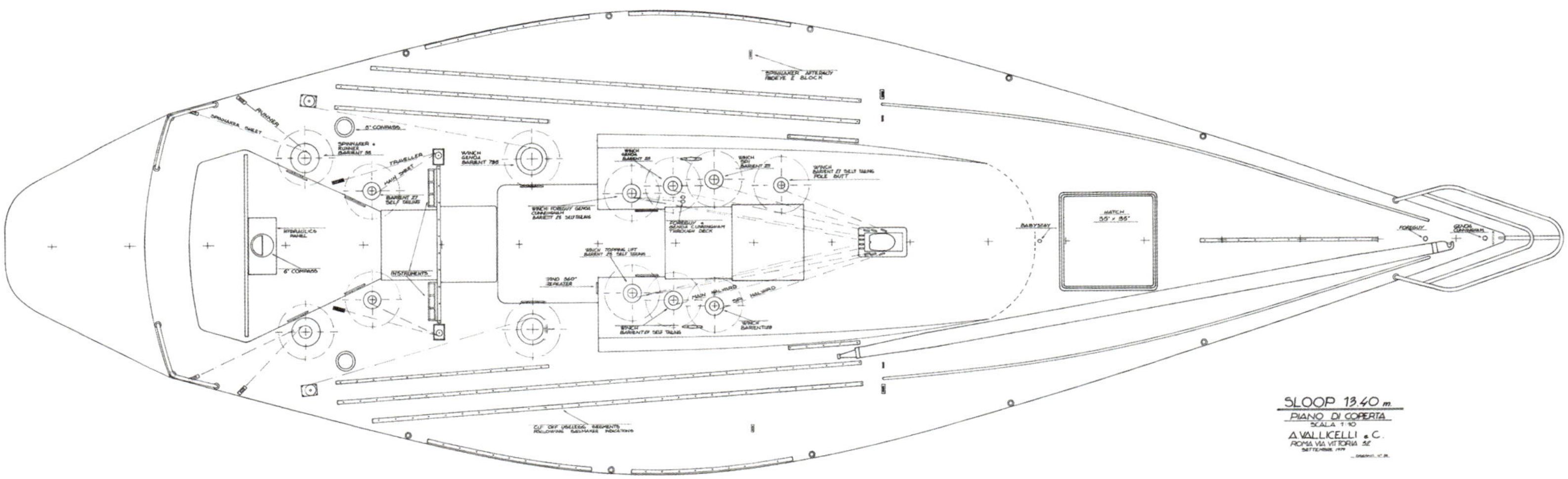

Brava 13.40 during a race.

Deck plan.

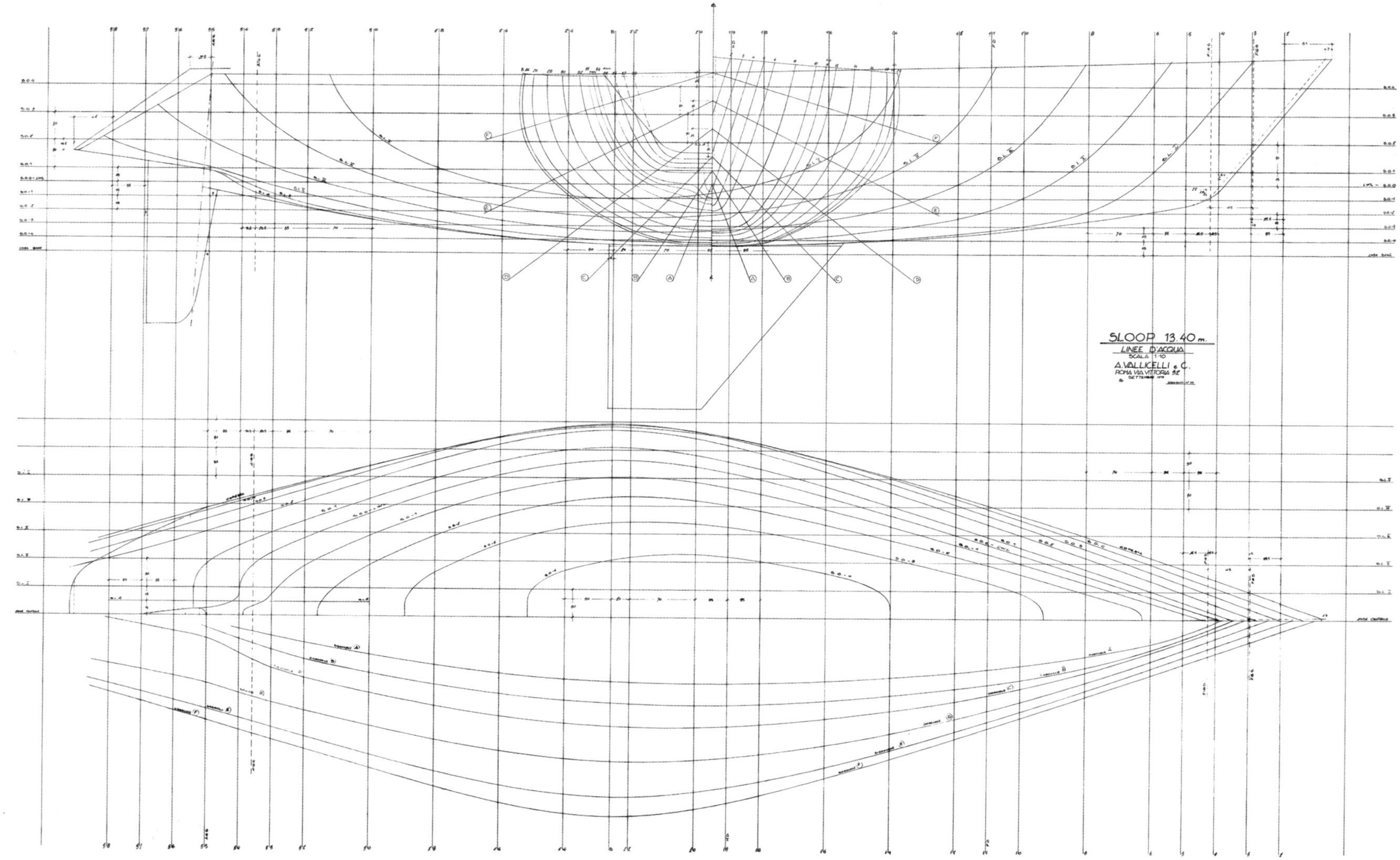
SLOOP 13.40 m.
LINEE D'ACQUA
SCALA 1:10
A. VALLICELLI e C.
ROMA VIA VITTORIA 52

Lines plan (stations, waterlines, buttocks, diagonal sections).

Under construction at the Minneford yard in New York, 1979.

Brava 13.40 under sail.

Profile and general arrangement. Ink drawing with halftone screen on tracing paper.

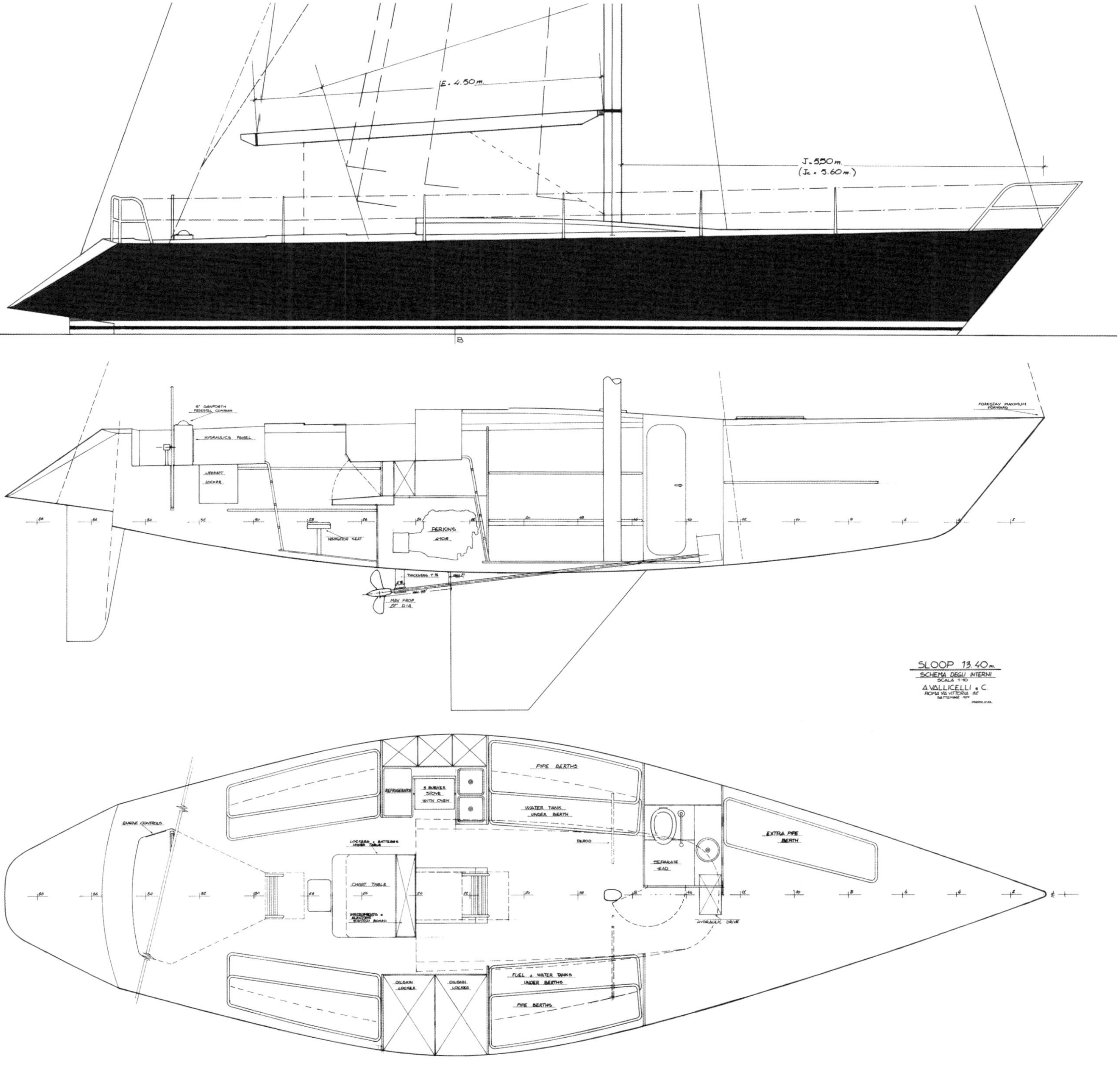
E. 4.50 m.
J. 5.50 m.
(Jc. 5.60 m.)
B
HYDRAULICS PANEL
NAVIGATOR SEAT
PERKINS
4.108
FORESTAY MAXIMUM FORWARD
SLOOP 13.40 m.
SCHEMA DEGLI INTERNI
SCALA 1:40
A.VALLICELLI e C.
ROMA VIA VITTORIA
SETTEMBRE
PIPE BERTHS
REFRIGERATOR
3 BURNER STOVE WITH OVEN
WATER TANK UNDER BERTH
EXTRA PIPE BERTH
ENGINE CONTROLS
CHART TABLE
SEPARATE HEAD
HYDRAULIC DRIVE
OILSKIN LOCKER
OILSKIN LOCKER
FUEL & WATER TANKS UNDER BERTHS
PIPE BERTHS

Blu Show

Blu Show is a medium-sized two tonner designed shortly after *Filo da torcere* and *Brava*, built in cold-molded wood at the Pezzini shipyard in Viareggio using special construction techniques. Indeed, for *Blu Show*, we decided to use longitudinal elements for the structure. A large number of red cedar stringers were laid over rather widely spaced frames, and the diagonally crossed planking was laid over this structure, using two layers of spruce and okumé for the inner plies and two mahogany layers for the external ones. Compared to *Brava* and *Filo da torcere*, *Blu Show* had slightly larger volumes forward and a more vertical stem. The hull was designed with the stern sections having a rather sharp turn of the bilge, as was visible in the squarish transom. *Blu Show* proved to be a good all round performer: there were no holes in performance and she performed especially well in medium to light winds. With Dick Deaver at the helm, she contributed to the Italian team's second place in the Sardinia Cup in 1980. The following year the project became a series model: the Show 42'. For the occasion, we changed the stern sections, softening and extending them. In 1986 the boat underwent a restyling marked by a new transom design, a new deckhouse, and interiors.

Year 1979
Builder Pezzini Viareggio (LU)
Construction Cold molded wood
LOA 12.495 m
B MAX 3.965 m
D 2.357 m
DSPL 7.7 t
Mainsail 35.5 mq
Headsail 69 mq
Spi 139 mq

Blu Show at the Sardinia Cup, 1980.

Construction plan.

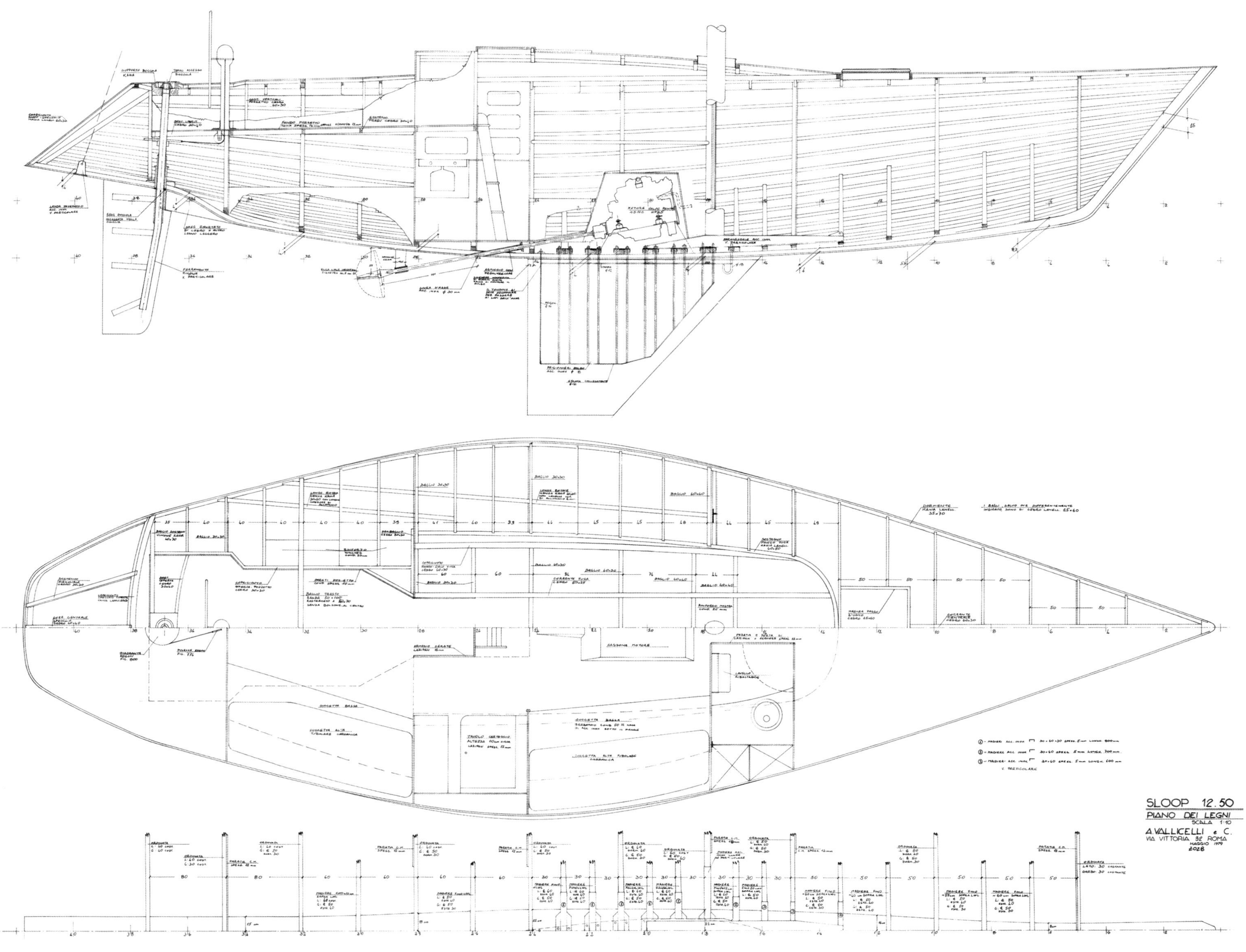
SLOOP 12.50
PIANO DEI LEGNI
SCALA 1:10
A.VALLICELLI e C.
VIA VITTORIA 32 ROMA
MAGGIO 1979
2028

Primadonna

A year after the success at the One Ton Cup in Naples, we designed a two tonner inspired in design by *Filo da Torcere*. Two years later, *Primadonna* was evidence of how our work always tended to be a compromise between the New Zealand school of light displacement, with great form stability, and the American school of medium-heavy displacement with a high prismatic coefficient. The boat was built in composite materials by the Mannini yard in Liguria.

In some ways, she recalled *Nonno Gigi*, although the hull did not plane as much and was, in general, less extreme. Her performance was not bad, but she presented some balance and trim problems, mainly attributable to a rig that was too heavy. *Primadonna* was a member of the Italian team at the 1983 Admiral's Cup, contributing to their second place in the overall standings.

Year 1980
Builder Mannini San Remo (IM)
Construction Composite
LOA 12.66 m
B MAX 4.05 m
D 2.315 m
DSPL 7.1 t
Mainsail 50 mq
Headsail 53 mq
Spi 115 mq

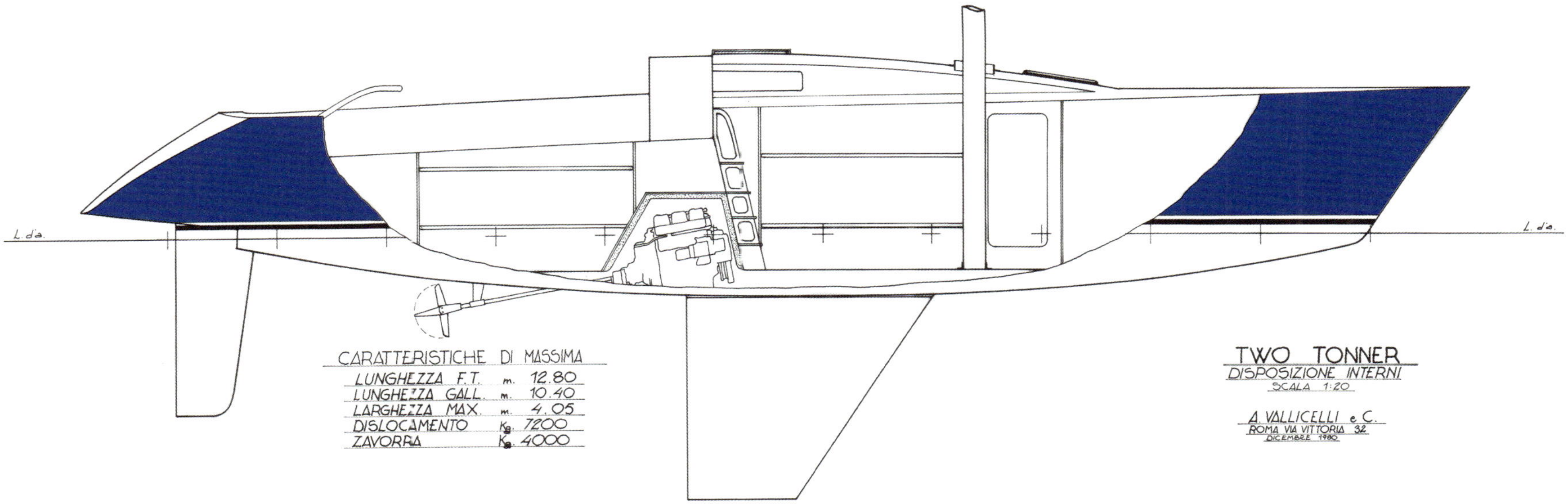

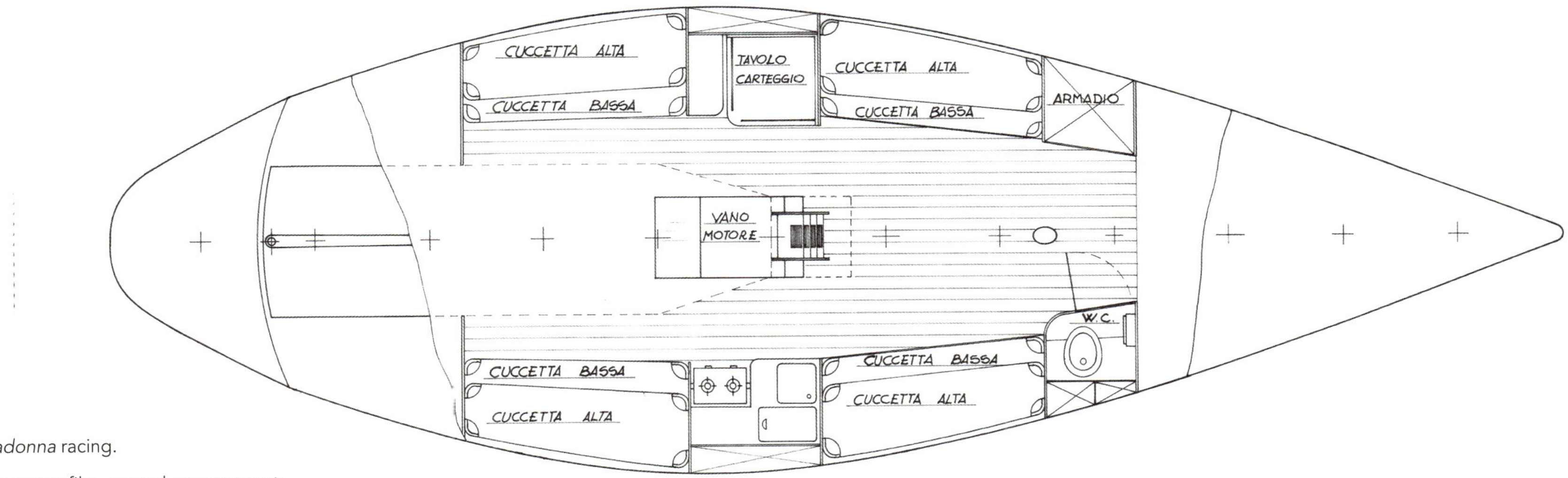

Primadonna racing.

Cut-away profile, general arrangement, deck plan.

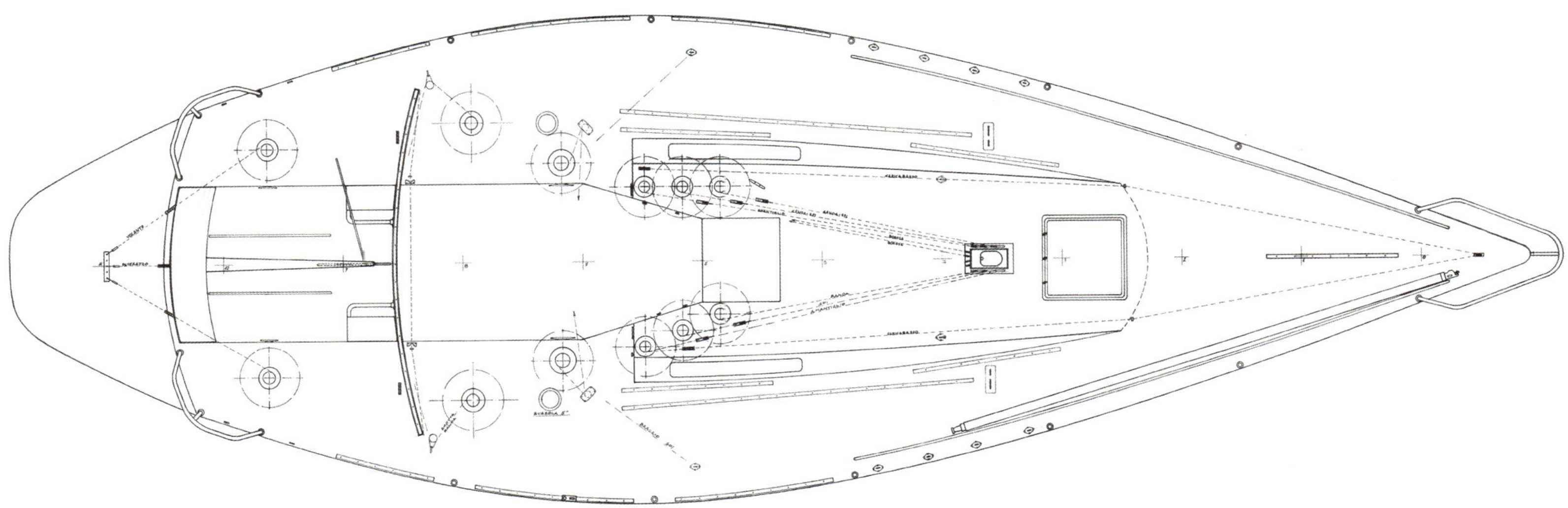

Azzurra

In the autumn of 1981, we began work on the project for the first Italian 12 Meter International Rule yacht. She was to compete in the challenge for America's Cup to be held in Newport in 1983 and which would subsequently go down in history for the first American defeat. For the first time, we were faced with the need to organize a major project research program to build a boat able to compete in this top sailing event. It was also the first time that an Italian yacht club (the Yacht Club Costa Smeralda), supported by a consortium of twenty companies, launched a challenge for America's Cup. The rules required the design, construction, and research to be performed in the challenger's country. The naval architecture was therefore carried out at the tank testing facility in Rome (INSEAN), which designed and calibrated the equipment necessary to perform the tests. Three 1:4 scale models were produced for towing tests in the tank and to visualize the flow dynamics of the circulation channel. The experiments lasted a few months, at the end of which we drew the final hull design, taking into account not only the test results but also our ideas and experience. Compared to the other, classic 12 Meter yachts, such as those designed by Olin Stephens (*Enterprise*, *Courageous*, *Freedom*), *Azzurra* was slightly shorter on the waterline, therefore lighter, and with less sail area. On the other hand, she had wider stern sections: a choice that could have been a gamble, as this was heavily penalized by the Measurement Rule. She also had a keel fin with a very sharp leading edge, flattened in the middle sections. Ultimately *Azzurra* was designed with a far slimmer keel than her opponents. The rudder in its first version was made of aluminum, trapezoidal in shape, and with a fixed skeg, but during the racing in Newport, we decided to change it, giving the rudder an elliptical form, without a skeg, and with a fiberglass blade. The mast was made with a 7000 Series aluminum alloy; this alloy could not be welded, so it was assembled with two half-shells glued and riveted together.
Built in aluminum by the Yacht Officine shipyard

Year 1982
Builder Yacht Officine Pesaro
Construction Aluminium
LOA 19.68 m
B MAX 3.73 m
D 2.7 m
DSPL 24.9 t
Mainsail 104 mq
Headsail 115 mq
Spi 280 mq

Azzurra out practicing.

I-4

Patrizia Ferri's color palette for choosing the hull color. Tempera on a Schoellershammer board.

Ugo Nespolo's poster for the 1983 Italian America's Cup challenge.

Sketches for the hull graphics.

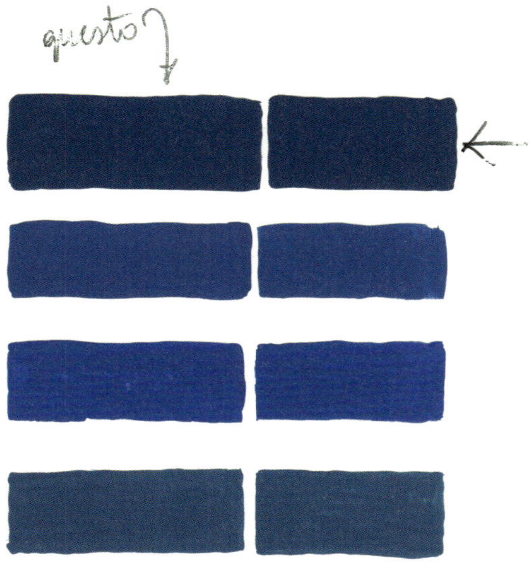

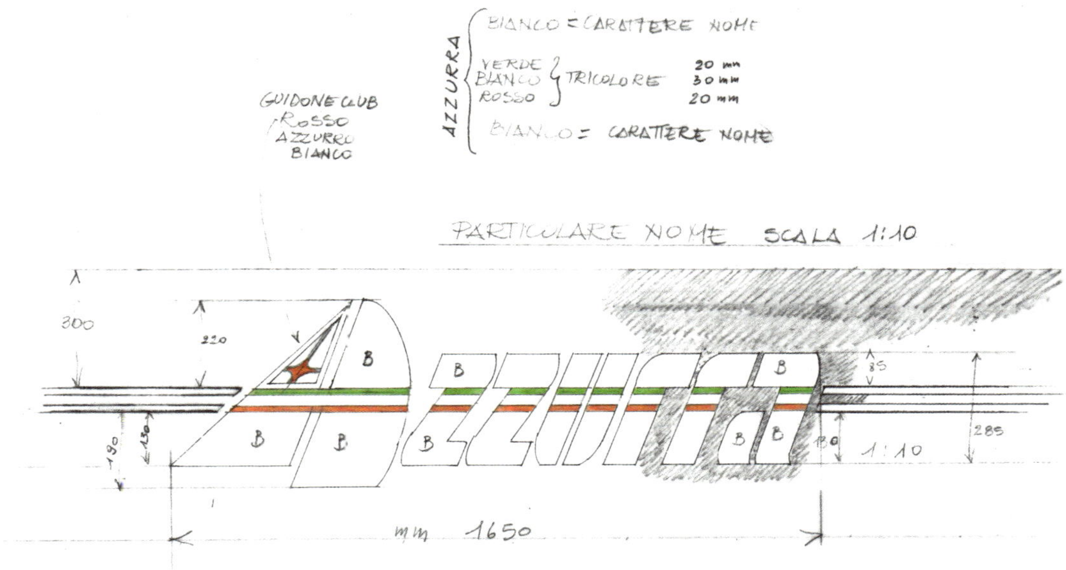

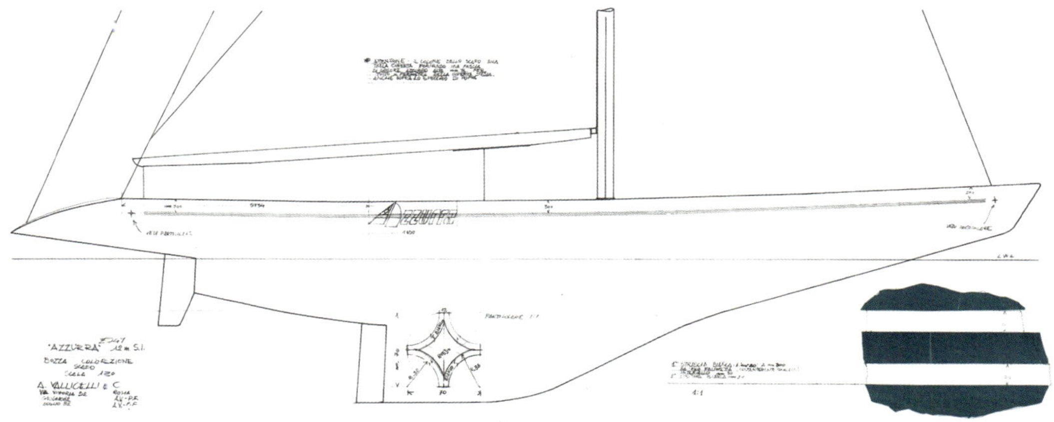

in Pesaro, *Azzurra* was launched a year before the start of the challenger series in Newport. The sea trials began immediately, based on comparisons with the excellent trial horse *Enterprise*, which in a short time proved slower than our new boat, to the surprise of everyone.

Mauro Pelaschier took the helm of *Azzurra*, and in Newport, the boat proved to be among the fastest, especially off the wind, benefiting from the wider stern sections and her unusual thin keel. After performing well in the selection trials, beating the revelation of the year *Australia II* on one occasion, she lost the semifinals of the Louis Vuitton Cup to the English aboard *Victory 83* because of a trivial spreader fitting failure. The following year, in 1984, *Azzurra* finished second in the 12 Meter World Championship in Porto Cervo.

Tank testing and construction.

Construction, launch and first sea trails.

Rudder shapes. Left: drawing of the first rudder, approved by Lloyd's Register. Right: drawing of the second rudder, made in Newport of aluminum and fiberglass during the 1983 selection trails.

Keel profile. Ink and pencil drawing on polyester paper.

Central cockpit coffee grinder winches.

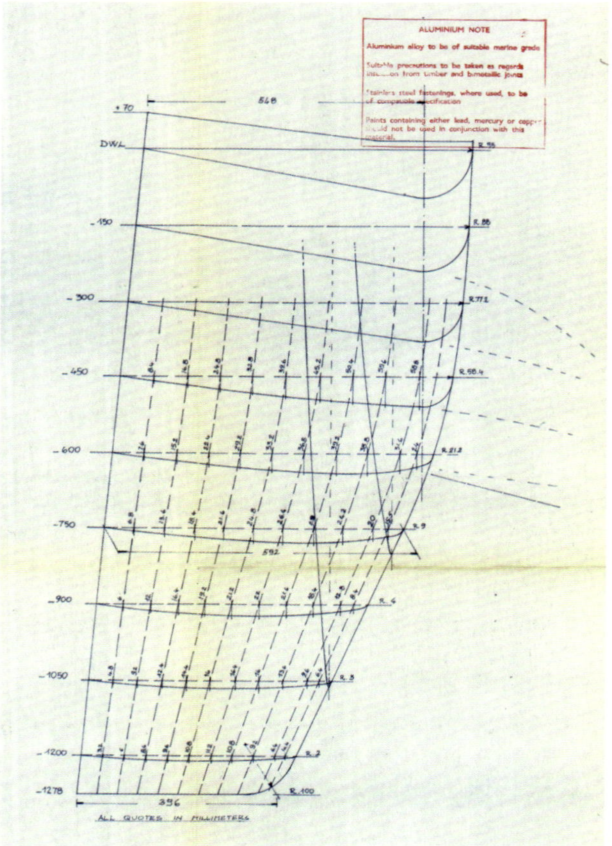

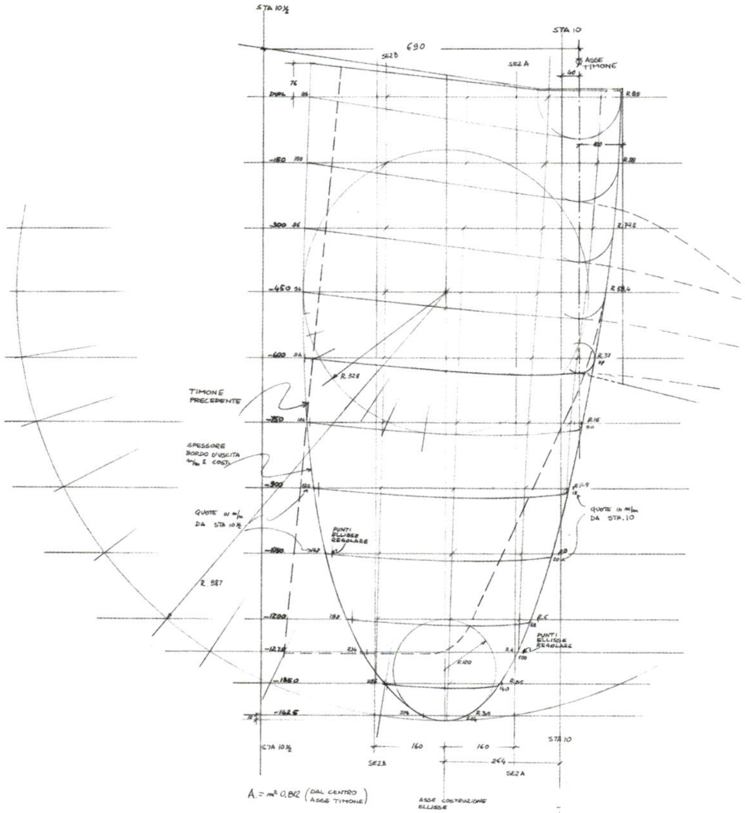

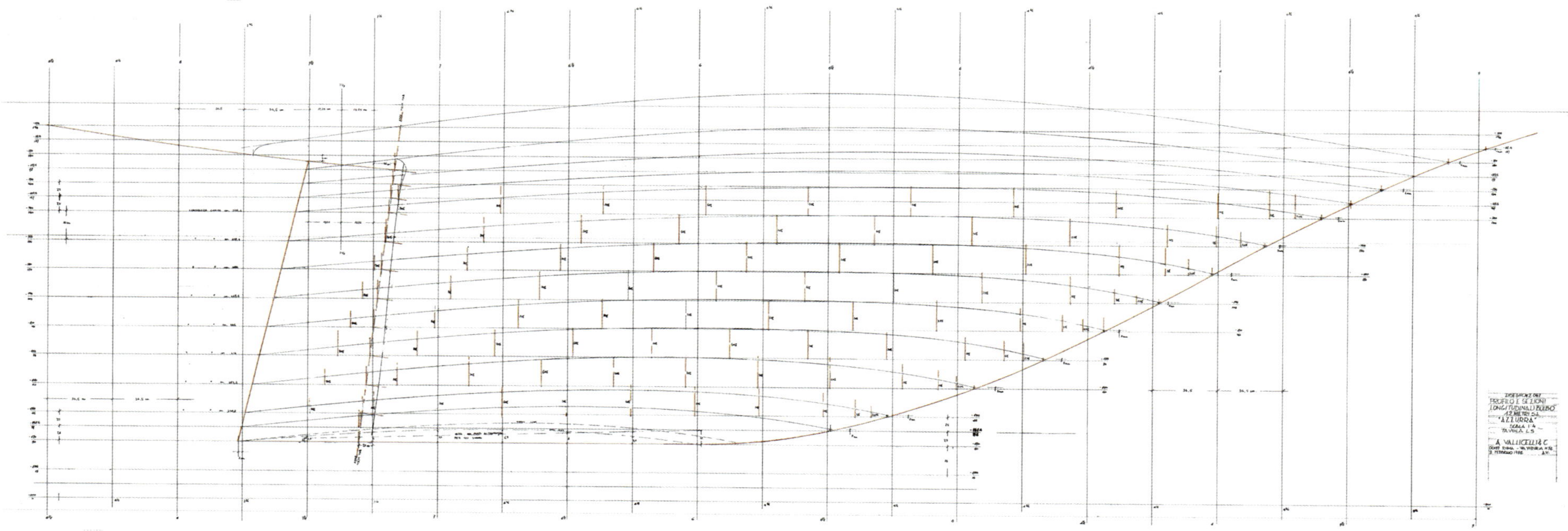

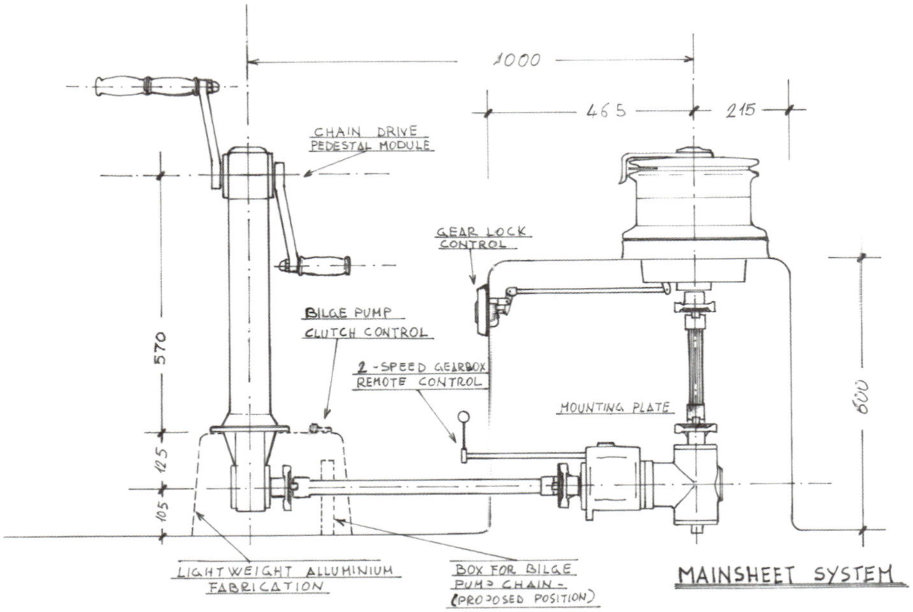
1000
465
215
CHAIN DRIVE PEDESTAL MODULE
GEAR LOCK CONTROL
BILGE PUMP CLUTCH CONTROL
2-SPEED GEARBOX REMOTE CONTROL
MOUNTING PLATE
570
125
105
600
LIGHTWEIGHT ALLUMINIUM FABRICATION
BOX FOR BILGE PUMP CHAIN (PROPOSED POSITION)
MAINSHEET SYSTEM

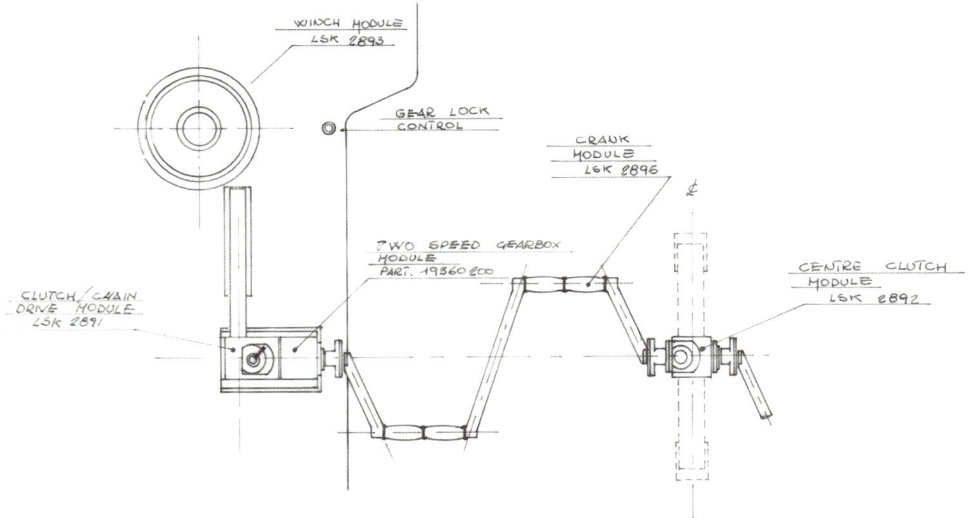
WINCH MODULE LSK 2893
GEAR LOCK CONTROL
CRANK MODULE LSK 2896
TWO SPEED GEARBOX MODULE
CLUTCH/CHAIN DRIVE MODULE LSK 2891
CENTRE CLUTCH MODULE LSK 2892

Top: The launch in Pesaro, 1982. Bottom left: Marco Cobau, Andrea Vallicelli and Cino Ricci in the yard during construction.

Bottom right: One of the wooden models for tank testing on a 1:4 scale.

Top: *Azzurra* during the selection trail race won against *Australia II*. Bottom left: (L to R) Stefano Roberti, Riccardo Bonadeo, Shakè Bonadeo, Andrea Vallicelli (at the helm) in a moment of relaxation. Bottom right: Leaving for New York from Naples in the spring of 1983.

Azzurra II

The second *Azzurra* we designed, this time for the 1987 America's Cup challenge in Australia, had a basically similar hull to *Azzurra I*, but with a slightly wider stern and less bow overhang. The main new feature was a keel fin with almost horizontal wings protruding out from the bulb, exploiting the breakthrough by the winners of the last Cup, *Australia II*. However, the geometry of our bulb differed in the form of the contour lines. Ours were not straight, but elliptical, as was the shape of the winglets. The same choice was made by the eventual 1987 America's Cup winner, *Stars & Stripes*. *Azzurra II* participated at the 12 Meter World Championship in Australia in 1986, revealing some problems: she did not perform very well in strong winds and tended to pitch too much in the steep waves of Perth.

Year 1985
Builder Scardellato Treviso
Construction Aluminum
LOA 19.875 m
B MAX 3.754 m
D 2.7 m
DSPL 25 t
Mainsail 105 mq
Headsail 115 mq
Spi 280 mq

Azzurra II off Fremantle (Western Australia) during the 1986 12 Meter World Championships.

Construction plan. Ink drawing on polyester paper.

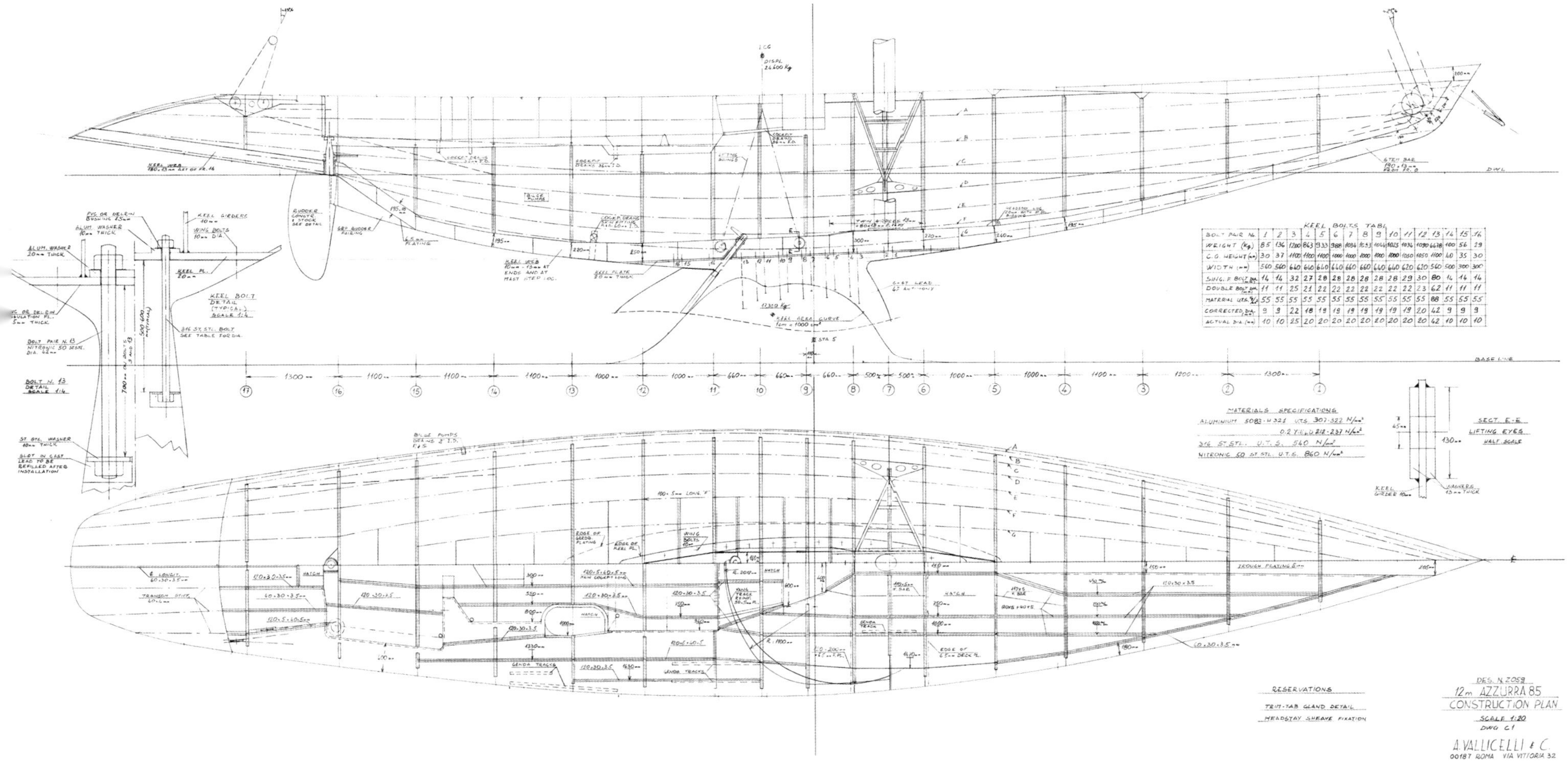
KEEL BOLTS TABL
BOLT PAIR No. 1 2 3 4 5 6 7 8 9 10 11 12 13 14 15 16
WEIGHT (Kg) 85 136 1200 863 933 988 1034 1053 1044 1023 1034 1090 6678 100 56 29
C.G. HEIGHT (mm) 30 37 1100 1100 1100 1000 1000 1000 1000 1000 1050 1050 1100 40 35 30
WIDTH (mm) 560 560 640 640 640 640 660 660 640 640 620 620 560 500 300 300
SINGLE BOLT DIA. (mm) 14 14 32 27 28 28 28 28 28 28 29 30 80 14 14 14
DOUBLE BOLT DIA. (mm) 11 11 25 21 22 22 22 22 22 22 22 23 62 11 11 11
MATERIAL U.T.S. 55 55 55 55 55 55 55 55 55 55 55 55 88 55 55 55
CORRECTED DIA. (mm) 9 9 22 18 19 19 19 19 19 19 19 20 42 9 9 9
ACTUAL DIA. (mm) 10 10 25 20 20 20 20 20 20 20 20 20 42 10 10 10
STEM BAR
190 x 13 mm
FROM FR. 0
DWL
BASE LINE
RESERVATIONS
TRIM-TAB GLAND DETAIL
HEADSTAY SHEAVE FIXATION
DES. N. 2059
12m AZZURRA 85
CONSTRUCTION PLAN
SCALE 1:20
DWG C1
A. VALLICELLI & C.
00187 ROMA VIA VITTORIA 32

Azzurra III

Azzurra III was the last boat we designed for America's Cup. She was a 12 Meter with a slightly shorter waterline than her predecessor, therefore lighter (approximately a ton less in displacement). Her keel was now a delta in shape, with larger winglets than those of *Azzurra II* and set further aft. One of the new features of *Azzurra III* was her hull structure, designed with the Finzi Studio in Milan. It was created with finite element calculation and was composed of elements placed longitudinally: this was a solution that we found in many 12 Meters boats present in Perth. *Azzurra III*'s design was influenced by two light displacement boats, *Australia III* and *French Kiss*, which showed impressive performance in the 12 Meter World Championship held in the waters of Fremantle, the same location as the 1987 America's Cup. Although shared by other designers, our decision to follow the approach of the Australian and French boats would prove to be wrong. The Cup was, in fact won by *Stars & Stripes*, the American boat designed by Britton Chance Jr., Bruce Nelson and David Pedrick, designed according to completely different criteria: quite long on the waterline, with a heavy displacement and less sail area. These choices proved optimal for the conditions of the match, with strong winds that caused steep waves. During the selection regattas *Azzurra III* was modified: we changed the keel, providing it with much longer and thinner winglets; it was made heavier and lengthened in the after section. These proved significant improvements, but not enough to make the boat rank higher in the final standings.

Year 1986
Builder SAI Ambrosini Passignano (PG)
Construction Aluminum
LOA 19.45 m
B MAX 3.7 m
D 2.7 m
DSPL 23.5 t
Mainsail 100 mq
Headsail 110 mq
Spi 270 mq

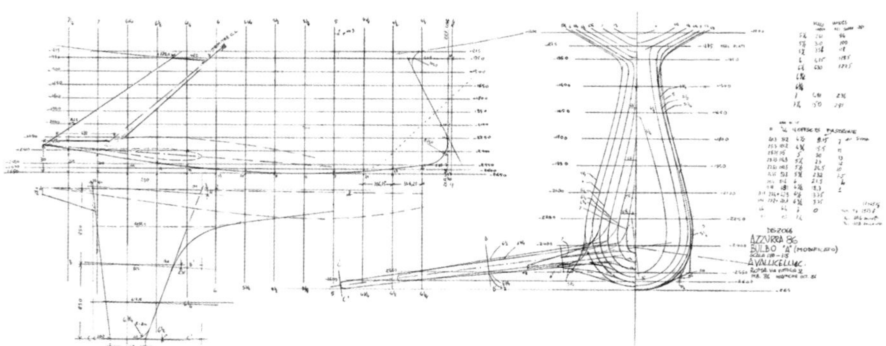

Drawing of the new keel made for *Azzurra III* during the Challenger series for the 1987 America's Cup in Perth.

During construction in the yard.

III under spinnaker off Fremantle.

Close up of the keel with winglets.

Deck plan and a longitudinal section detailing deck hardware. Ink drawing on polyester paper.

12
I-10

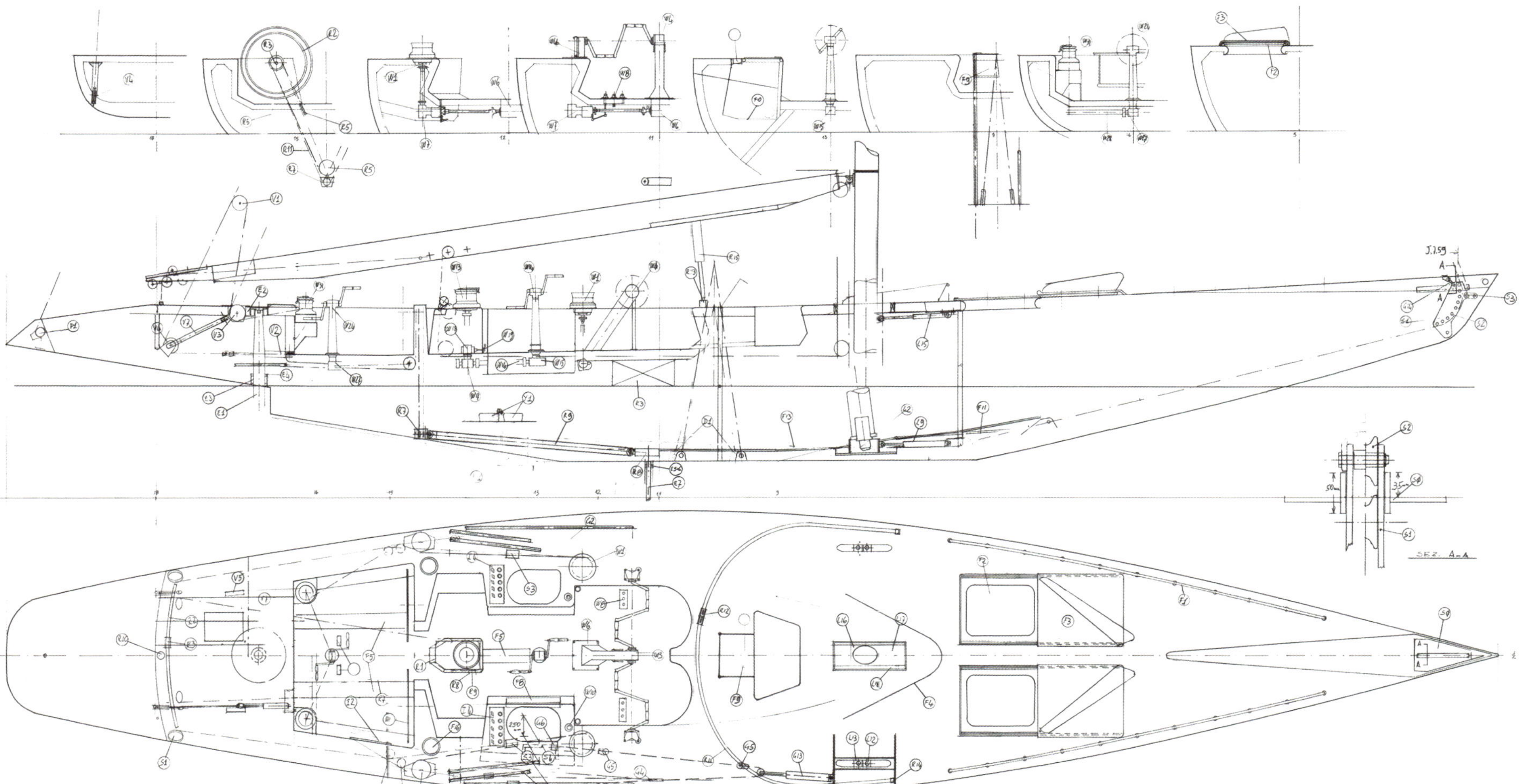
SEZ. A-A

Entreprise

Entreprise was a 50′ yacht designed in 1983 for a French buyer and built at the C.M. de Normandie yard, which he owned. A few years had passed since *Brava*, *Gemini* and *Fantasque*, our first generation of 44-footers, and together with other 50-footers like *Sirena* and *Springbok*, *Entreprise* expressed a change in our design approach. Compared to *Brava*, *Entreprise* had a fuller bow and a smoother transition towards the maximum beam. The canoe body had fair lines in the diagonal sections and was less influenced by the IOR measurements. The construction was the result of the yard's experience in the hybridization of materials and the techniques of resin polymerization. *Entreprise* was produced by coupling an aluminum hull to a composite deck, made with a special lamination technique. Metallic wires were inserted inside the laminate: these acted as electrical resistances, which, when heated, made it possible to attain the optimal curing temperature. The rudder was manufactured in composite by Green Marine, specialized in composite constructions. The twin cockpit deck, with the after one in direct communication with the navigation station belowdecks, was similar to that of Brava, although equipped with coffee grinders. The coachroof was low and streamlined, almost a stylization of our previous designs. Helmed by one of the most important sailors of the time, Ron Love, the boat won the Cowes-Dynard in 1984, setting a new record for the crossing, and participated in the Sardinia Cup of the same year under the Irish flag. She also won the important North American Championship in 1985 and entered the SORC circuit in the same year.

Year 1983
Builder CMN Cherbourg (France)
Construction Aluminum – Composite deck
LOA 15.4 m
B MAX 4.5 m
D 2.8 m
DSPL 13.5 t
Mainsail 52 mq
Headsail 62 mq
Spi 150 mq

Sail plan, general arrangement, deck plan.

Construction in the yard.

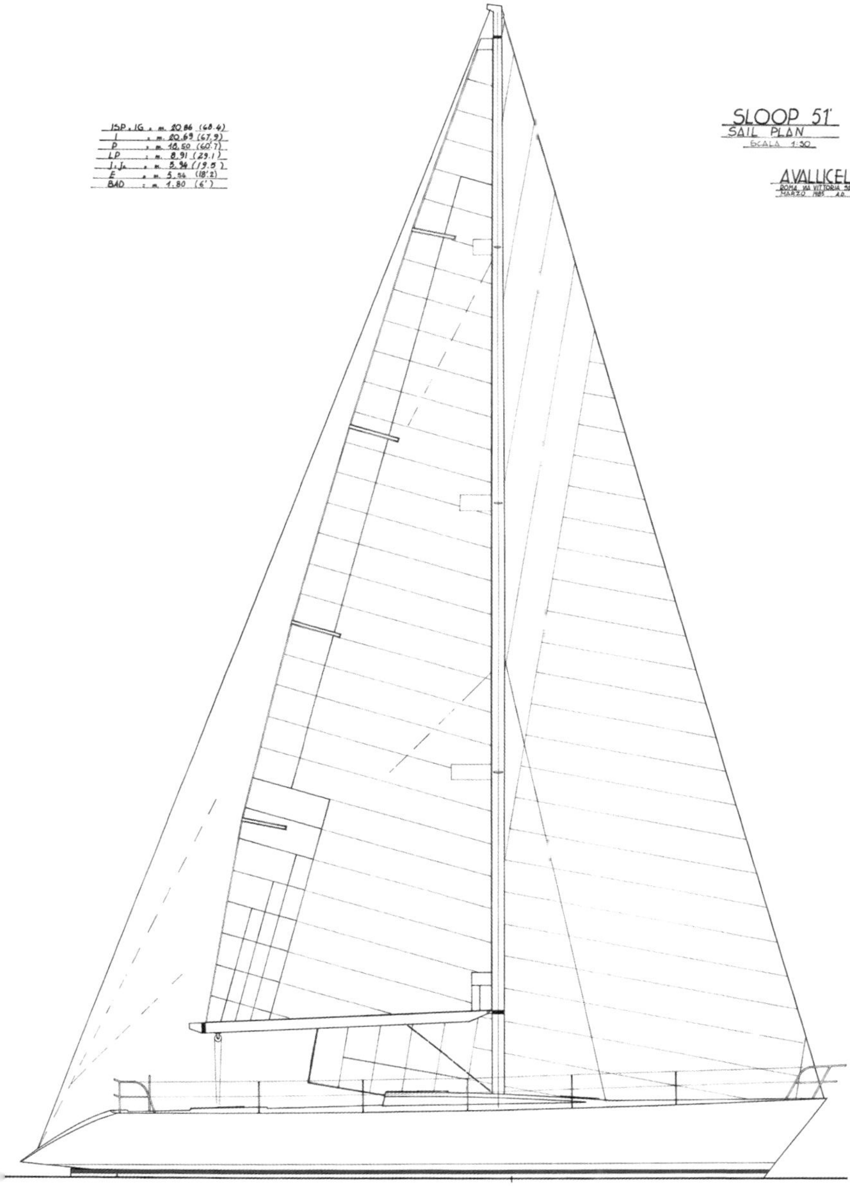
ISP, IG = m. 20.86 (68.4)
I = m. 20.69 (67.9)
P = m. 18.50 (60.7)
LP = m. 8.91 (29.1)
J, Ja = m. 5.94 (19.5)
E = m. 5.54 (18.2)
BAD = m. 1.80 (6')
SLOOP 51'
SAIL PLAN
SCALA 1:30
A.VALLICELLI & C.
ROMA VIA VITTORIA 52 00187
MARZO 1985 A.D.

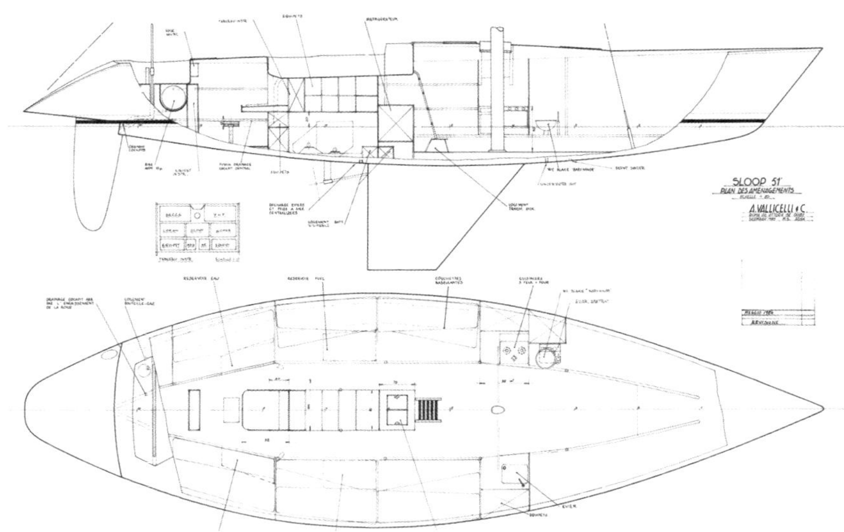
SLOOP 51'
PLAN DES AMENAGEMENTS
A.VALLICELLI & C.

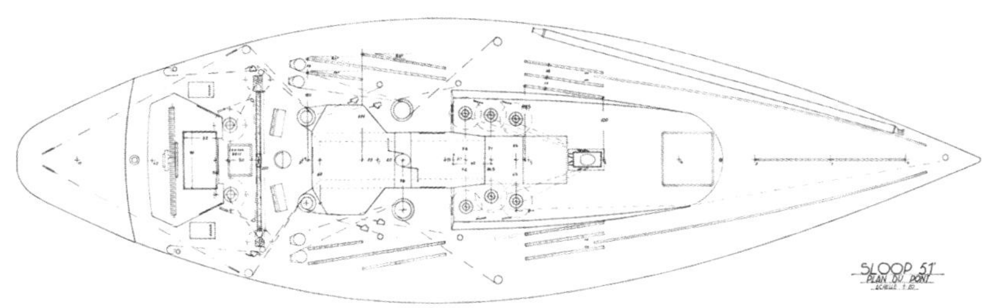
SLOOP 51'
PLAN DU PONT

Sketches for hull graphics
(logo, colors and graphics).

Entreprise racing.

Construction plan.

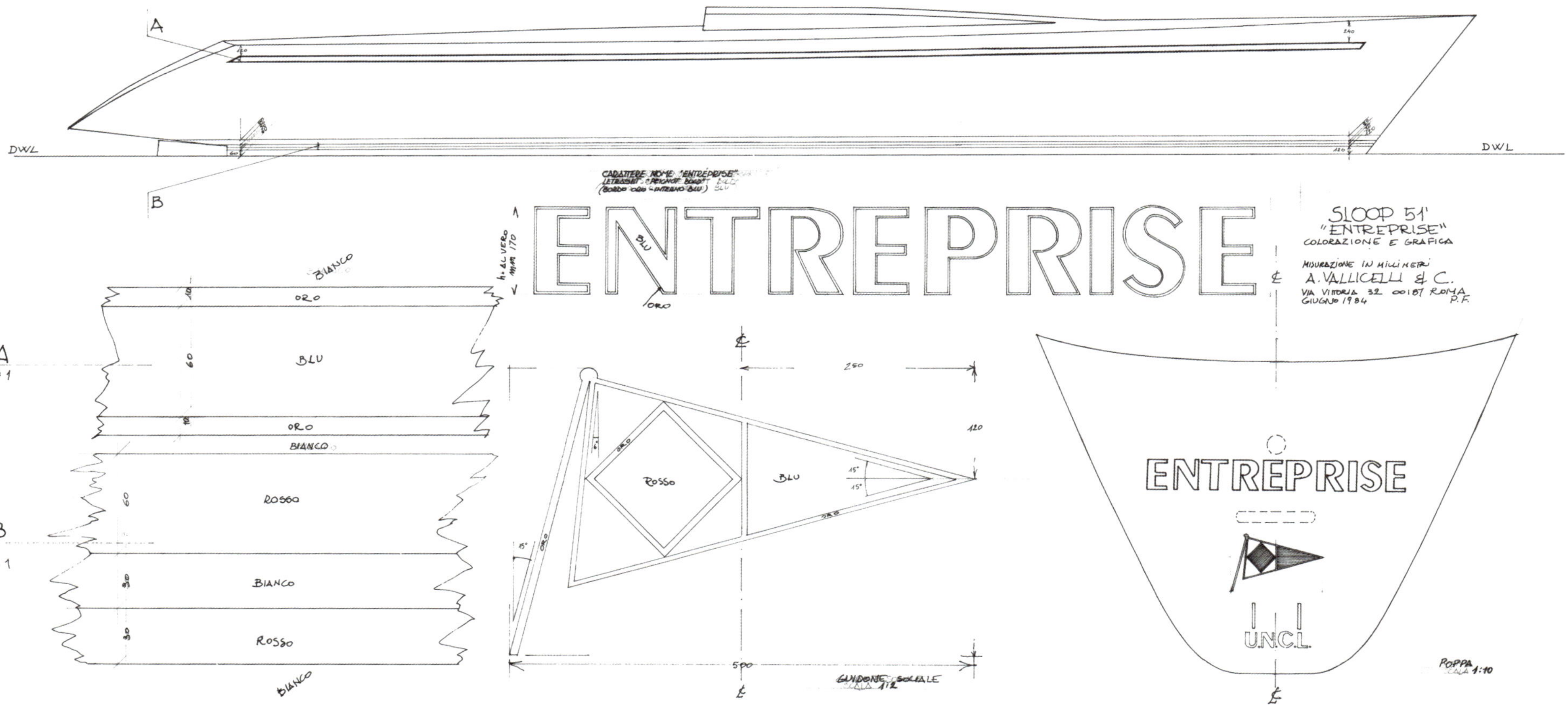

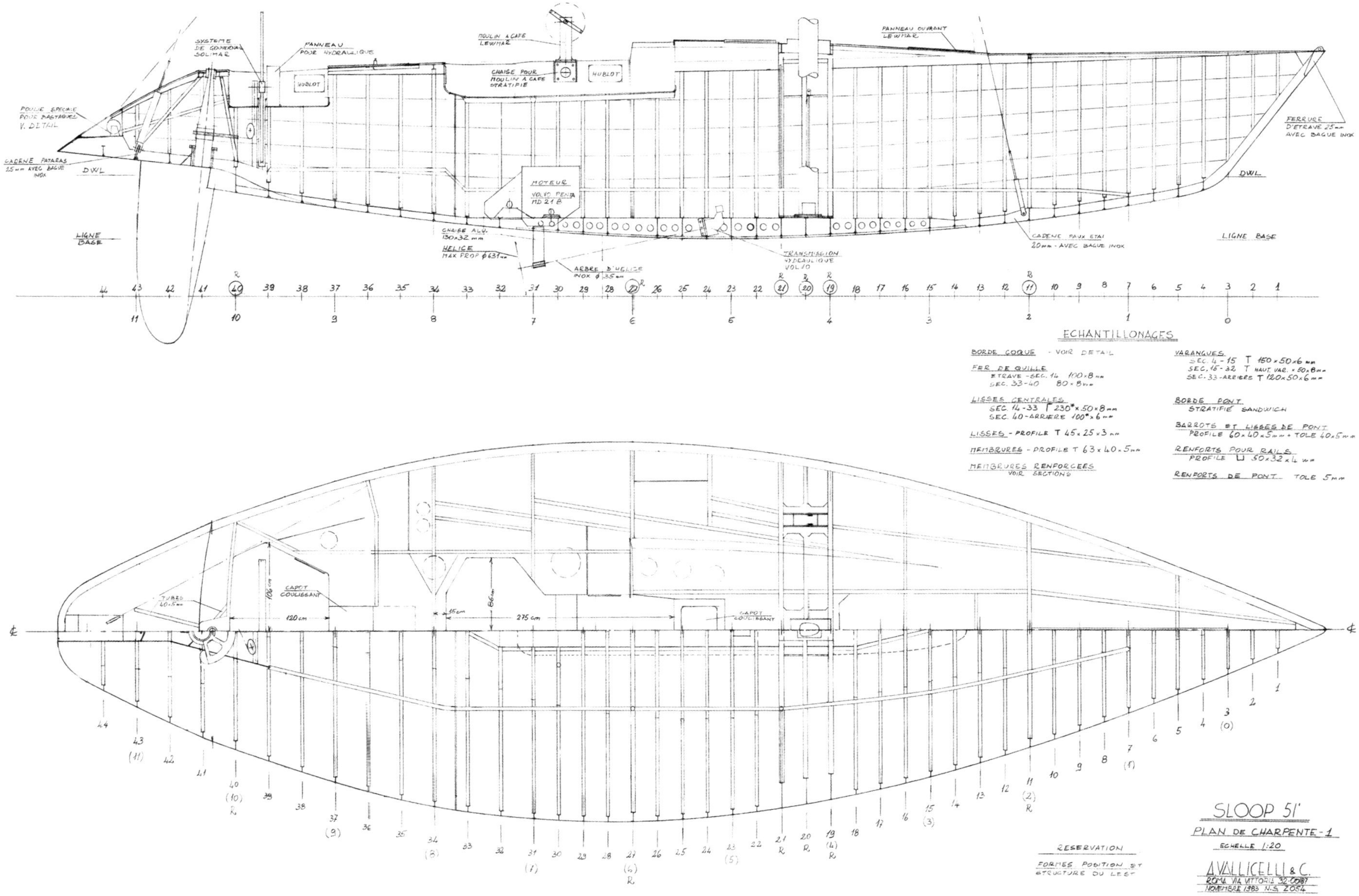
SYSTEME DE GOUVERNAIL SOLIMAR
PANNEAU POUR HYDRAULIQUE
HUBLOT
MOULIN A CAFE LEWMAR
CHAISE POUR MOULIN A CAFE STRATIFIE
HUBLOT
PANNEAU OUVRANT LEWMAR
POULIE SPECIALE POUR BASTAQUES V. DETAIL
CADENE PATARAS 25 mm AVEC BAGUE INOX
DWL
FERRURE D'ETRAVE 25 mm AVEC BAGUE INOX
DWL
MOTEUR VOLVO PENTA MD 21 B
LIGNE BASE
CHAISE ALU. 130x32 mm
HELICE MAX PROP ϕ631 mm
ARBRE D'HELICE INOX ϕ 35 mm
TRANSMISSION HYDRAULIQUE VOLVO
CADENE FAUX ETAI 20mm - AVEC BAGUE INOX
LIGNE BASE
ECHANTILLONAGES
BORDE COQUE - VOIR DETAIL
FER DE QUILLE
ETRAVE - SEC. 14 100×8 mm
SEC. 33-40 80×8 mm
LISSES CENTRALES
SEC. 14-33 ⎾ 230*×50×8 mm
SEC. 40-ARRIERE 100*×6 mm
LISSES - PROFILE T 45×25×3 mm
MEMBRURES - PROFILE T 63×40×5 mm
MEMBRURES RENFORCEES VOIR SECTIONS
VARANGUES
SEC. 4-15 T 150×50×6 mm
SEC. 15-32 T HAUT. VAR. ×50×8 mm
SEC. 33-ARRIERE T 120×50×6 mm
BORDE PONT STRATIFIE SANDWICH
BARROTS ET LISSES DE PONT PROFILE 60×40×5 mm + TOLE 40×5 mm
RENFORTS POUR RAILS PROFILE ⊔ 50×32×4 mm
RENFORTS DE PONT TOLE 5 mm
CAPOT COULISSANT
120 cm
275 cm
CAPOT COULISSANT
SLOOP 51'
PLAN DE CHARPENTE - 1
ECHELLE 1:20
RESERVATION
FORMES POSITION ET STRUCTURE DU LEST
A.VALLICELLI & C.
ROMA VIA VITTORIA 32-00187
NOVEMBRE 1983 N.S. 2054

Brava 30.5

In 1983 the IOR rating for the One Ton Class was increased from 27.5' to 30.5' to encourage the participation of this type of boat in team races such as the Admiral's and the Sardinia Cup. After the successes of *Brava*, Pasquale Landolfi instructed us to design a new generation one tonner. We produced a less extreme boat than the first one, more harmonized and all round. The yard chosen for the construction was that of Luigi Morri and Dino Para of Rimini, real artists of wooden construction. *Brava 30.5* was made of cold-molded laminate, glued with West System (a special epoxy resin for wooden constructions) and reinforced with unidirectional carbon fiber inserted between the frame laminates. We had already experimented this technique in a previous boat, *I vezzi mie*, built-in Liguria by the Maccianti yard. The deck of *Brava 30.5* was instead an Airex sandwich with Kevlar-carbon fibers. We also tested a very thin keel fin, which offered excellent performance when running downwind, but less so when beating. In 1984 the boat finished sixth in the One Ton Cup at La Trinité (the first edition with the 30.5' rating) and third overall in the individual ranking at the Sardinia Cup. In 1985 we decided to change the keel for a more traditional one; with the new set-up, the boat took part in the 1985 One Ton Cup with Paul Cayard at the helm. That same year she took part in the Admiral's Cup and was the only Italian boat to finish the Fastnet course in similar conditions to those of 1979. *Brava 30.5* represented the swan song of wooden construction for international-level racing boats; from that moment onwards, advanced composite materials definitively replaced not only wood but also aluminum.

Year 1983
Builder Morri & Para Rimini
Construction Cold molded wood – Composite deck
LOA 12.19 m
B MAX 3.91 m
D 2.35 m
DSPL 6 t
Mainsail 49 mq
Headsail 48 mq
Spi 97 mq

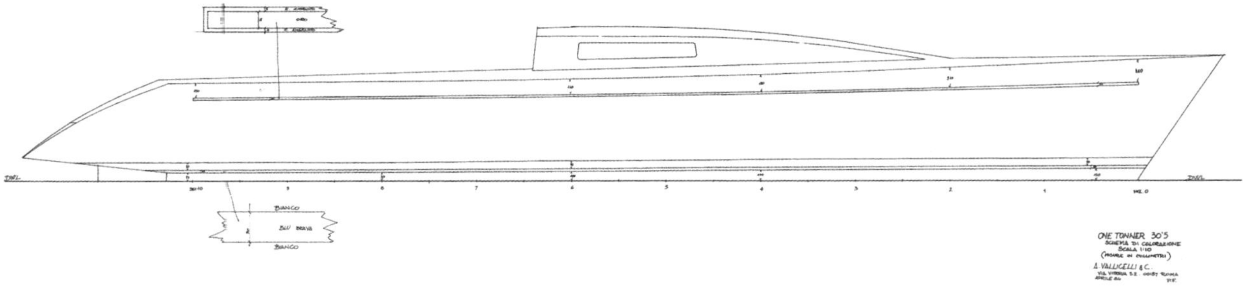

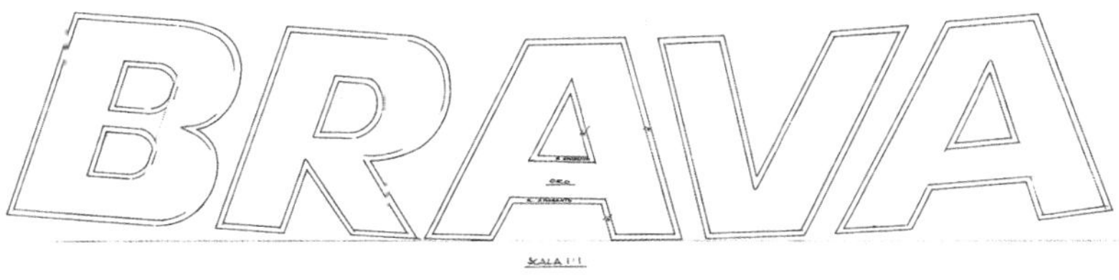

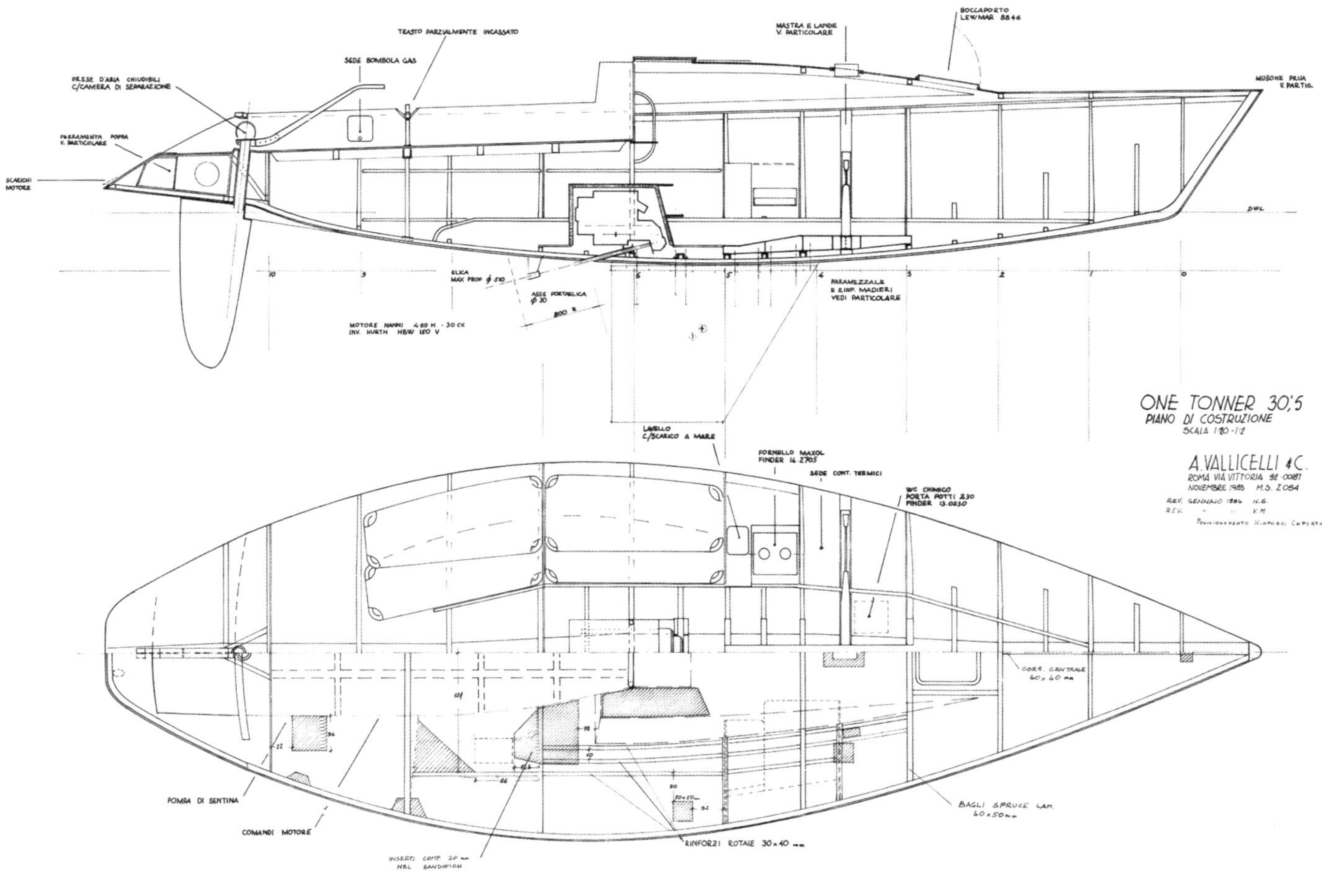

Brava 30.5 racing.

Graphics and logo. Construction plan.

Springbok

Springbok was a 50′ boat built in the United States in 1986 by Geraghty Marine of San Diego using advanced composite materials. The hull was a sandwich made up of a variable-density PVC core and a laminate reinforcement layer consisting of unidirectional carbon and Kevlar fibers, impregnated with epoxy resin. *Springbok's* design was an evolution of *Entreprise*, but with variations aimed at improving her performance in medium to light winds. Compared to the French boat, she had a higher ratio between sail surface and displacement, a smaller wetted surface, and higher aspect ratio appendages. The twin steering wheels were located rather far forward on deck, to give the helmsman a better sight of incoming waves and the boat as she rode them. Mainsail trimmer and tactician had their place aft of the helmsman. *Spingbok* also had coffee grinder pedestals for the genoa, which were very advanced for the time. The boat raced in U.S. waters with Dennis Conner at the helm. In 1986 she won the Onion Patch Regatta and the Bermuda Race.

Year 1986
Builder Geraghty Marine San Diego (California)
Construction Composite
LOA 15.086 m
B MAX 4.54 m
D 2.7 m
DSPL 11.4 t
Mainsail 55 mq
Headsail 95 mq
Spi 190 mq

Finishing the graphics in the yard.

Under sail.

Deck lines and keel construction detail drawings.

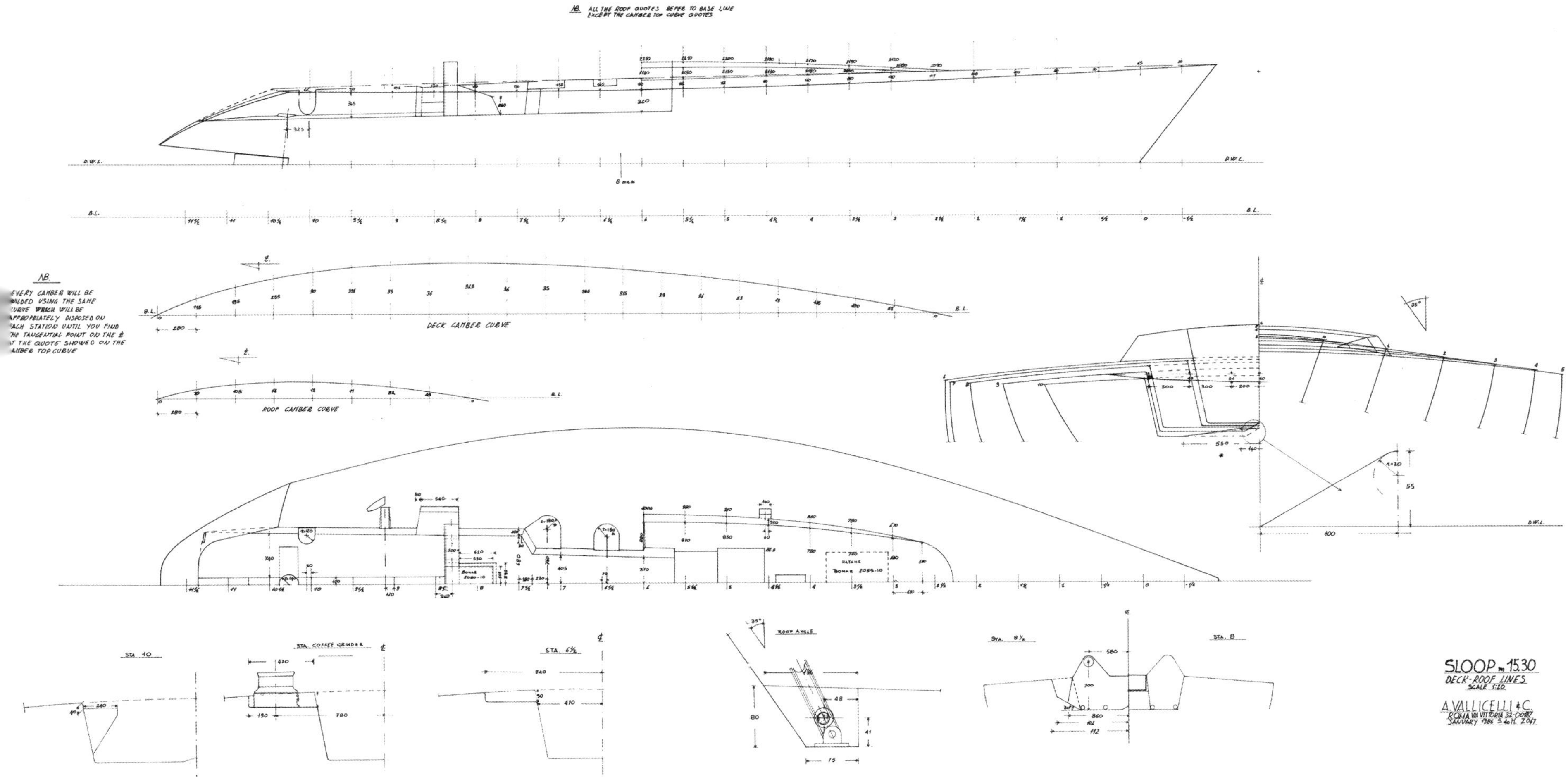

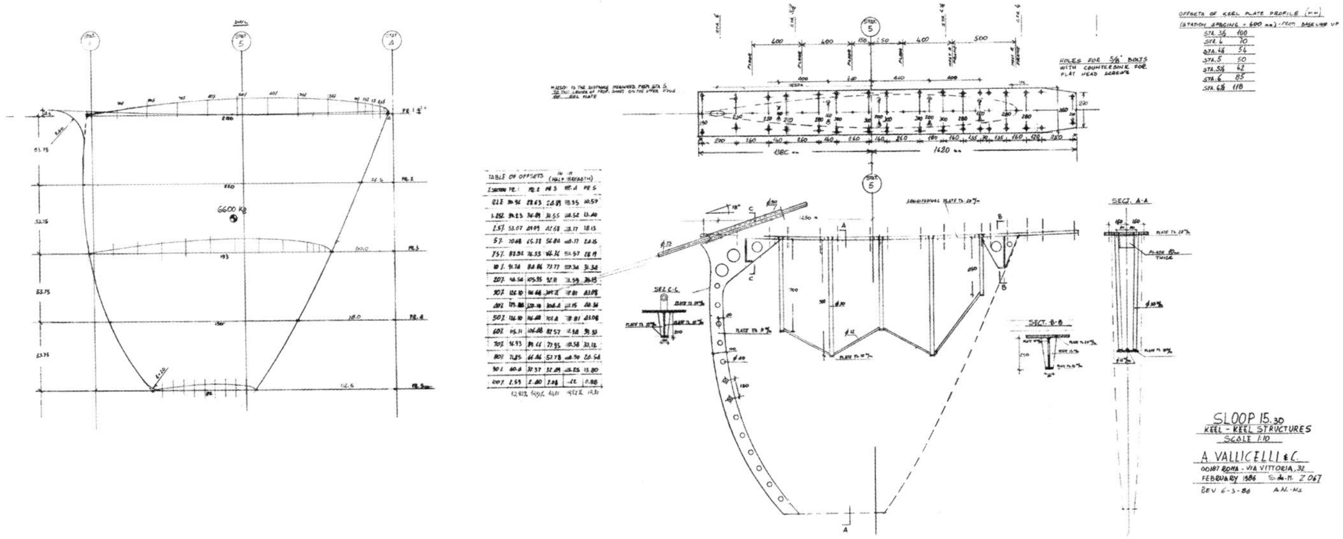

Sirena

After the first *Brava* and after *Entreprise*, in 1987, we designed another First Class: *Sirena*. She was commissioned by an Istanbul owner with the aim of building the first composite racing boat in Turkey, with which to participate in the 1988 Sardinia Cup. The boat recalled the design choices of *Springbok*: medium displacement, moderately broad stern, a balanced compromise between form stability, and hull slenderness. The design of the deck was essential but equipped with top-quality fittings, chosen to minimize weight and for ergonomic functionality. The interiors were minimized as in a racing boat but maintained some cruising comforts. The construction assured lightness and rigidity and was achieved thanks to the collaboration of SP System, at the time a leading company in composites, which made a team of technicians available on-site during construction. *Sirena* then participated in the 1988 Sardinia Cup, winning a race, the first victory for a Turkish boat in a high-level sailing competition.

Year 1986
Builder Regat Ltd Istambul
Construction Composite
LOA 15.2 m
B MAX 4.25 m
D 2.76 m
DSPL 11.4 t
Mainsail 55 mq
Headsail 95 mq
Spi 190 mq

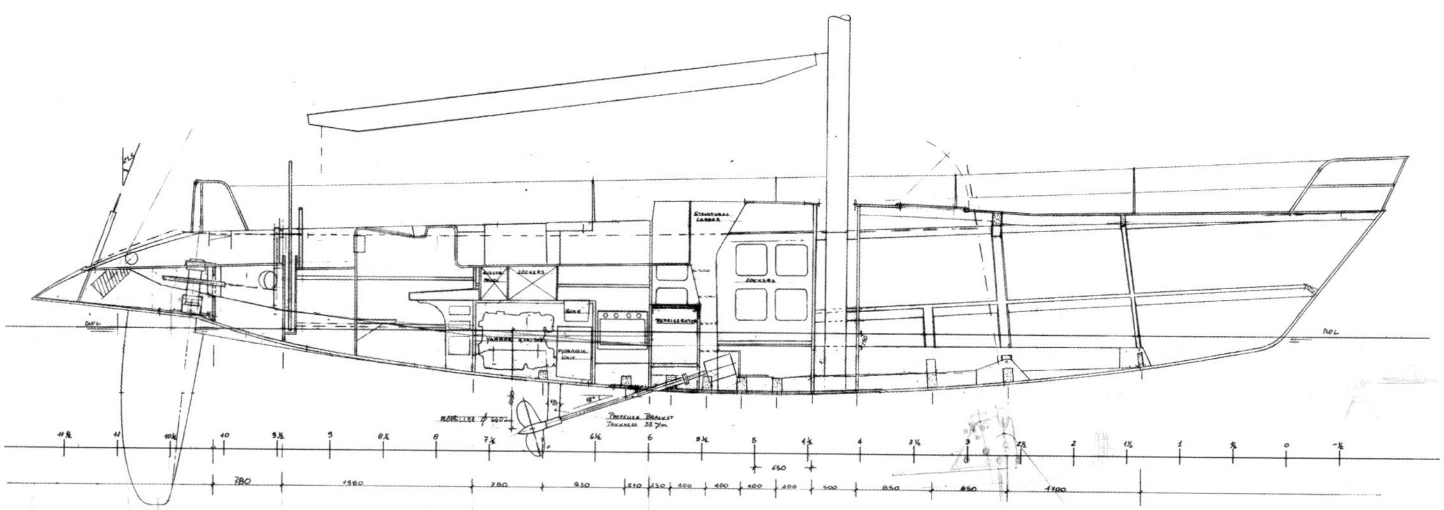

Longitudinal section. Black and blue ink on tracing paper.

Mast and rigging scheme. Black and blue ink on tracing paper.

Sirena sailing in the Bosphorus.

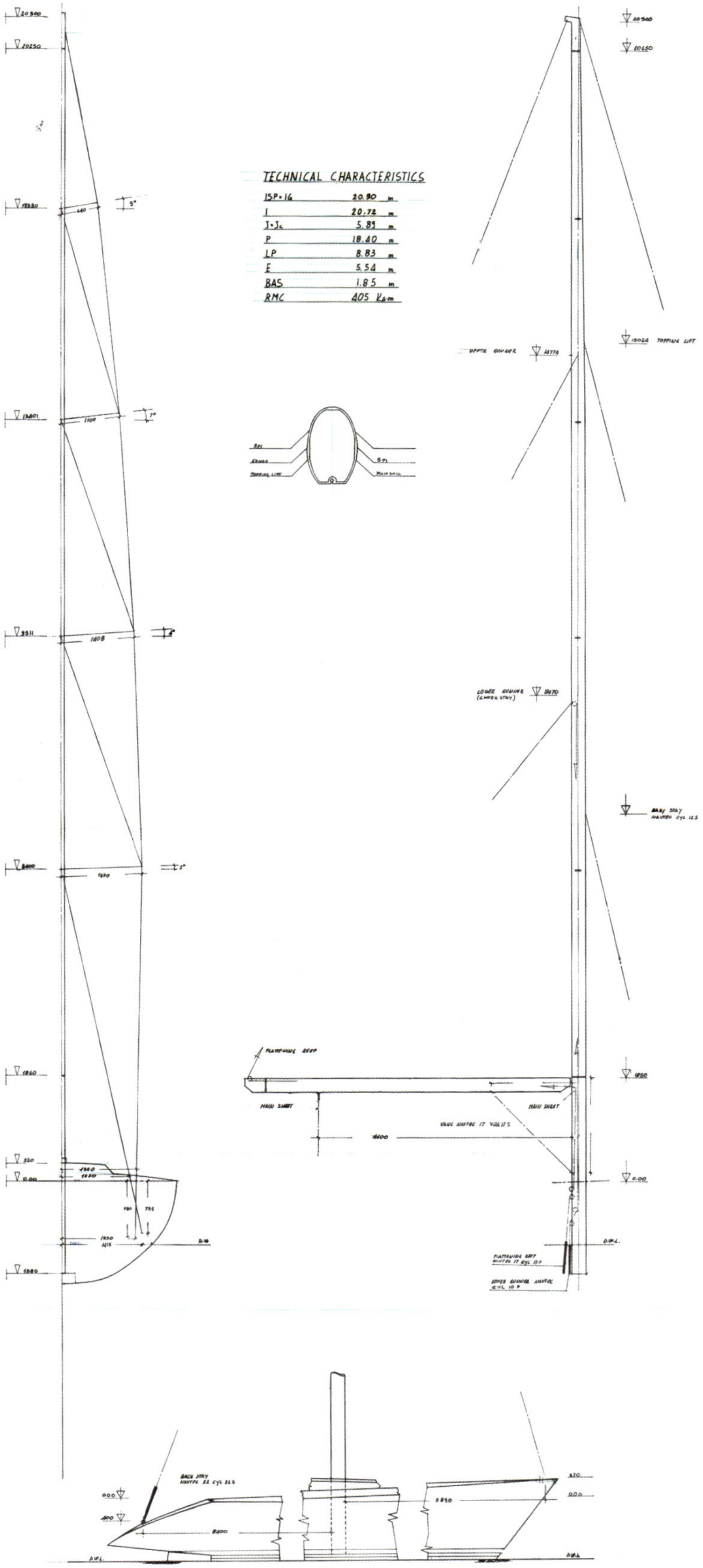
TECHNICAL CHARACTERISTICS
ISP=IG 20.90 m
I 20.72 m
J=Jc 5.89 m
P 18.40 m
LP 8.83 m
E 5.54 m
BAS 1.85 m
RMC 405 Kg·m

Brava Les Copains

Brava Les Copains was the last boat we designed for Pasquale Landolfi. Built in 1987 in the UK by the well-known Green Marine yard, she was very similar in design to *Merope*, built at the same time and also made of composite materials but in Italy, by the Crosato yard. Her design incorporated that streamlining of the hull and appendages of elliptical shape and higher aspect ratio that were emerging at the end of the '80s. Although she always performed well in races, *Brava Les Copains* never won any major trophies. Probably her fault was to have been conceived as too much of an all-round boat, capable of working well in all sea and wind conditions, but without achieving excellence in any specific situation. In the second half of the 1980s, indeed, there emerged an increasingly marked radicalization of design choices: a trend that would contribute significantly to the decline of the IOR Rule.

Year 1986
Builder Green Marine Southampton (UK)
Construction GRP
LOA 12.07 m
B MAX 3.85 m
D 2.316 m
DSPL 5 t
Mainsail 48 mq
Headsail 47 mq
Spi 95 mq

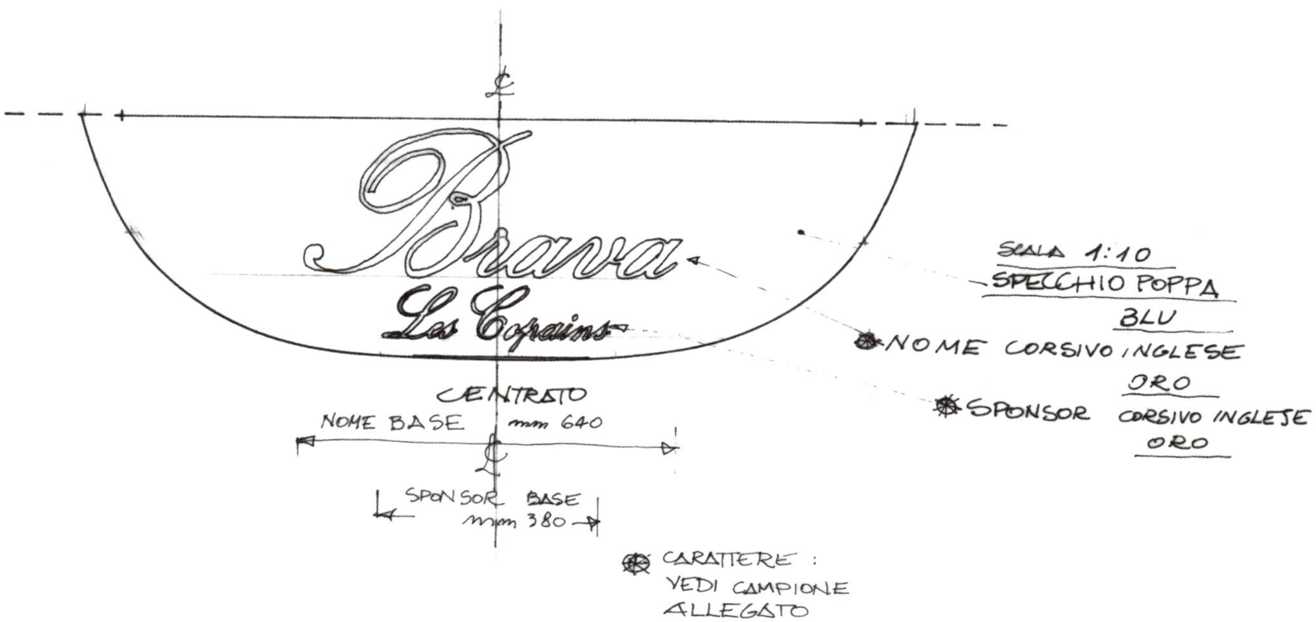

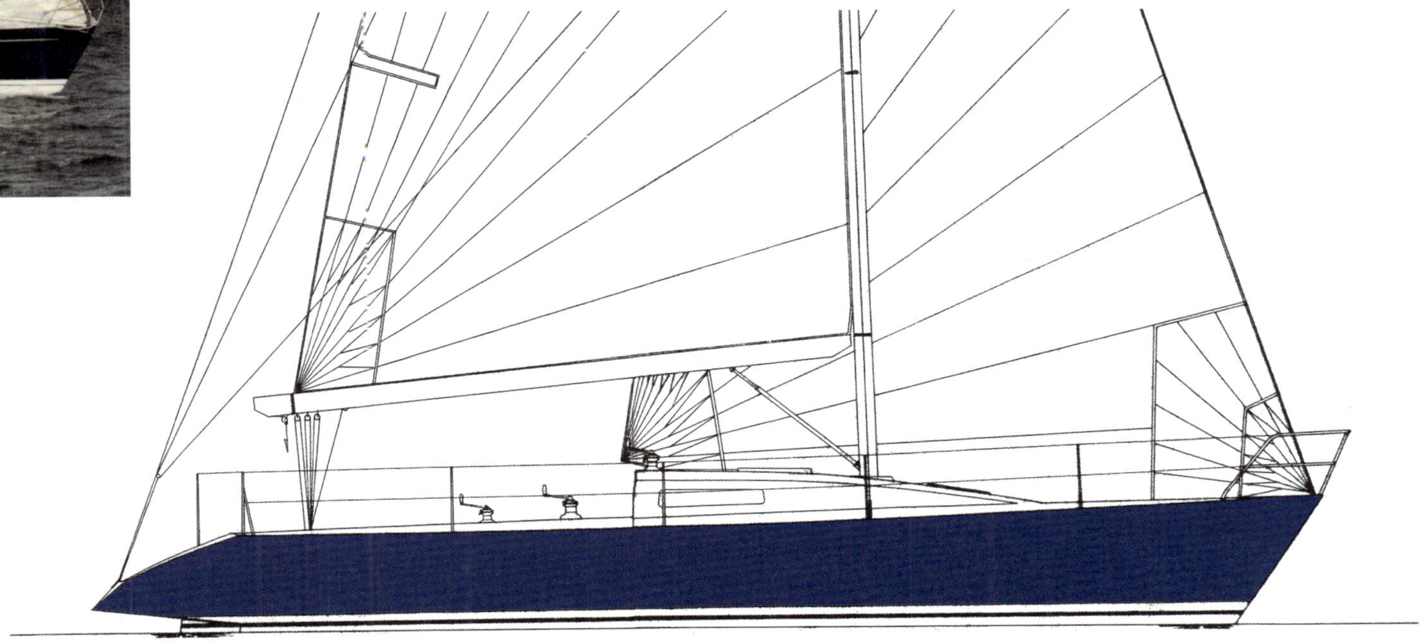

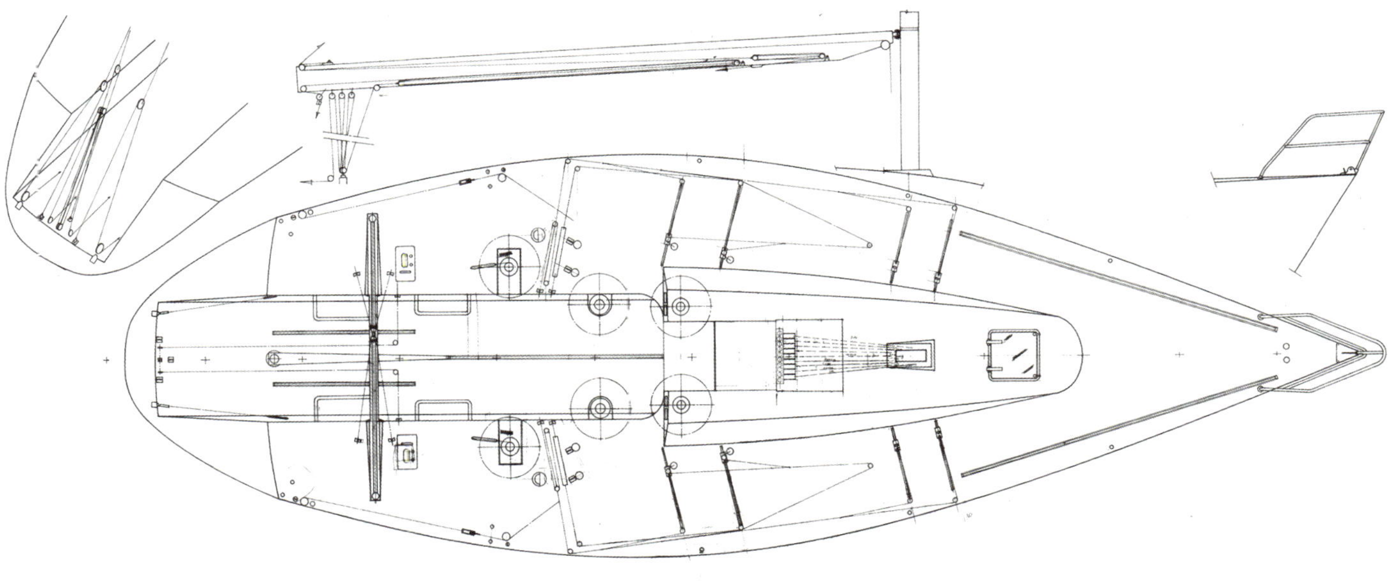

Sketch for the transom graphics. On the hard.

Brava Les Copains racing.

Deck plan with details of backstay, boom, and bow pulpit.

Merope

After the First Class *Gemini* and the half tonners *Castore* and *Polluce*, which had raced in the 1985 Half Ton Cup, the Italian Navy commissioned us to design a one tonner. The boat was built by the C4 yard in Treviso, with the assistance of SP System, a leading company in the composites sector at the time, and Deck Marine, an Italian company producing deck hardware. The hull and deck were of sandwich, with a differentiated-density PVC foam core; for lamination, unidirectional kevlar and carbon fibers were used. The curing was first done under vacuum, using the peel-ply technique to eliminate excess resin, then baked in an oven in order to increase the mechanical characteristics of the material. *Merope* had fair lines, without the forcing of the IOR typical of the past, with reduced stern volumes and high aspect ratio appendages, in line with other boats of the time. Like *Merope*, *Brava Les Copains* participated with the Italian team in the Sardinia Cup of 1986. *Merope* won the Italian Championship in her class in the same year, while *Brava Les Copains* finished in second place: the same result was repeated the following year.

Year 1986
Builder Nautica C4 Treviso
Construction GRP
LOA 12.062 m
B MAX 3.93 m
D 2.165 m
DSPL 5.6 t
Mainsail 48 mq
Headsail 46 mq
Spi 93 mq

Working on the hull in the yard.

Merope close hauled.

Construction plans, color scheme for hull, deckhouse and cockpit. Black and blue ink on tracing paper.

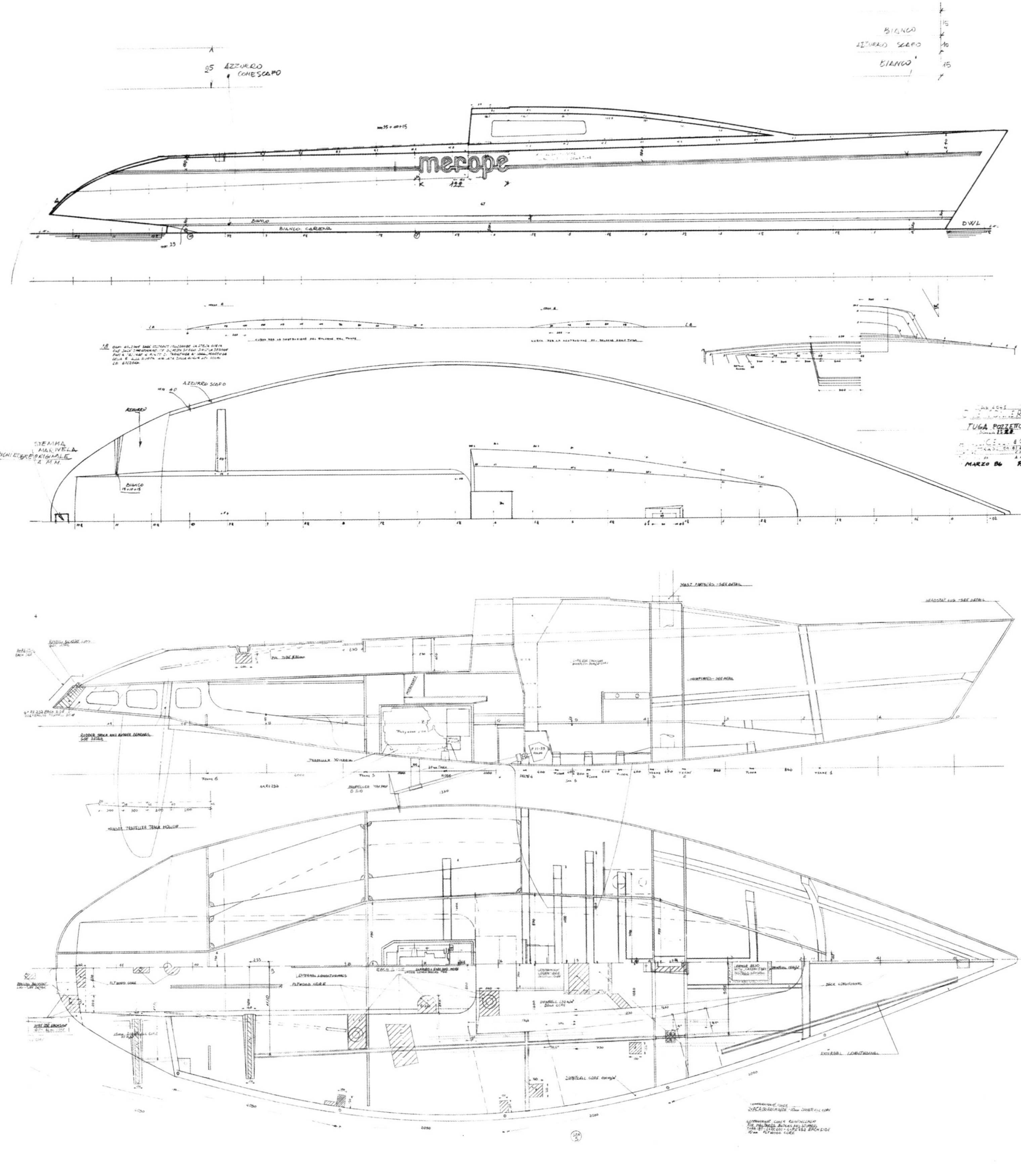
25 AZZURRO CONESCAFO
BIANCO
AZZURRO SCAFO
BIANCO
merope
BIANCO
BIANCO CARENA
DWL
AZZURRO SCAFO
AZZURRO
STEMMA MANOVELA RICHIEDERE ORIGINALE A M.M.
BIANCO
TUGA POZZETTO-COLORAZIONE
MARZO 86 RF.

Brancaleone Open

Five years after the Three Quarter Tonner *Brancaleone*, we designed a boat for the same owner to race in the Adriatic in the Open 10.67 category, a class regulated only the maximum length (in this case precisely 10.67 meters), with no other restrictions except those imposed by safety standards. The hull was designed with wide stern sections, a good form stability, a limited wetted surface and a relatively slender shape, notwithstanding the rather flat bottom. In this case, too, we used a structure reinforced with carbon fiber in both the hull's skin and frames. The internal layers were arranged at various angles of incidence, while the outer were laid were laid longitudinally and then varnished. For the construction, we again returned to the yard of Luigi Morri and Dino Para, the *non plus ultra* as regards wooden construction: the result, as expected, was a beautifully built boat.

Year 1990
Builder Morri & Para Rimini
Construction Cold molded wood
LOA 10.67 m
B MAX 3.24 m
D 2.1 m
DSPL 3.2 t
Mainsail 36 mq
Headsail 33 mq
Spi 72 mq

Carbon reinforcement
of the laminated wooden hull.

Construction plan.

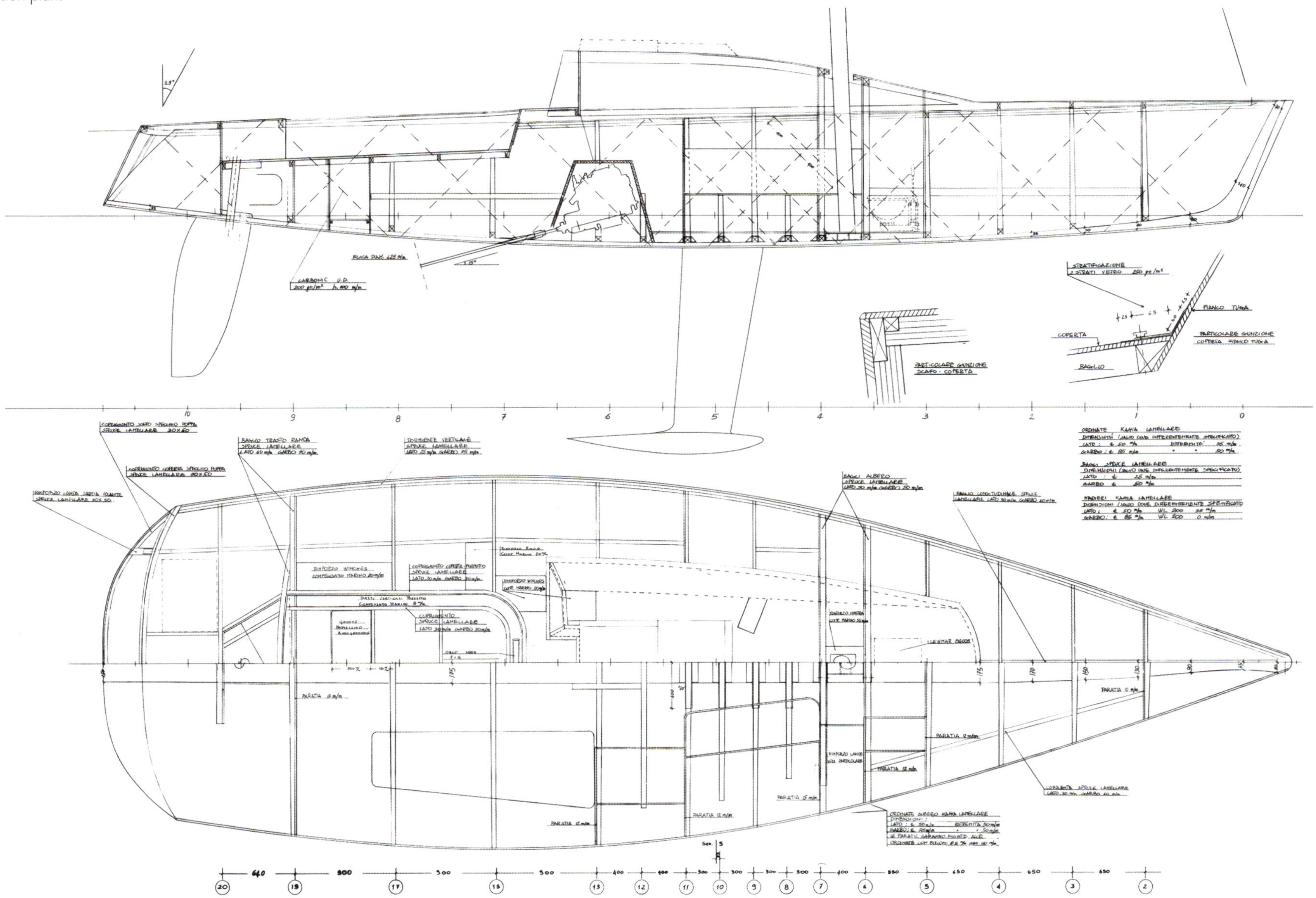

Stradivaria

This was the first of our projects for the Lake Garda "Libera," or Open Class (here in the cruising category), a regulation that set the maximum length limit at 12.7 meters. *Stradivaria* was built in 1992 in composite materials by the Soleri yard with a fractional rig and the possibility to hoist a masthead spinnaker and a large light headsail. With *Stradivaria*, we developed our design strategy for open classes, which we subsequently proposed in boats such as *Gaia Cube* and *Lisa*: avoid excessive beam and flat sections in the counter (and therefore not pushing stability of form to excess); and limit the wetted surface to assure good performance with moderate wind, especially upwind. For the hull livery below the waterline, we decided to replicate the motif of black and white rectangles that is customarily used in tank testing models, a design that resembles a chessboard and is used to evaluate better the shape of the waves generated by the boat's motion. In the topsides, instead, there was a dark red stripe from bow to stern. *Stradivaria* achieved excellent results, winning the Centomiglia del Garda in its class in three consecutive years, from 1992 to 1994.

Year 1992
Builder Soleri Ravenna
Construction Composite
LOA 12.55 m
B MAX 4.118 m
D 2.785 m
DSPL 2.95 t
Mainsail 49 mq
Headsail 47 mq
Spi 110 mq

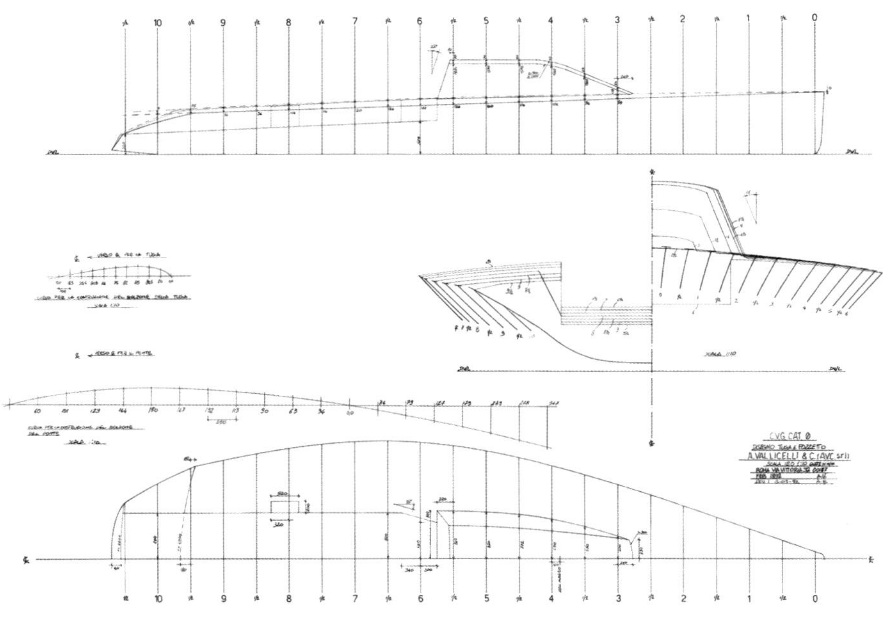

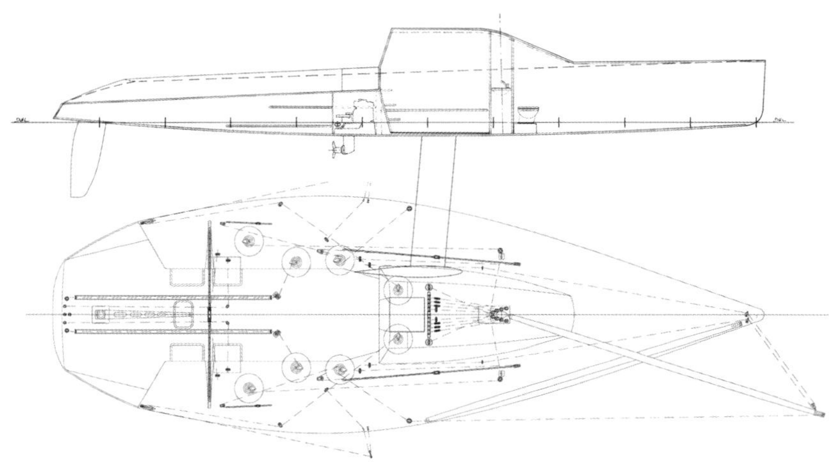

Deck and cabin lines, deck plan, longitudinal section.

Stradivaria on Lake Garda.

Sail plan.

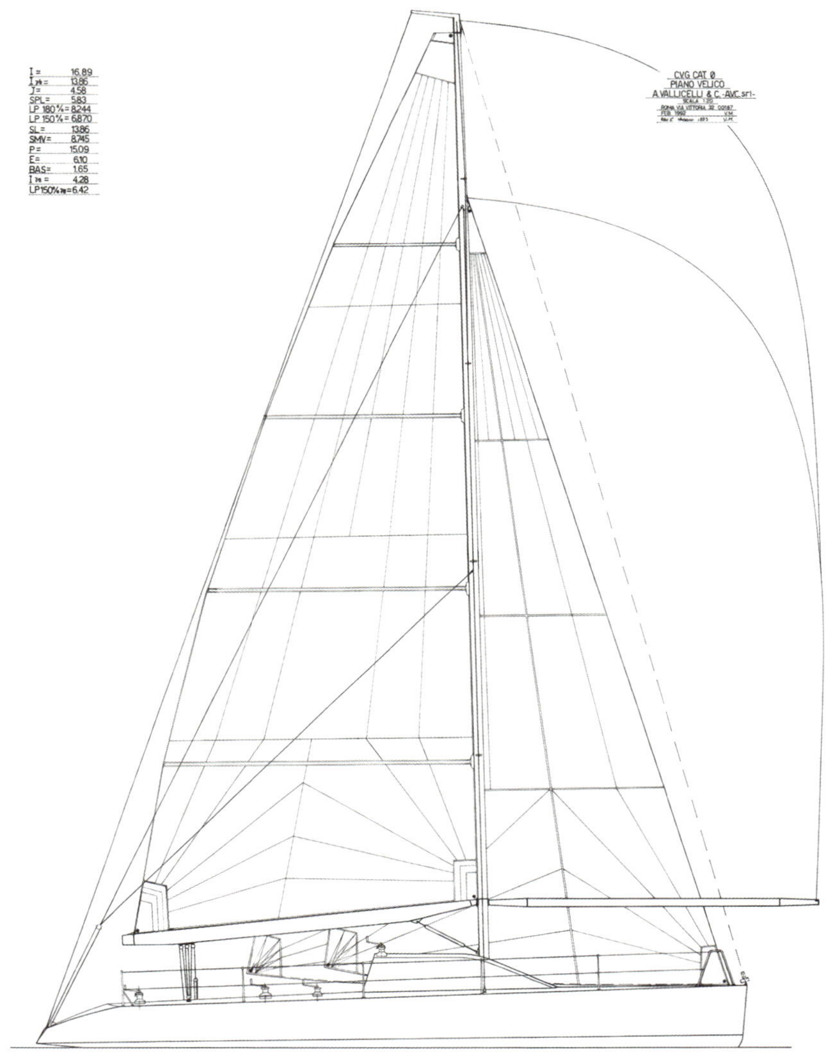

Gaia Legend

Gaia Legend, also known by the sponsor's name *Istra Benz*, was is an open class of 16.75 meters. She was designed in 1995, two years before *Riviera di Rimini*, and built by the Soleri yard using carbon and Kevlar prepreg laminate. These were the years in which regattas for open classes began to be popular also at sea, and so the project was conceived on the basis of wind and sea conditions. Compared to our designs of the period for open boats racing on lakes, *Gaia Legend* had a hull with rounder cross-sections and with less turn of the bilge, and waterlines that ran very gently from bow to stern, without forcing.

The intention was to reduce wetted surface, compared to the standards used previously, gaini ng something thereby also in terms of slender lines. We also equipped the boat with an appropriate sail plan to make her perform well in the most common sea and wind conditions in the Adriatic. The result was a boat that was particularly happy in medium to light winds and very agile in a chop. *Gaia Legend* won the Barcolana in her class three times in a row (1995, 1996 and 1997) and the Rimini-Corfu-Rimini twice, in 1996 and 1997.

Year 1995
Builder Soleri Ravenna
Construction Composite
LOA 16.75 m
B MAX 4.76 m
D 3.5 m
DSPL 6.2 t
Mainsail 102 mq
Headsail 76 mq
Spi 240 mq

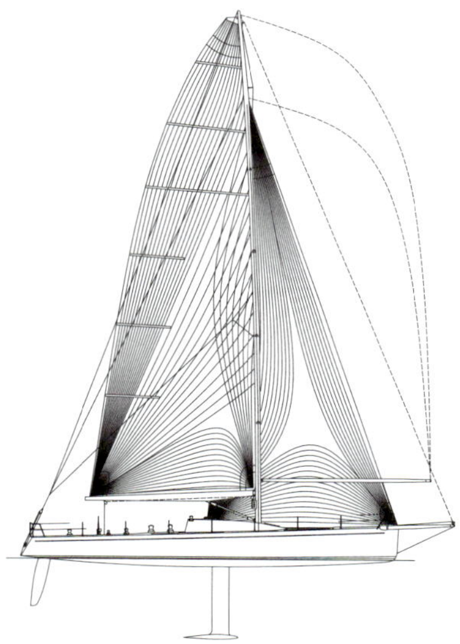

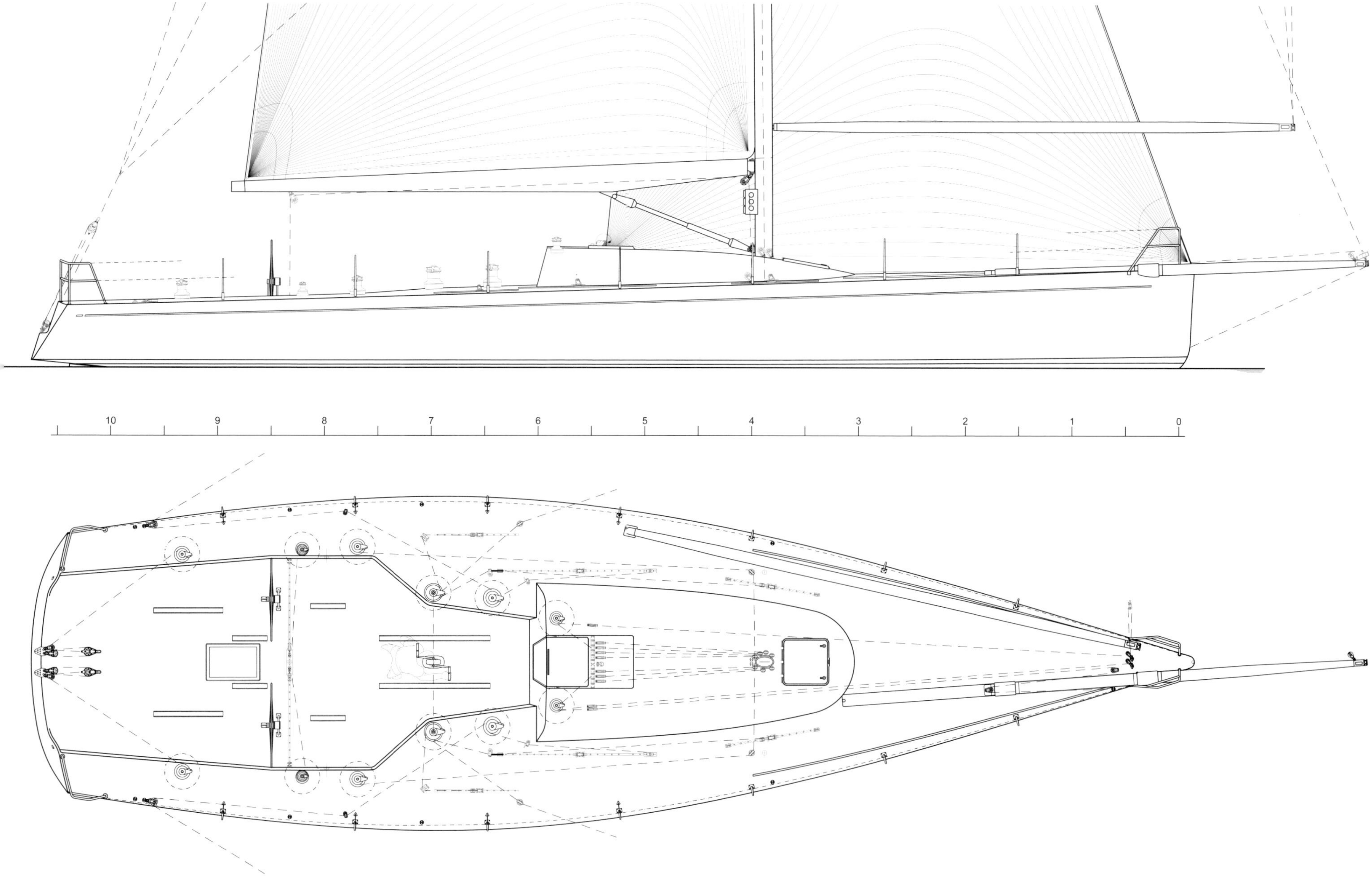

Digital deck plan and profile.

Riviera di Rimini

After the excellent performances of *Gaia Legend* we were asked to design another open class: *Riviera di Rimini*. The project was designed taking into account the few constraints imposed by the regulation of the Open 55 class, first of all, that of the maximum overall length of 16.75 meters. The hull was drawn with a fine bow, a limited wetted area, and a very low displacement (originally 5.2 tons, then 5.8 after the boat was extended). The maximum beam was broad and the sections flared to increase the righting moment generated by the weight of the crew on the rail and the water ballast. The stern sections were wide and became flat in the central area, to facilitate planing when sailing downwind. The sail plan was flexible: a basic fractional rig (85%), but on which it was possible to hoist a large genoa at the masthead (code zero) for light winds and mount a bowsprit for the gennaker.
The keel fin was a special steel narrow and deep blade, fitted with a lead bulb of approximately 1500 kg. Like our other boats of this type, *Riviera di Rimini* was built by the Soleri yard in composite material using vacuum bag technology. The skin was made of an epoxy resin sandwich with a Termanto core and unidirectional carbon, glass and Kevlar fabrics. In 1999, to adapt the boat to the Open 60 category, used in the Fastnet race and the Tour d'Europe, *Riviera di Rimini* returned to the yard to be extended by a meter and a half. In addition to the hull extension, the bulb was replaced with a heavier one to increase the righting moment, and two asymmetric canards were fitted forward of the mast. Moreover, the sail plan was increased by 20%, attaining an area of 310 sq.m. upwind, and 540 sq.m. downwind.
Riviera di Rimini has a rich palmarès of wins: in 1998, she was awarded the Boat of the Year prize after winning the Giraglia, Rimini-Corfù-Rimini, Barcolana and the Middle Sea Race, establishing some new course records (the Soleri website shows course records for the Giraglia and Rimini-Corfù-Rimini). In 1999 she won the Rimini-Corfù-Rimini again and came second in the Fastnet and the Tour d'Europe. In 2000 she won first place in the Roma per Tutti and in 2001 the Roma per Due.

Year 1997
Builder Soleri Ravenna
Construction Composite
LOA 16.75 m
B MAX 4.76 m
D 4.2 m
DSPL 6.2 t
Mainsail 109 mq
Headsail 62 mq
Spi 235 mq

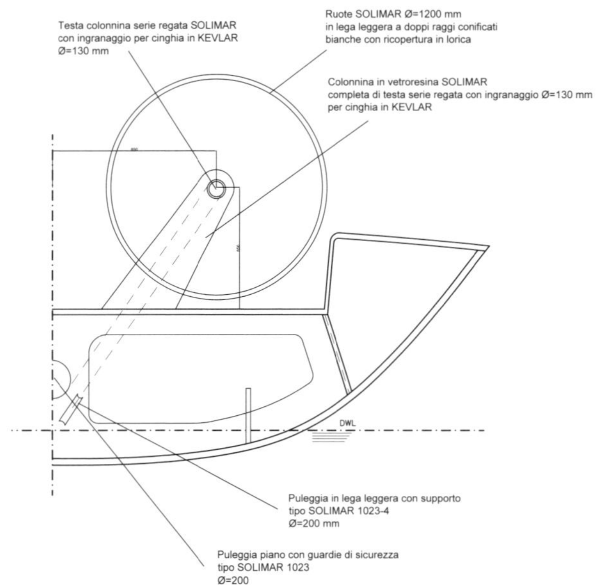

Wheel section details.

Riviera di Rimini at the Giraglia.

Deck plan and profile.

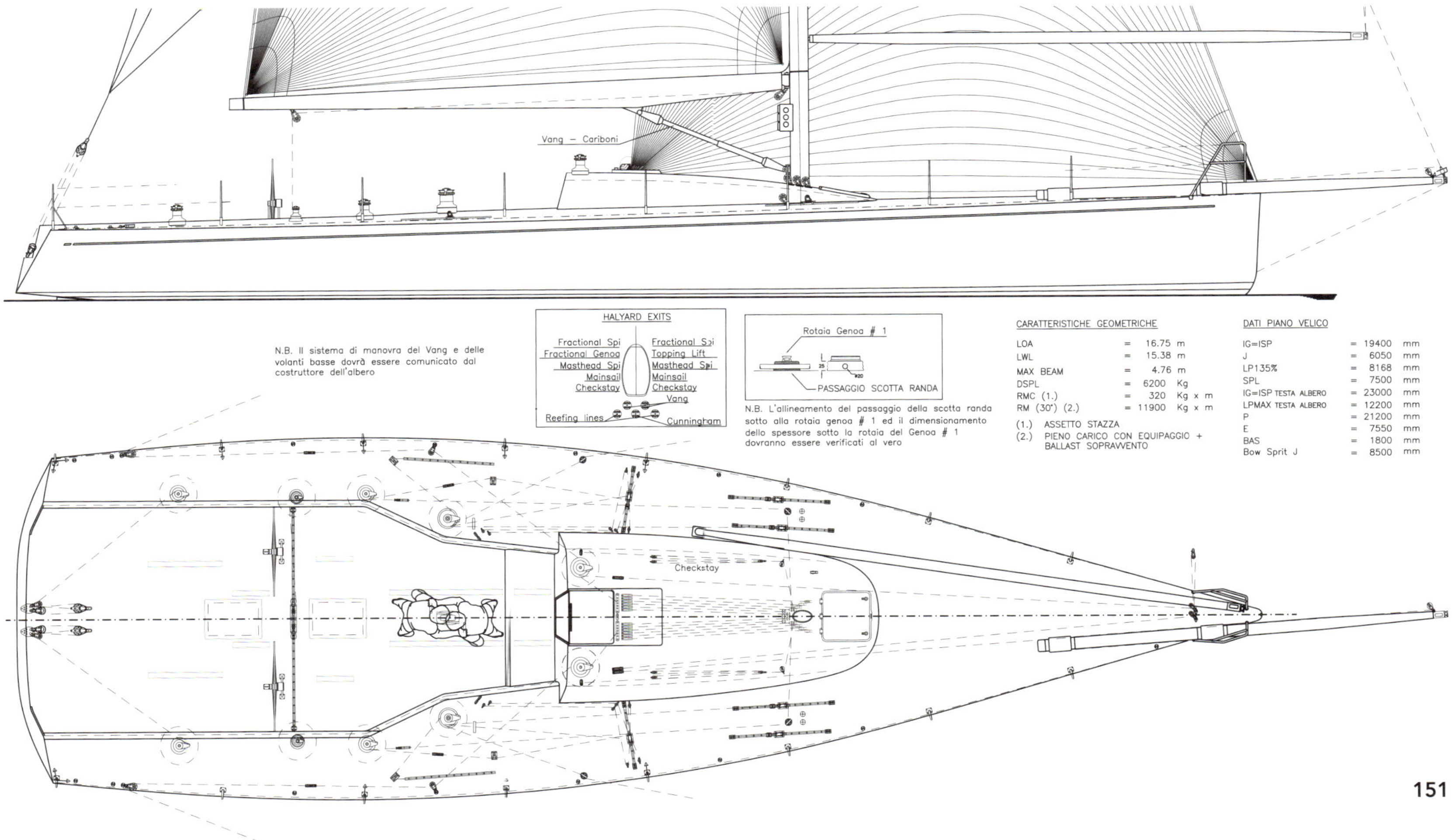

Hull lamination construction plan.

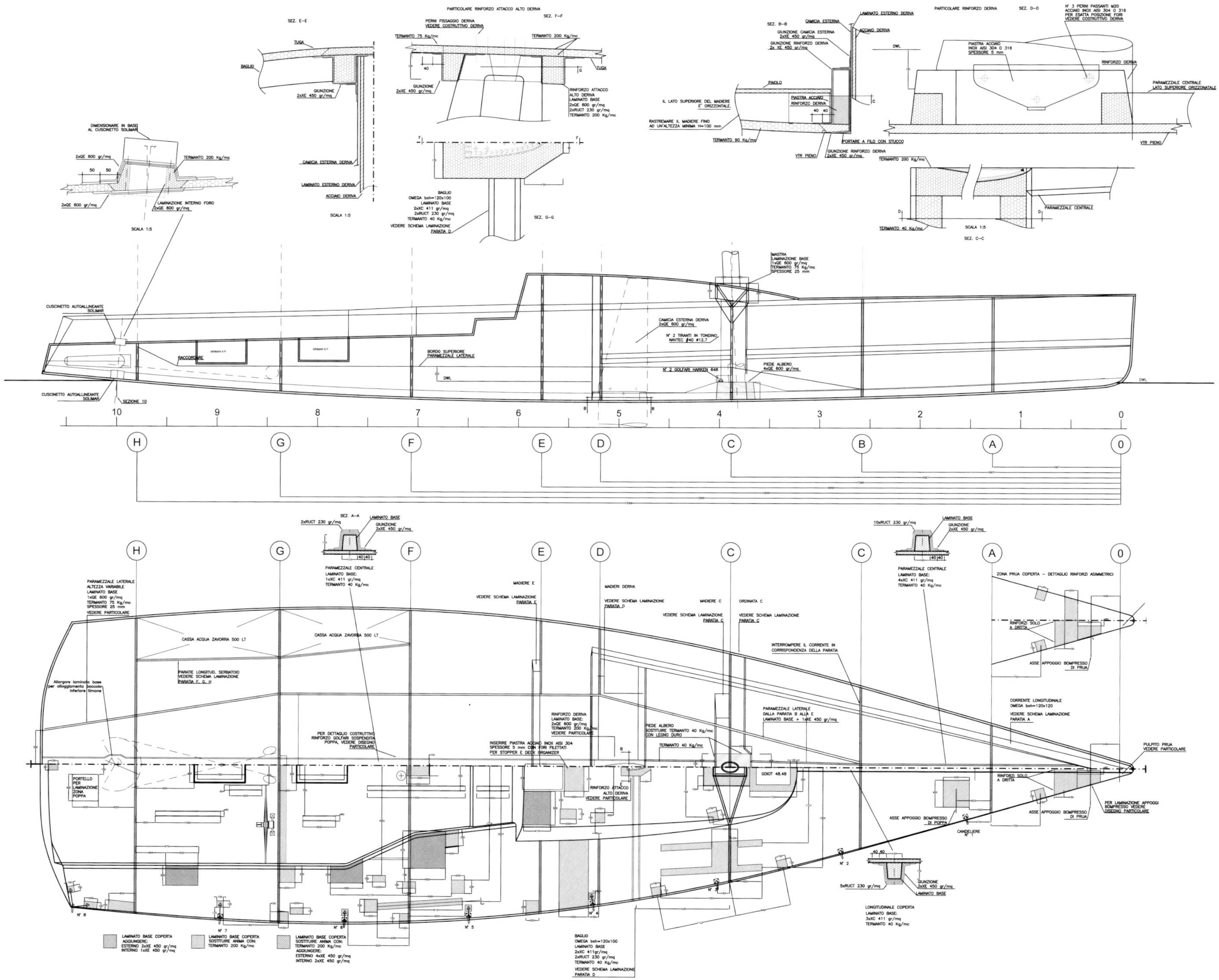

WWW.PORTOSANROCCO.IT
RIVIERA DI

Libertine

We were unsure whether to include *Libertine* among the racing or cruising boats since she was built using the molds of a series model, the Comet 45, designed for recreational use and therefore drawn with a high freeboard and a rather generous coachroof volume. Some changes were made on *Libertine*, for the purposes of competition, which mainly concerned the appendages and sail plan. Thanks to good management of their racing program, the boat proved much faster than we expected, so much to win the Giraglia in 2007 and the ORC World Championship in Athens in 2008. The Comet 45 is a medium-light displacement boat, with a very fine bow and a full, round stern. The coachroof is quite long, and near the entrance, hatch joins up with a large coaming where the halyard winches are located. The cockpit coamings, which form a continuum with the deckhouse, taper near the transom, which is open like that of the Comet 51. The cockpit seats extend transversally in the front part of the cockpit, to provide full height in the aft cabins. The solution of placing the head under the entrance ladder was borrowed from *Genesi*, making it accessible from the two twin cabins arranged under the cockpit. Another peculiarity is the chart table, placed opposite to the horseshoe settee and fitted with two comfortable facing seats. In 2003 the boat was named European Boat of the Year by "Yacht magazine" of Dusseldorf and the Boat of the Year 2007 by "Yacht & Sail Magazine."

Year 2003
Builder Comar Fiumicino (RM)
Construction GRP
LOA 13.782 m
B MAX 4.2 m
D 2.63 m
DSPL 9.5 t
Mainsail 66 mq
Headsail 55 mq
Spi 146 mq

Libertine racing.

Line plan for hull and appendages.

LIBERTINE

LIBERTINE

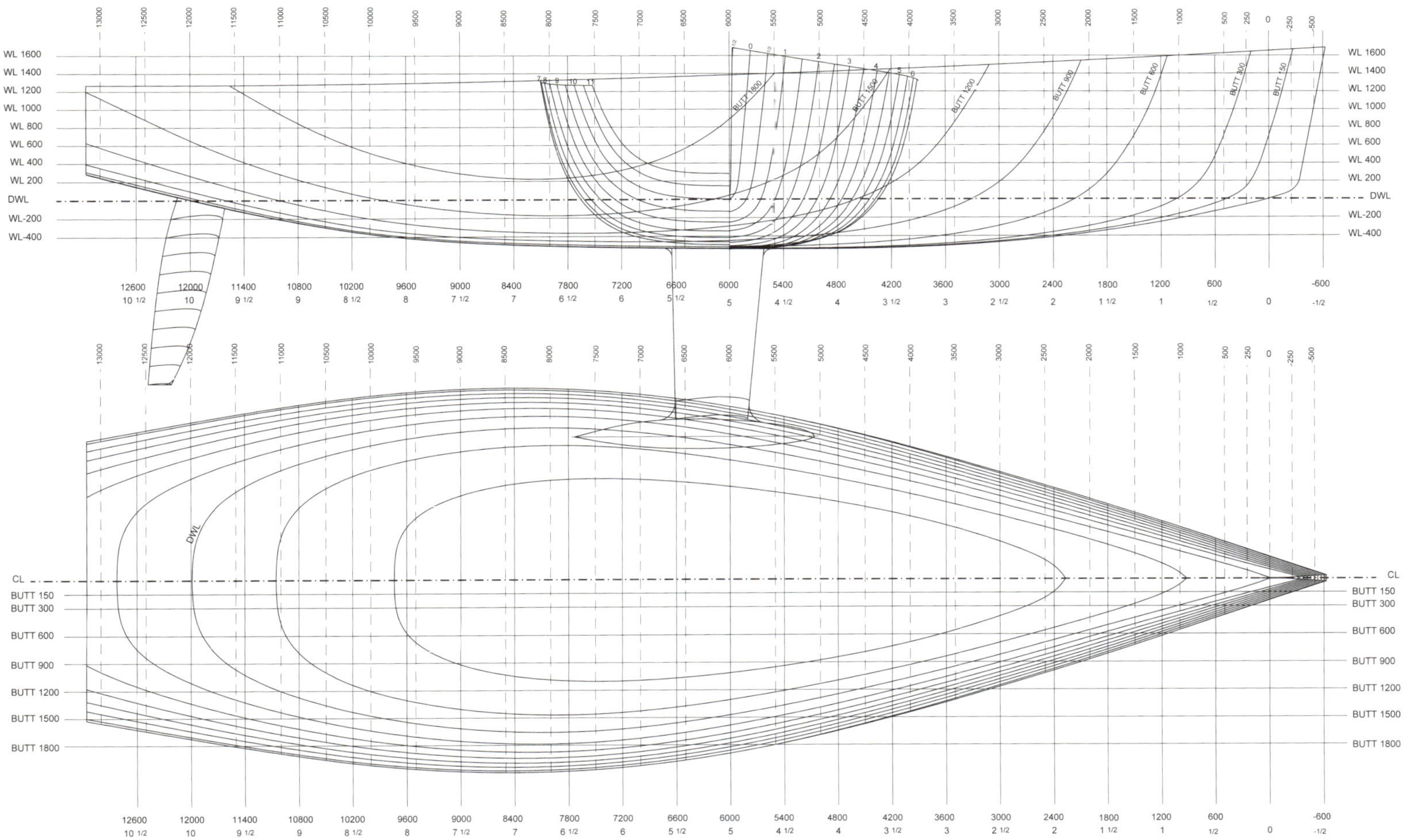
WL 1600
WL 1400
WL 1200
WL 1000
WL 800
WL 600
WL 400
WL 200
DWL
WL-200
WL-400
BUTT 1800
BUTT 1500
BUTT 1200
BUTT 900
BUTT 600
BUTT 300
BUTT 150
CL
DWL
12600 12000 11400 10800 10200 9600 9000 8400 7800 7200 6600 6000 5400 4800 4200 3600 3000 2400 1800 1200 600 -600
10 1/2 10 9 1/2 9 8 1/2 8 7 1/2 7 6 1/2 6 5 1/2 5 4 1/2 4 3 1/2 3 2 1/2 2 1 1/2 1 1/2 0 -1/2
13000 12500 12000 11500 11000 10500 10000 9500 9000 8500 8000 7500 7000 6500 6000 5500 5000 4500 4000 3500 3000 2500 2000 1500 1000 500 250 0 -250 -500

Verve

Designed to replace the Comet 51 in the Comar yard production, the Comet 50 Sport was a cruiser-racer designed for the ORC International formula, but also IRC, given the hull shape and the high stability. *Verve* was the prototype of the new production and was made entirely of advanced composite materials, with an expanded PVC sandwich, using unidirectional and biaxial carbon fibers with epoxy resins, and post-cured in an oven. The longitudinal stringers and the floors were of carbon, and the composite bulkheads were also reinforced with carbon fibers. The chosen displacement places it in the range of medium-light boats in this category. The heavy keel had a lead bulb and a special cast iron blade. The sail plan had a 9/10 fractional rig with a square-top mainsail and 105% jibs with asymmetrical spinnakers that can also be flown from the spinnaker pole. The mast, without running backstays, was made of carbon with two levels of boomerang-shaped spreaders, swept aft 20°. The deck plan featured twin helms and a removable coffee grinder in the cockpit. The full cruising interiors included three cabins, in addition to the galley, the chart table and services.

Year 2011
Builder Comar Fiumicino (RM)
Construction Composite
LOA 15.1 m
B MAX 4.09 m
D 3.45 m
DSPL 9.5 t
Mainsail 80 mq
Headsail 69 mq
Spi 190 mq

Construction at the Comar yard in Fiumicino.

Sail plan and rigging scheme.

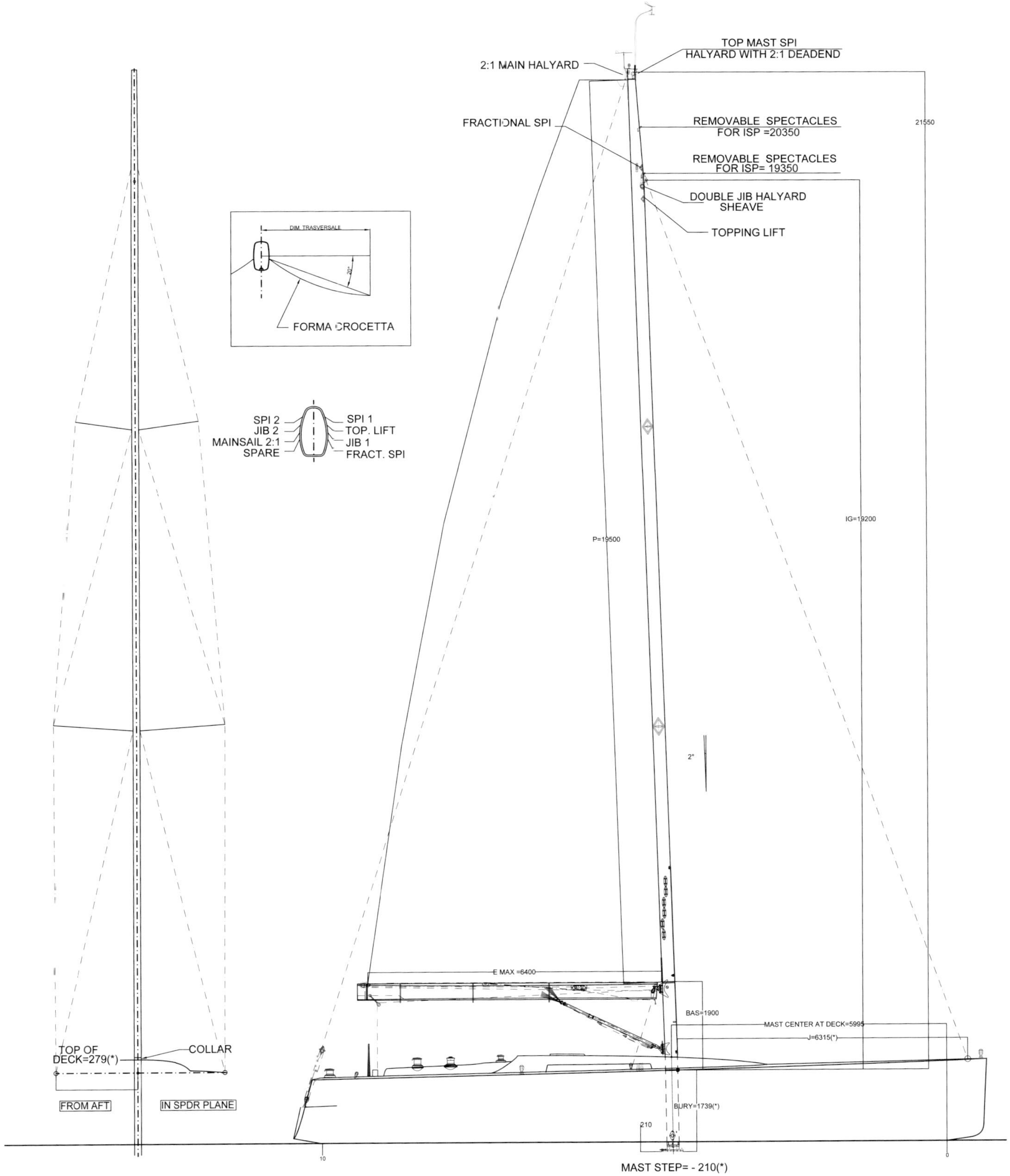

Verve racing.

Deck plan and profile.

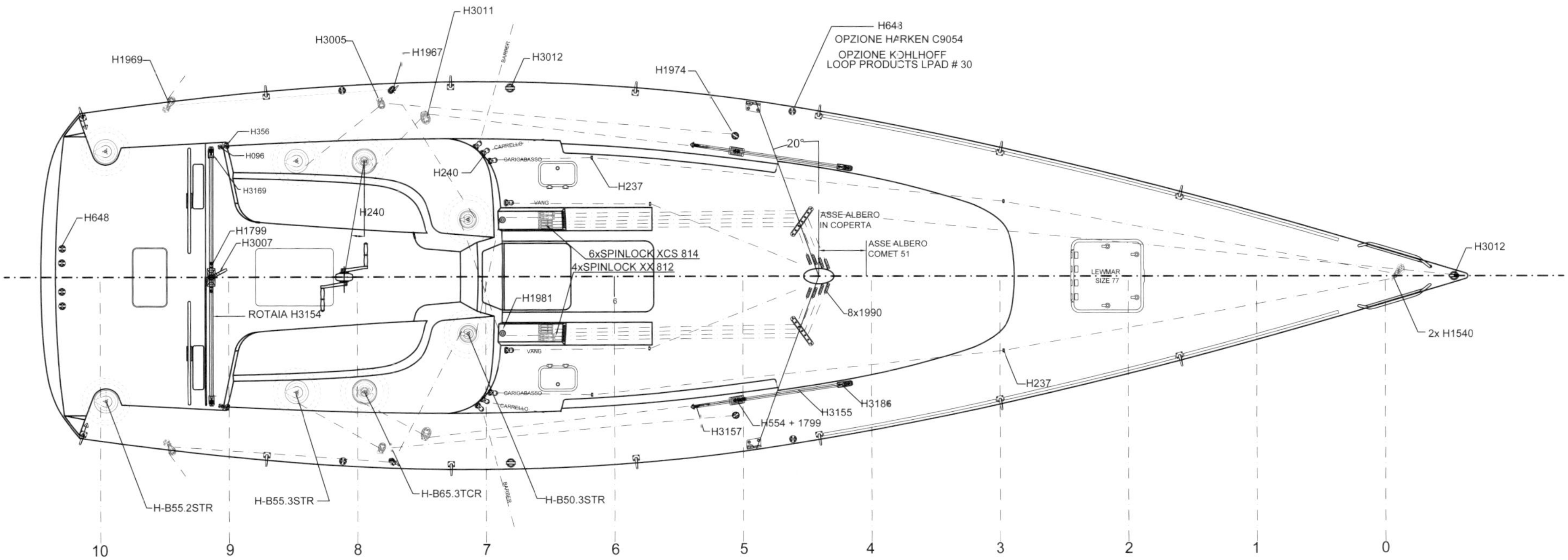
H3011
H3005
H1967
H3012
H1969
H1974
H648
OPZIONE HARKEN C9054
OPZIONE KOHLHOFF
LOOP PRODUCTS LPAD # 30
H356
H096
H3169
H240
H237
20°
ASSE ALBERO
IN COPERTA
ASSE ALBERO
COMET 51
H648
H1799
H3007
6xSPINLOCK XCS 814
4xSPINLOCK XX 812
H1981
ROTAIA H3154
8x1990
LEWMAR
SIZE 77
H3012
2x H1540
H237
H3186
H3155
H554 + 1799
H3157
H-B55.2STR
H-B55.3STR
H-B65.3TCR
H-B50.3STR
10
9
8
7
6
5
4
3
2
1
0

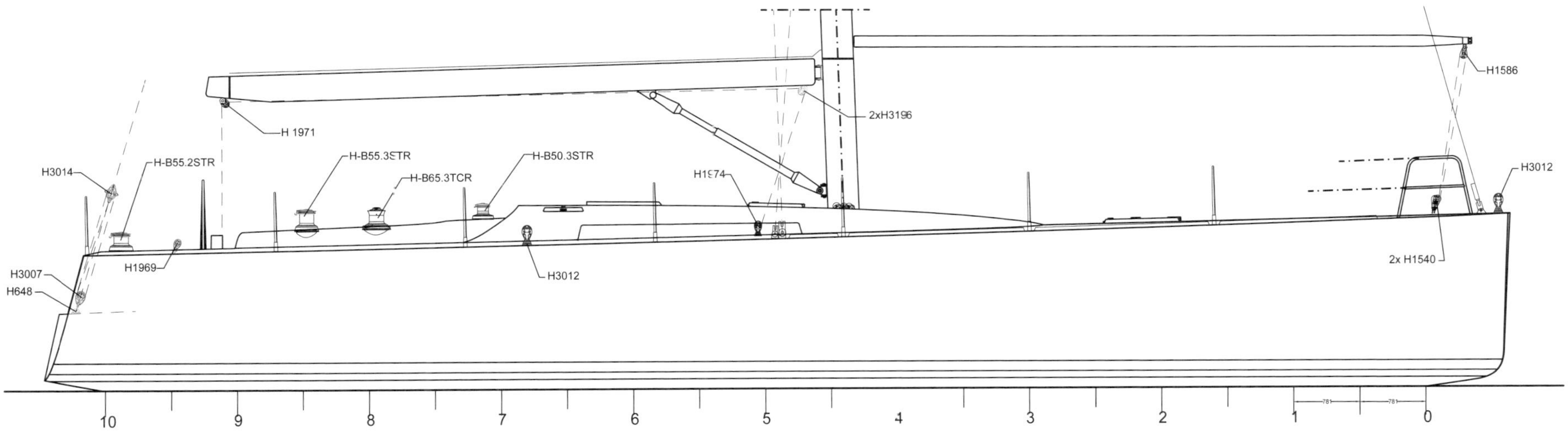
H1586
2xH3196
H 1971
H3014
H-B55.2STR
H-B55.3STR
H-B65.3TCR
H-B50.3STR
H1974
H3012
H3007
H648
H1969
H3012
2x H1540
10
9
8
7
6
5
4
3
2
1
0

CRUISERS

One of the first cruisers the Studio designed was *Manua* in 1978, a medium-heavy displacement yacht made out of cold molded wood, and *Koala V*, a one-off cruiser designed that same year, for which architect Pierluigi Spadolini designed the interiors. Continuing with the same desire to combine fast hulls with comfortable fittings that catered to owners' needs, the practice went on to design boats such as *Dioscura*, *Koala VI*, *La Cardinala* and *Fetch III*. In the early '90s, the school ship *Orsa Maggiore* was launched, a very special almost thirty meter boat designed for the Italian Navy for long ocean cruises with a large crew of students. The Studio had previously designed a number of boats for the Navy, with *Gemini*, *Castore*, *Polluce* and *Merope*. At the end of the '90s, our Studio teamed up with famous French designer Philippe Starck for *Virtuelle*, an experimental fast cruiser with an inimitable style. This yacht showcased a whole series of innovative solutions, including twin rudders and streamlined, rounded hull shapes that were to return in many of our later designs. Fast cruisers such as *H_2O* and *Shadow* may be considered evolutions of *Virtuelle:* despite being designed for fast cruising, they have always acquitted themselves well in racing – *H_2O* has won the Rolex Cup in 2015, 2017, 2018 and 2019.

Manua

Manua was one of the first large boats we designed, two years after the studio's debut with *Gattone* and *Ziggurat* (in those days a 12.7 meter boat was considered rather large). It was a Second Class boat of medium-heavy displacement, in cold-molded construction built by the Trieste shipyard of Arrigo Petronio and Sergio Pecarich. The construction, of remarkable crafted refinement, is a beautiful example of the quality achieved by the North Adriatic school following in the steps of their maestro, Craglietto.

Among the special details in wood construction, which we had the chance to witness very closely, I would like to mention what appears to be a merely technical detail: to reduce weight in the upper topsides of the hull, the two masters from Trieste gradually reduced the thickness of the wood planking, decreasing it from the keel up to the sheer. *Manua* has a hull similar to *Koala V*, but features a streamlined roof characteristic of our early designs and has a more classic look, with a more pronounced sheer.

Year 1978
Builder Petronio & Pecarich Trieste
Construction Cold molded wood
LOA 12.77 m
B MAX 3.82 m
D 2.1 m
DSPL 8 t
Mainsail 36 mq
Headsail 68 mq
Spi 135 mq

Manua on a close reach.

Construction plan and general arrangement. Ink drawing on tracing paper.

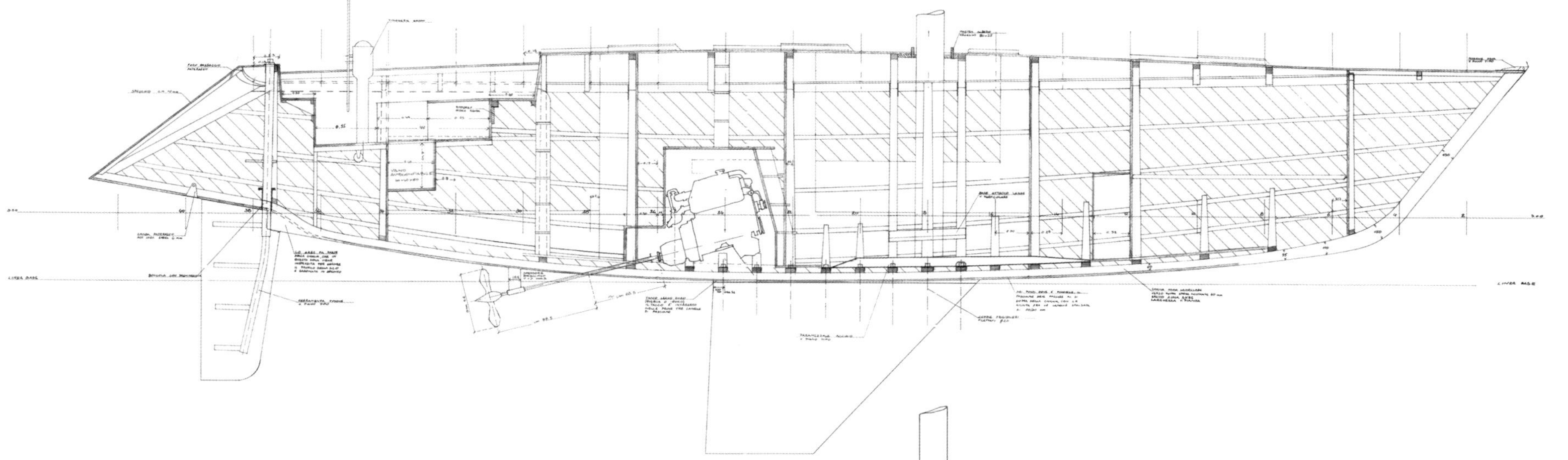

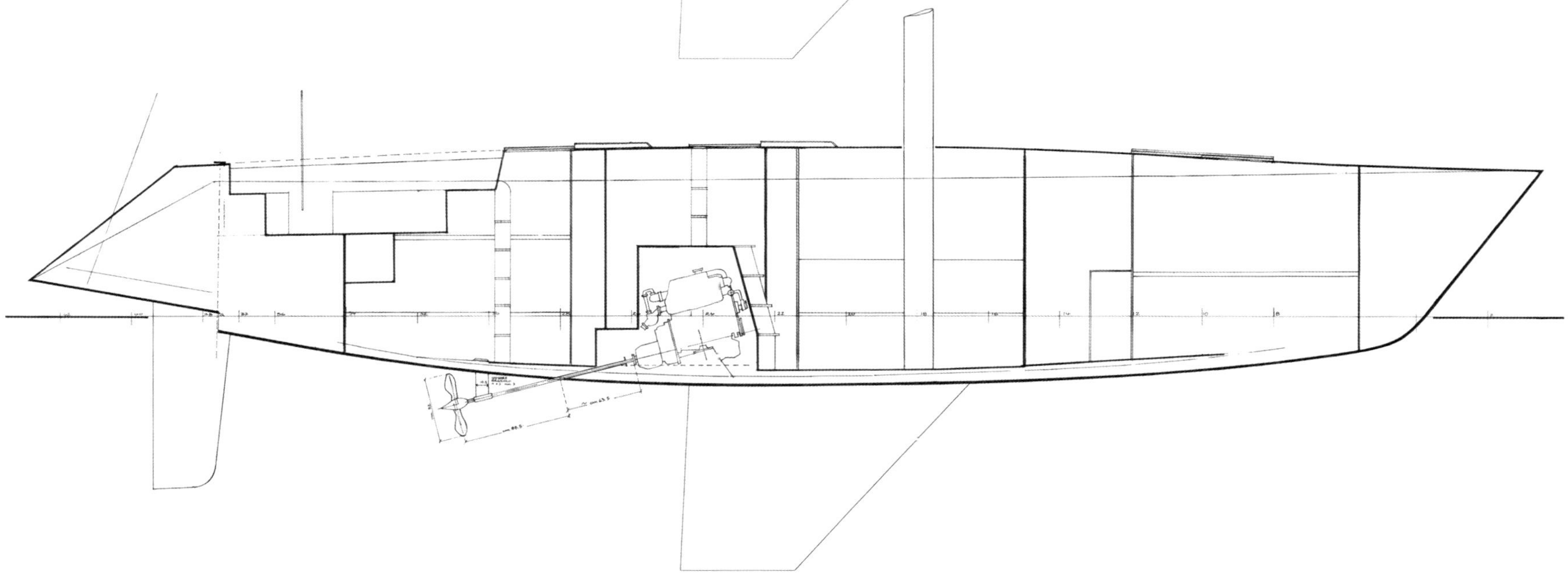

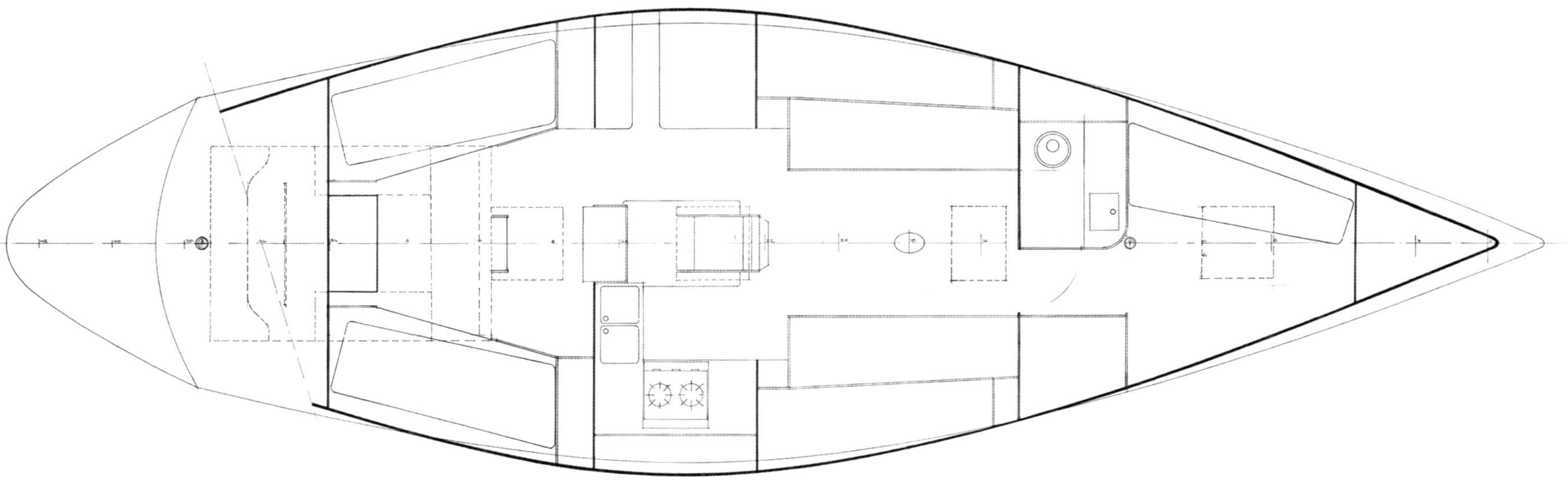

Koala V

For the 1978 season, we designed *Koala V* for the well-known designer and architect Pierluigi Spadolini (founder of the first university faculty of nautical design in Italy), who subsequently became a friend and mentor for our studio. The boat was built in cold-molded wood by the De Cesari yard using a characteristic technique he used for several of our designs of a mold with false frames, where the keel and longitudinal stringers were attached. The real frames were laminated within the mold, then smoothed and painted externally and only then placed over the planking, setting them on top of the stringers. For the interiors, Spadolini proposed an innovative layout, with two cabins in the bow from which one could access a shared head via two separate doors. From these two cabins, one entered the main saloon directly, without intermediate spaces. The aft cabin, set up with separate berths, was in direct communication with the cockpit where the helm was located via a convenient companionway. The design of the boat was deliberately essential, with a flush deck and an almost straight sheer.
In *Koala*, we used titanium for the first time for the stanchions and the rudder post: light and strong material, but fragile and difficult to work (in particular as regards welding) and therefore later replaced by more reliable composite materials.

Year 1978
Builder De Cesari Cervia (RA)
Construction Cold molded wood
LOA 12.76 m
B MAX 3.8 m
D 2.2 m
DSPL 8 t
Mainsail 31 mq
Headsail 70 mq
Spi 141 mq

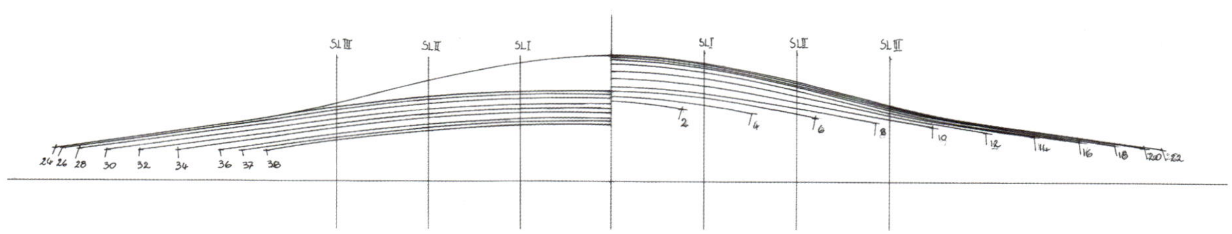

Deck sections.

Pierluigi Spadolini at the helm of *Koala V*.

Koala V with spinnaker and staysail.

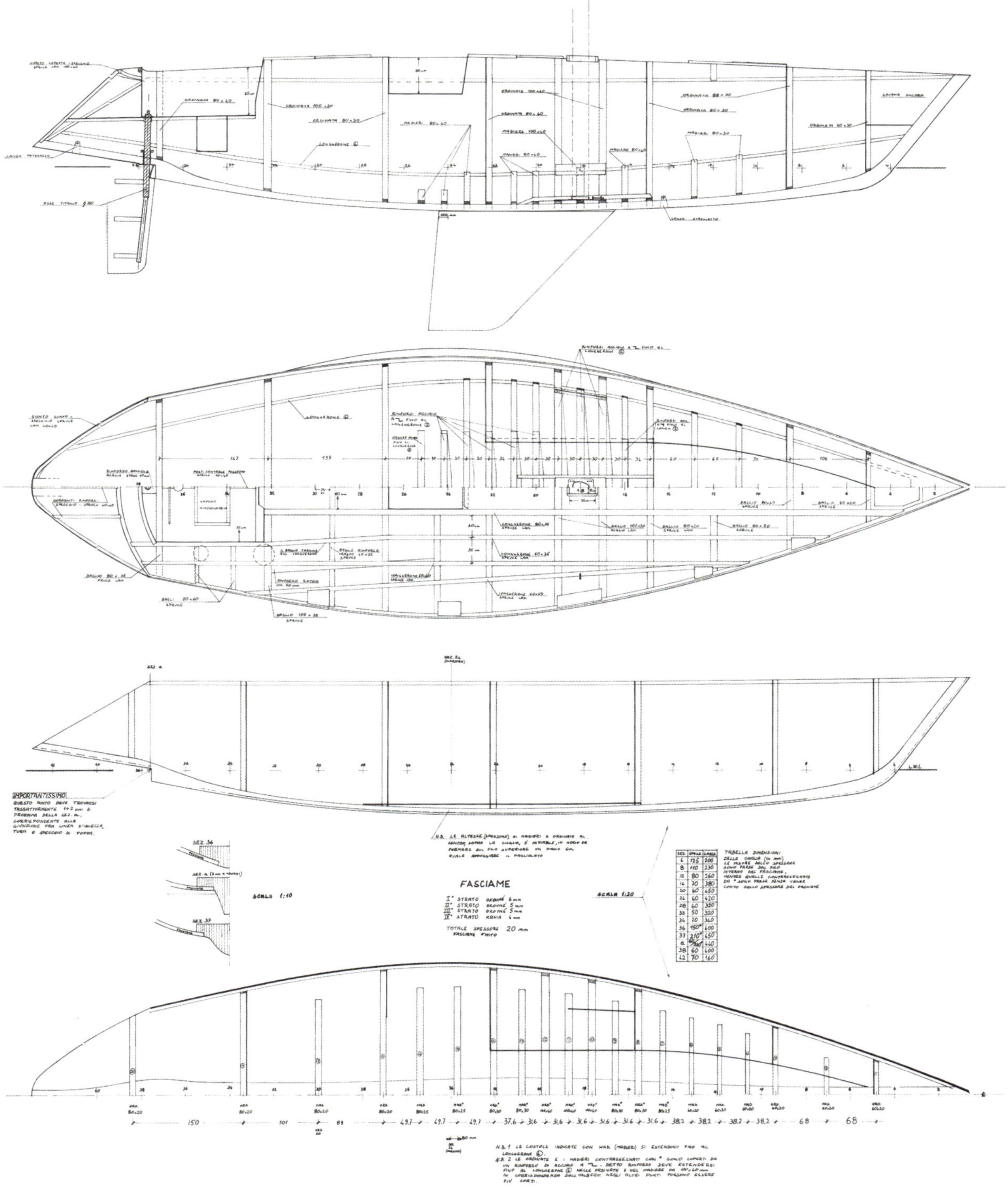

Construction plan and detail of hull stringers.

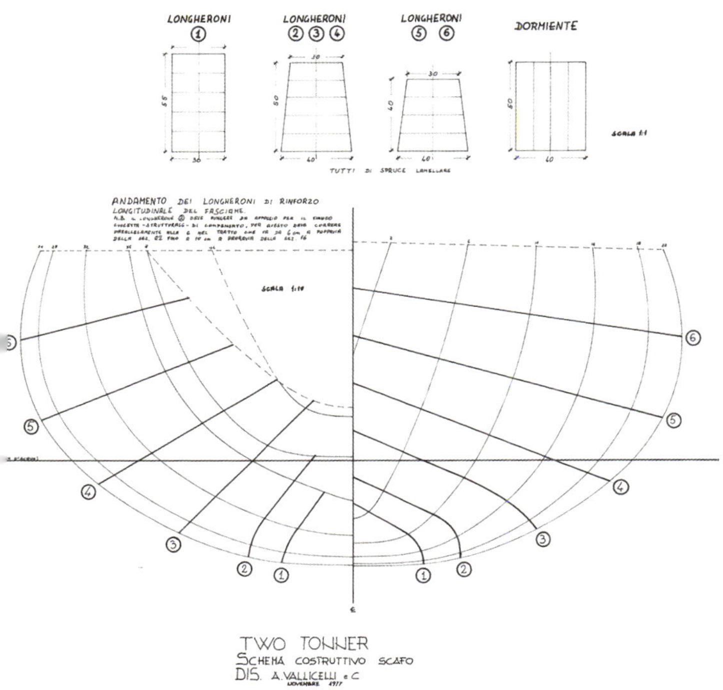

Gemini

After the first *Brava*, we designed another First Class, this time for the Italian Navy: *Gemini* was intended for student training, including participation in inshore and offshore races. The boat was built by the De Cesari yard in cold-molded wood, with carbon fiber reinforcements inserted in the hull skin lamination. Compared to *Brava*, *Gemini* had a hull with a less extreme shape: its lines were more streamlined, with a maximum beam that receded more gently fore and aft. She also had a broader bow, with higher volumes, allowing better sailing upwind with waves, but above all, to increase stability when running in heavy wind, avoiding dangerous broaches under spinnaker. This means that the boat was not as unbeatable as *Brava* in light winds and flat sea conditions, but she suffered less in rough seas and strong winds. With *Gemini*, we directed our design philosophy to all-round performance, capabilities that would later characterize our other boats, such as *Sirena* and *Springbok*. *Gemini* performed very well in many national and international races in the Mediterranean. In the early 1990s, we designed another boat for the Italian Navy, the *Orsa Maggiore* training ship.

Year 1982
Builder De Cesari Cervia (RA)
Construction Cold molded wood
LOA 13.73 m
B MAX 4.112 m
D 2.6 m
DSPL 9.2 t
Mainsail 78 mq
Headsail 41 mq
Spi 155 mq

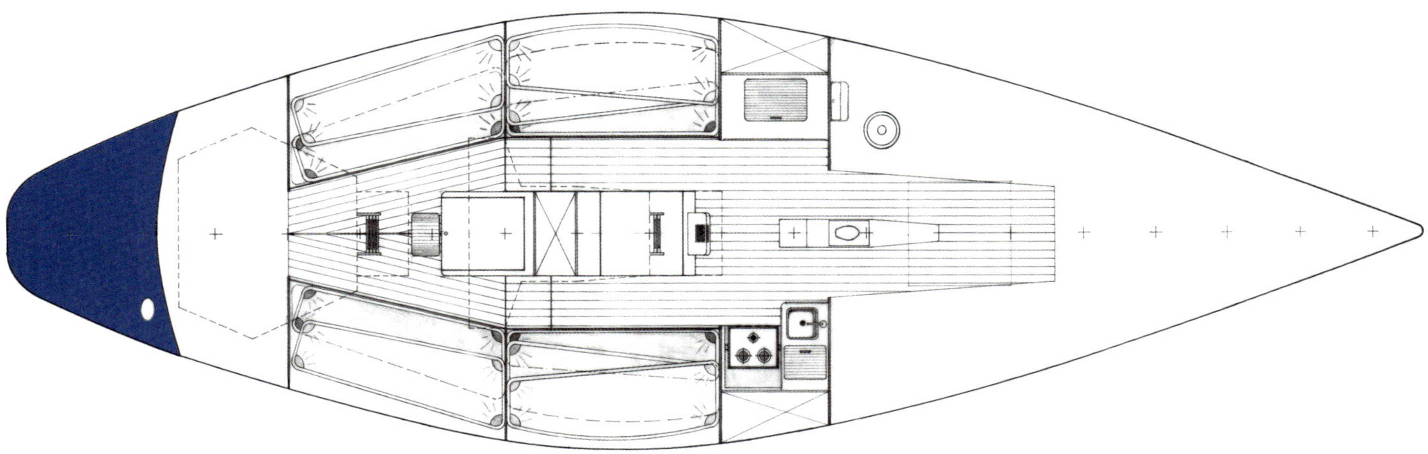

General arrangement.

Construction of the laminated wooden hull at the De Cesari yard in Cervia.

Gemini is coming off a wave.

Cut-away hull profile.

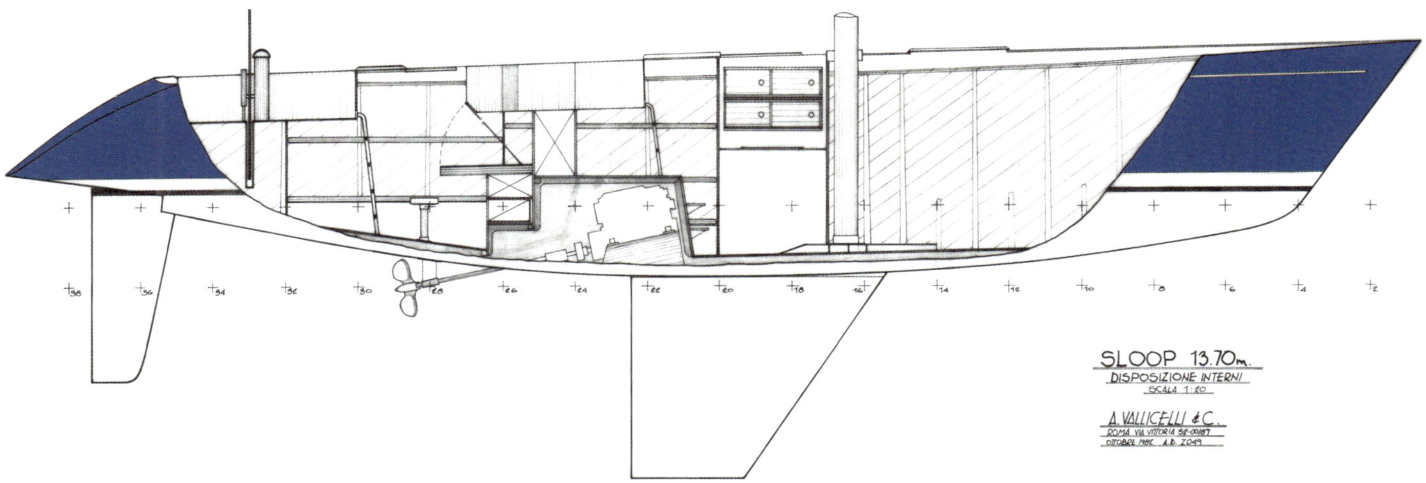
SLOOP 13.70m.
DISPOSIZIONE INTERNI
SCALA 1:20
A. VALLICELLI & C.

Fetch III

Fetch, along with *Mia*, *La Cardinala* and *Atlantis*, typified the performance cruisers inspired by the IOR formula of the early 1990s. These designs already had some features of the fast cruisers but boasted a classic elegance without the minimalistic component that would be a sort of manifesto for us at the end of the 1990s. *Fetch* had fair lines, with a large maximum beam and a generous volume aft. In contrast to earlier models, she had a wedge-shaped cabin top faired aft with the coamings to protect the cockpit. She had a double cockpit allowing full standing room under the bridge deck. The deck layout was designed with particular attention to solutions giving maximum functionality and correct ergonomics. The construction in cold-molded wood and epoxy resin was undertaken by the De Cesari yard with their customary excellence.

Year 1990
Builder De Cesari Cervia (RA)
Construction Cold molded wood
LOA 22.35 m
B MAX 5.87 m
D 3 m
DSPL 32.5 t
Mainsail 112 mq
Headsail 175 mq
Spi 350 mq

Construction and launch.

Sail plan, cabin and deck shapes,
color scheme and general arrangement.

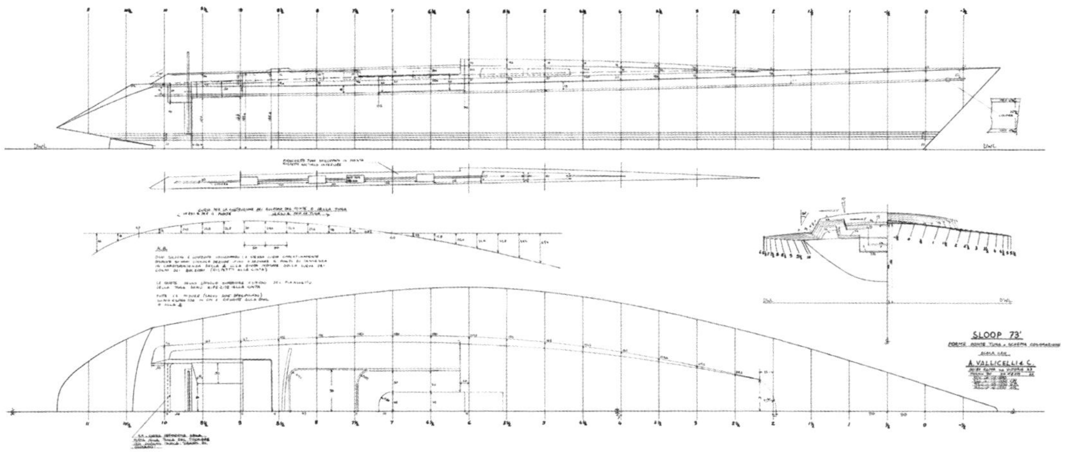

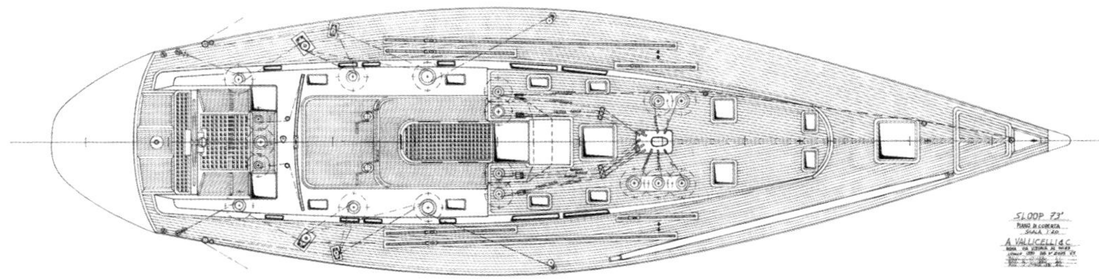

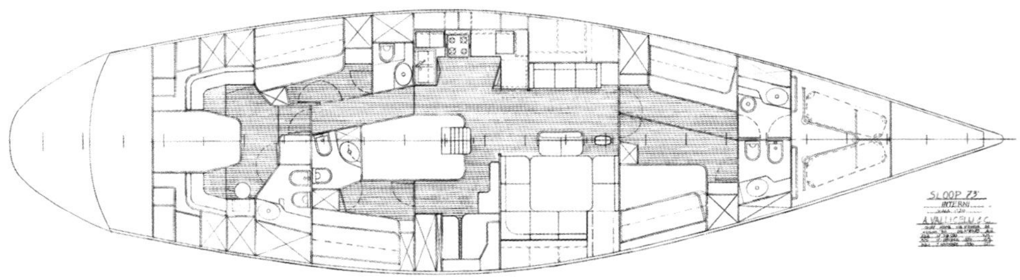

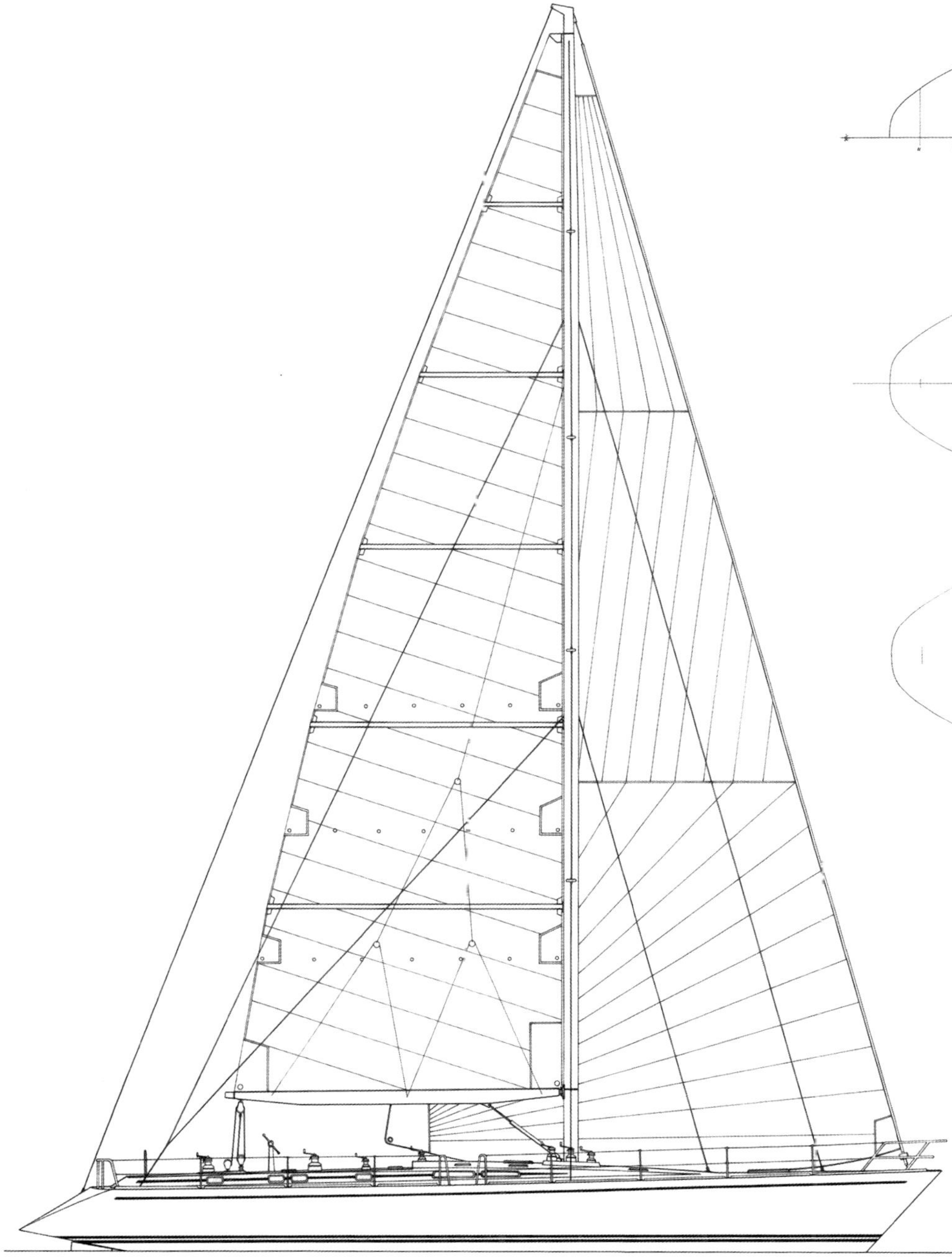

A portrait of *Fetch III*.

Turning the hull over at the De Cesari yard in Cervia.

Construction plan.

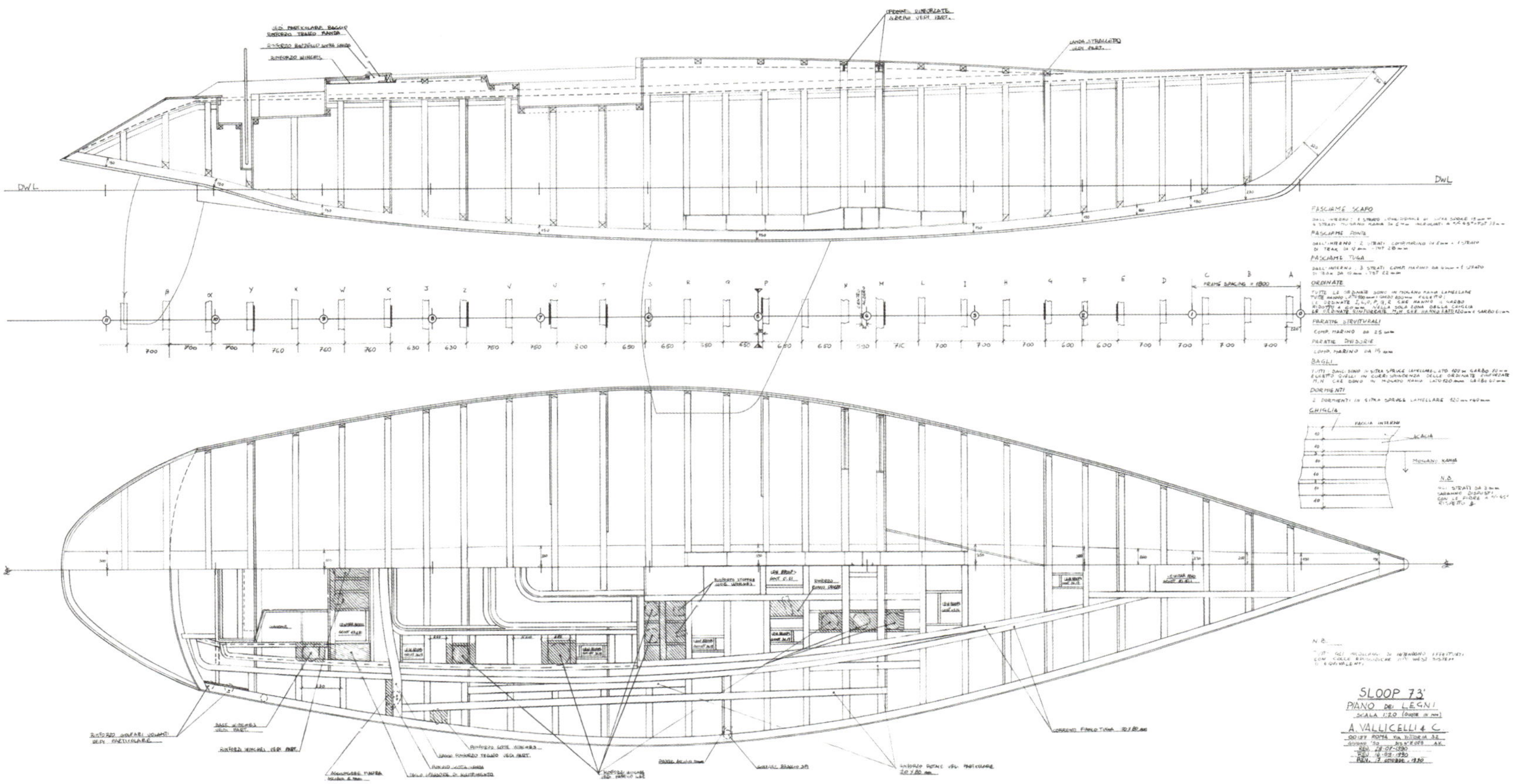

Orsa Maggiore

Orsa Maggiore was a 28.3 meter training ship designed for the Italian Navy and intended for long, non-stop ocean sailing. She was therefore conceived to accommodate a large crew of students while ensuring maximum safety in all wind and sea conditions. For the hull design, we used the CIRCE 3D program, which enabled us to study and optimize the hull geometries in relation to different wave configurations. Even though this was not a real seakeeping study, this type of analysis allowed us to minimize the asymmetries of the wetted hull while heeling. Our goal was to design a fast, balanced and stable boat that wouldn't pound. The keel and rudder planes were designed to assure directional stability while minimizing leeway and maximizing maneuverability. The hull, equipped with an independent engine room, was divided into watertight compartments to provide maximum safety in the event of a hull failure. *Orsa Maggiore* was rigged as a ketch, with a particularly tall mizzen mast structurally independent from the mainmast. The element characterizing the layout of the deck was the deckhouse, equipped with a large polycarbonate window to provide maximum visibility to the internal wheelhouse; its lenticular design was born with the aim of evoking both organic marine forms and the domes of fighter planes. With different variations, we made use of this design in other boats, such as the VR 54 and Solaris 55 Open. Another characteristic of *Orsa Maggiore* was the stowage of the tender in a special self-emptying cockpit under the main boom, while the boom itself was used for launching and retrieving the tender. The boat was built by the Tencara yard in a PVC foam sandwich with different densities and thicknesses pre-impregnated with glass, Kevlar and carbon fibers.

Year 1993
Builder Tencara Marghera (VE)
Construction Composite
LOA 28.3 m
B MAX 6.53 m
D 4.2 m
DSPL 75 t
Mainsail Maestra 140 mq
Mezzana 68 mq
Headsail 180 mq
Spi 425 mq

Orsa Maggiore close hauled.

Sail plan.

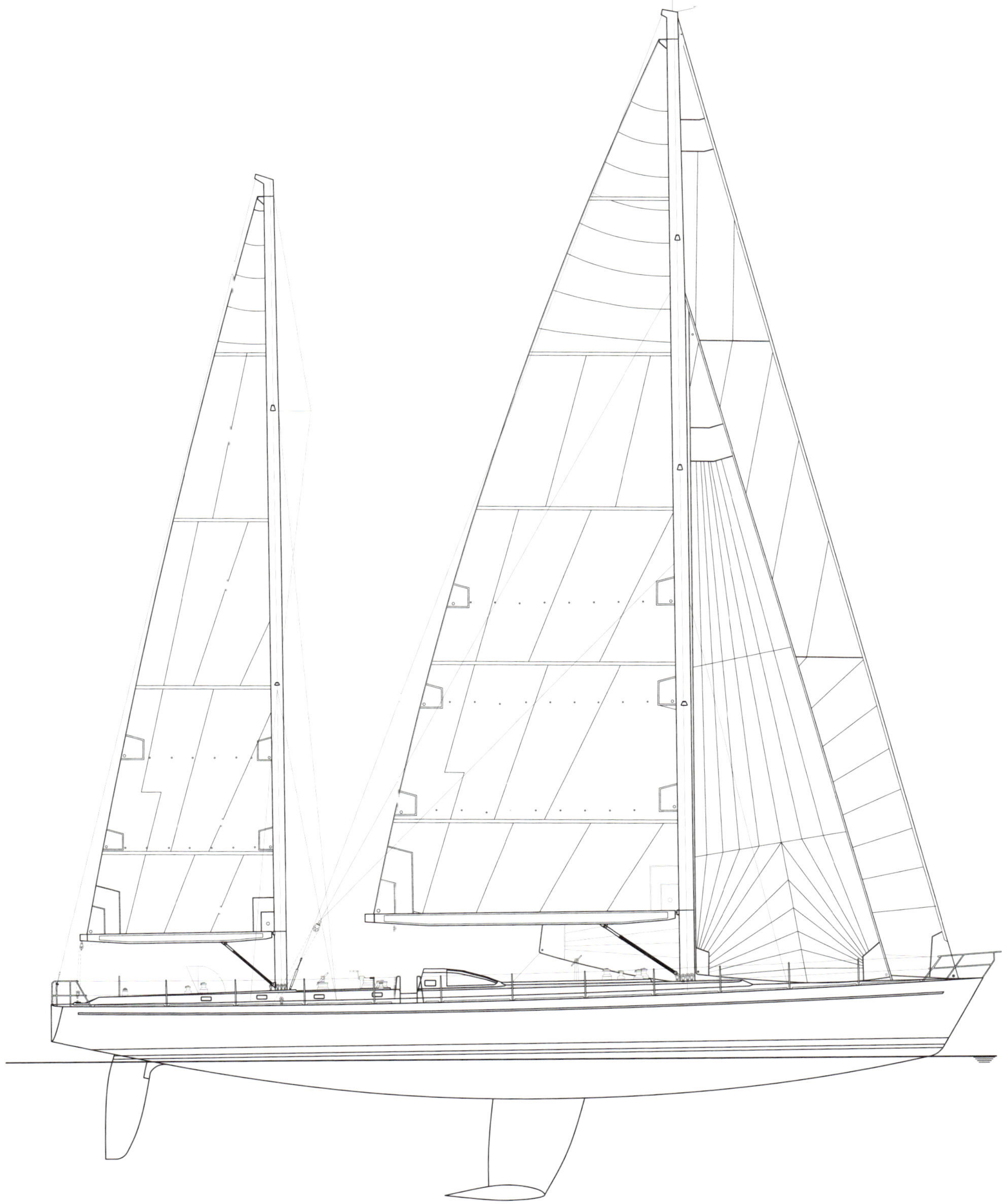

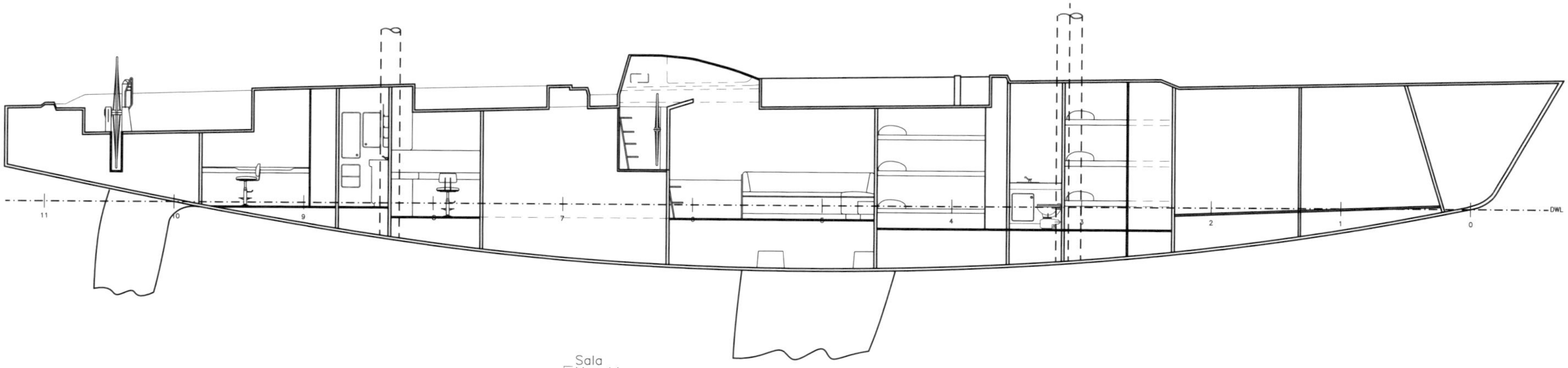

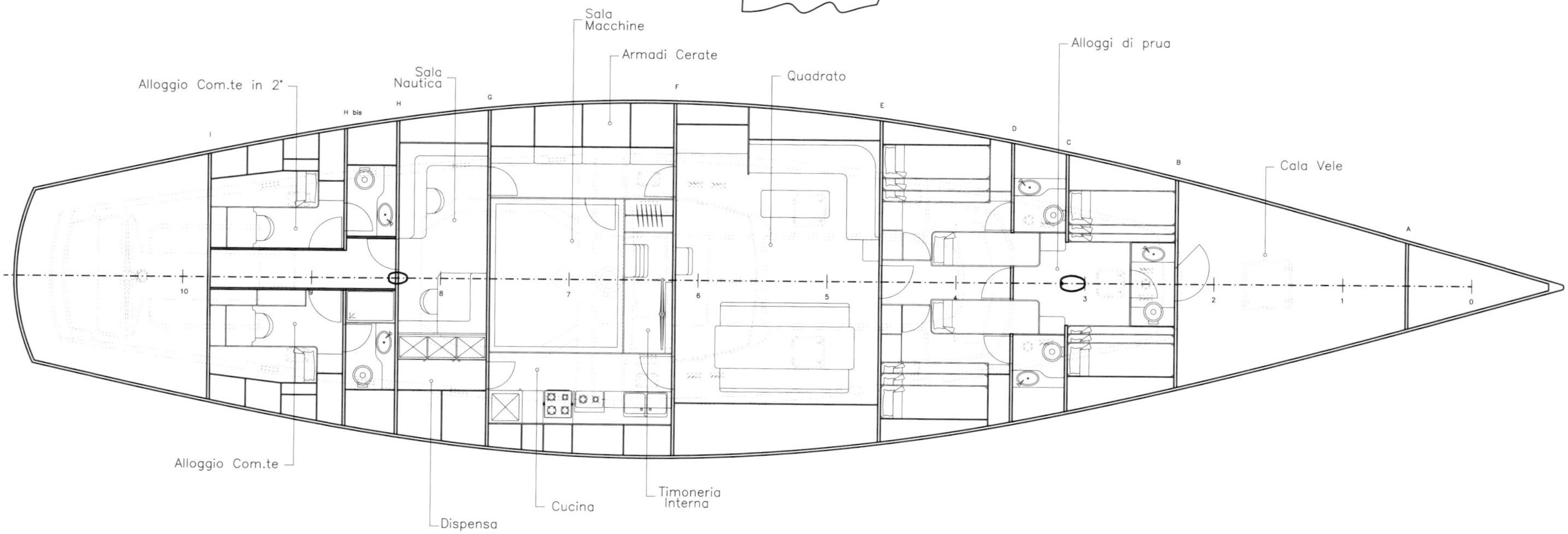

Sala Macchine
Armadi Cerate
Sala Nautica
Alloggio Com.te in 2°
Quadrato
Alloggi di prua
Cala Vele
Alloggio Com.te
Timoneria Interna
Cucina
Dispensa

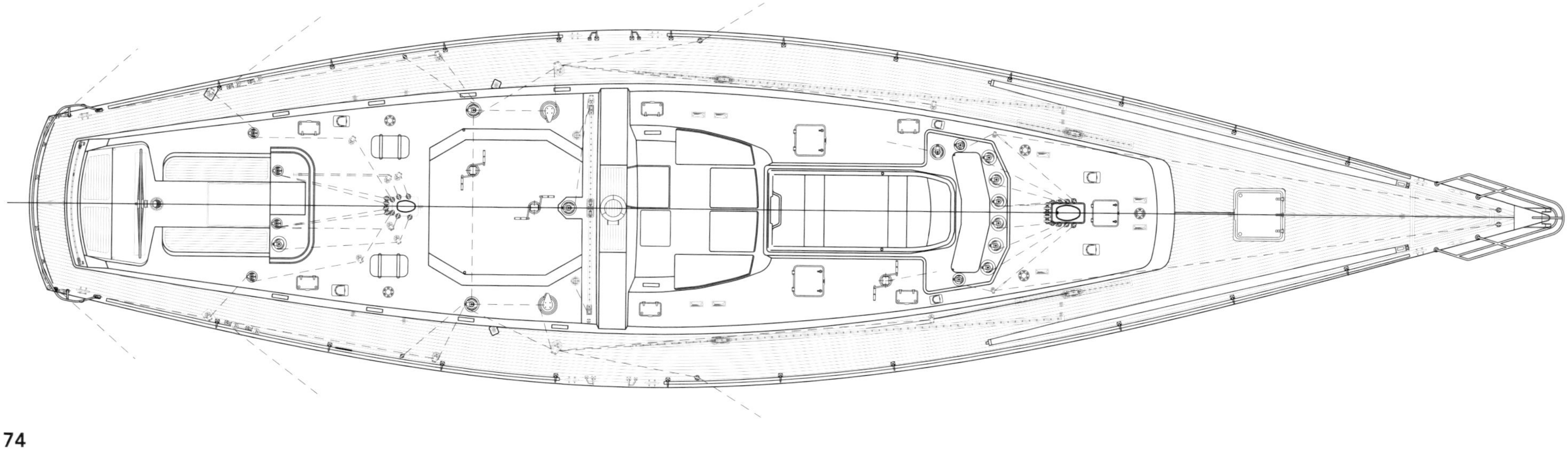

Deck plan and general arrangement.

Building the wooden male plug for composite hull lamination.

Star Cat

This was the smallest sailboat we ever designed. She was an elegant day-cruiser, 6.85 meters long, cat-rigged and intended for small series production in cold-molded wood or fiberglass. The idea was to offer a trailerable boat, easy to maneuver even single-handed; an idea that offers a design with a retro flavor, inspired by American cat-rigged boats, from which she differs in not having a coachroof. The hull of the wooden version was made of three layers of kahia mahogany, two of which are crossed at 45 degrees and one, the external one, arranged longitudinally and varnished. The deck is made of marine plywood, covered with teak strips set fore and aft. The hull structure, formed of bulkheads of mahogany plywood, divides the hull into four compartments, thus making the boat extremely safe. The hull is fitted with a daggerboard with a ballast bulb, while the rudder is mounted outboard, with the blade hinged via an easily operated lever system. The inboard engine was fitted with a sail-drive and protected in shallow waters by the daggerboard's bulb when in the raised position. The rigging includes a large roach full-battened mainsail and a gennaker that can be set from the bowsprit. *Star Cat* was deliberately characterized by a considerable displacement in relation to her size; this to make her not only stiff and comfortable while underway, but also to offer the helmsman the pleasant sensation of controlling a classic, seaworthy boat, and not a dinghy.

Year 1995
Builder Latini Roma
Construction Cold molded wood
LOA 6.85 m
B MAX 2.45 m
D 1.2 m
DSPL 0.9 t
Mainsail 24 mq
Spi 18 mq

Cockpit. Under full sail.

Construction plan and hull sections with sail plan illustration. Pencil and watercolor on board.

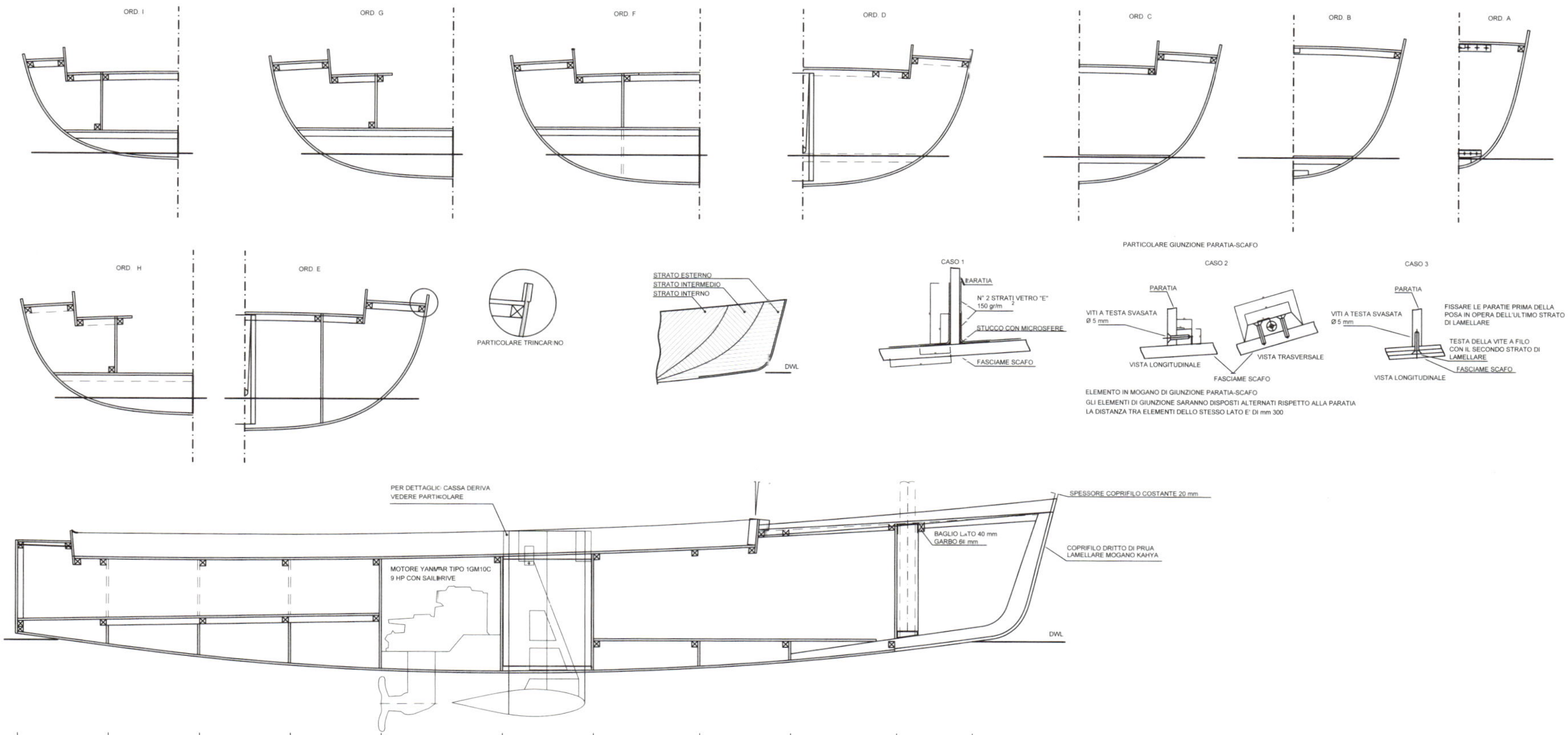
ORD. I
ORD. G
ORD. F
ORD. D
ORD. C
ORD. B
ORD. A
ORD. H
ORD. E
PARTICOLARE TRINCARINO
STRATO ESTERNO
STRATO INTERMEDIO
STRATO INTERNO
DWL
PARTICOLARE GIUNZIONE PARATIA-SCAFO
CASO 1
PARATIA
N° 2 STRATI VETRO "E" 150 gr/m2
STUCCO CON MICROSFERE
FASCIAME SCAFO
CASO 2
PARATIA
VITI A TESTA SVASATA Ø 5 mm
VISTA LONGITUDINALE
VISTA TRASVERSALE
FASCIAME SCAFO
CASO 3
PARATIA
VITI A TESTA SVASATA Ø 5 mm
FISSARE LE PARATIE PRIMA DELLA POSA IN OPERA DELL'ULTIMO STRATO DI LAMELLARE
TESTA DELLA VITE A FILO CON IL SECONDO STRATO DI LAMELLARE
FASCIAME SCAFO
VISTA LONGITUDINALE
ELEMENTO IN MOGANO DI GIUNZIONE PARATIA-SCAFO
GLI ELEMENTI DI GIUNZIONE SARANNO DISPOSTI ALTERNATI RISPETTO ALLA PARATIA
LA DISTANZA TRA ELEMENTI DELLO STESSO LATO E' DI mm 300
PER DETTAGLIO CASSA DERIVA VEDERE PARTICOLARE
SPESSORE COPRIFILO COSTANTE 20 mm
BAGLIO LATO 40 mm
COPRIFILO DRITTO DI PRUA LAMELLARE MOGANO KAHYA
MOTORE YANMAR TIPO 1GM10C 9 HP CON SAILDRIVE
DWL

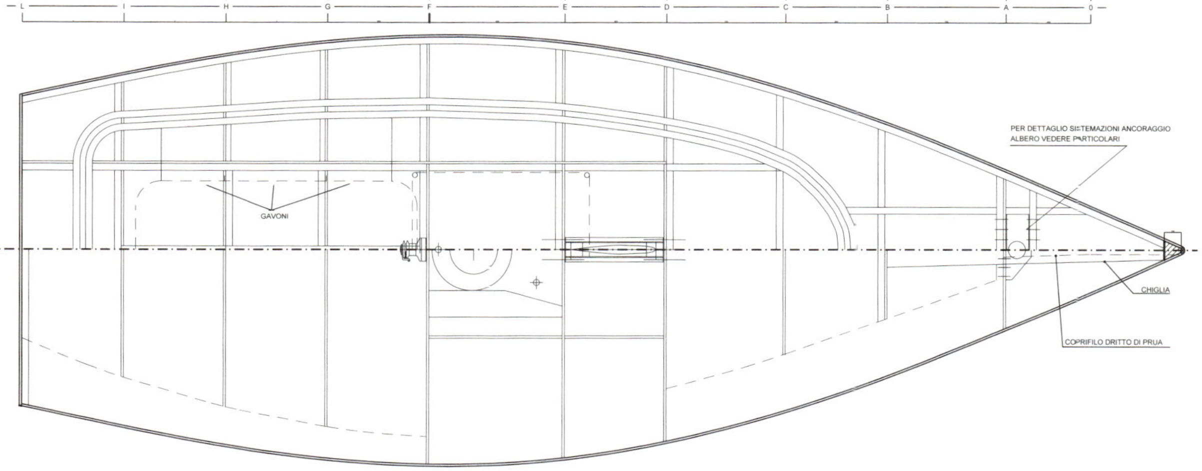
PER DETTAGLIO SISTEMAZIONI ANCORAGGIO ALBERO VEDERE PARTICOLARI
GAVONI
CHIGLIA
COPRIFILO DRITTO DI PRUA

Conchita

Built of cold-molded mahogany by the historic Carlini yard in Rimini, *Conchita IV* was a pilot boat for long and comfortable cruises. In this case, too, as in the Ocean 54, the general layout was marked by an intermediate space between the cockpit and the interior, which were located in the chart table and a sofa. In this area, next to the large polycarbonate windows of the deckhouse, we find the mahogany frames deliberately left in sight; this was to achieve an aesthetically balanced effect between tradition and new technologies. The cockpit was divided by the double steering position into two areas: aft for relaxation and forward for sail handling. Great attention was paid to the design of the deckhouse profile, which, although necessarily bulky, delicately links to the deck and harmonizes with the sheer.

Year 1996
Builder Carlini Rimini
Construction Cold molded wood
LOA 17.3 m
B MAX 5.27 m
D 2.95 m
DSPL 20.5 t
Mainsail 70 mq
Headsail 97 mq
Spi 210 mq

Conchita cruising.

Construction plan.

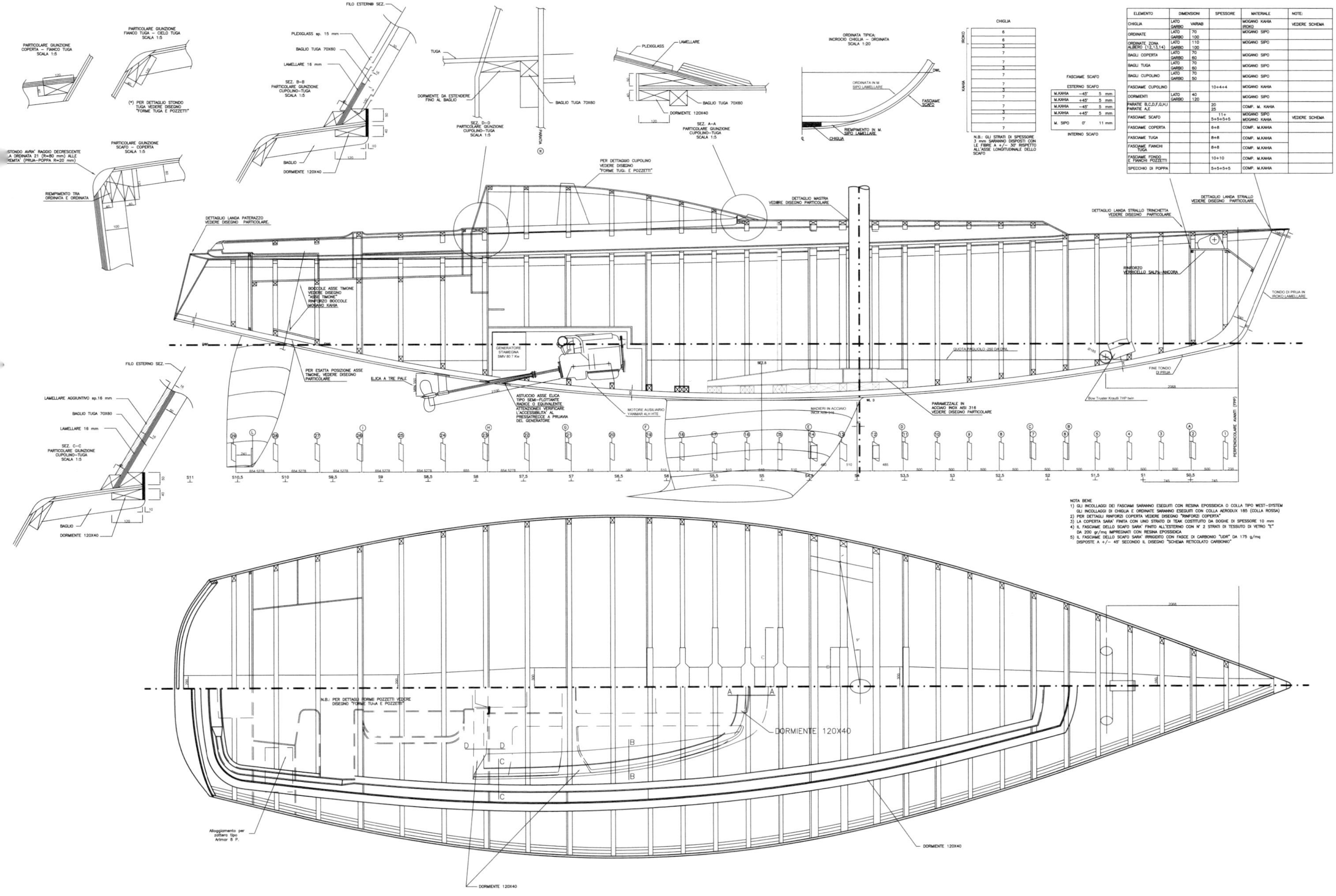

PARTICOLARE GIUNZIONE COPERTA – FIANCO TUGA SCALA 1:5
PARTICOLARE GIUNZIONE FIANCO TUGA – CIELO TUGA SCALA 1:5
(*) PER DETTAGLIO STONDO TUGA VEDERE DISEGNO "FORME TUGA E POZZETTI"
PARTICOLARE GIUNZIONE SCAFO – COPERTA SCALA 1:5
RIEMPIMENTO TRA ORDINATA E ORDINATA
FILO ESTERNO SEZ.
PLEXIGLASS sp. 15 mm
BAGLIO TUGA 70X60
LAMELLARE 18 mm
SEZ. B–B PARTICOLARE GIUNZIONE CUPOLINO–TUGA SCALA 1:5
BAGLIO
DORMIENTE 120X40
TUGA
DORMIENTE DA ESTENDERE FINO AL BAGLIO
SEZ. D–D PARTICOLARE GIUNZIONE CUPOLINO–TUGA SCALA 1:5
BAGLIO TUGA 70X60
PARATIA
PLEXIGLASS
LAMELLARE
SEZ. A–A PARTICOLARE GIUNZIONE CUPOLINO–TUGA SCALA 1:5
ORDINATA TIPICA: INCROCIO CHIGLIA – ORDINATA SCALA 1:20
ORDINATA IN M. SIPO LAMELLARE
FASCIAME SCAFO
RIEMPIMENTO IN M. SIPO LAMELLARE
CHIGLIA
IROKO
KAHIA
N.B.: GLI STRATI DI SPESSORE 3 mm SARANNO DISPOSTI CON LE FIBRE A +/– 30° RISPETTO ALL'ASSE LONGITUDINALE DELLO SCAFO
FASCIAME SCAFO
ESTERNO SCAFO
M.KAHIA –45° 5 mm
M.KAHIA +45° 5 mm
M.KAHIA –45° 5 mm
M.KAHIA +45° 5 mm
M. SIPO 0° 11 mm
INTERNO SCAFO

ELEMENTO	DIMENSIONI		SPESSORE	MATERIALE	NOTE:
CHIGLIA	LATO GARBO	VARIAB		MOGANO KAHIA IROKO	VEDERE SCHEMA
ORDINATE	LATO GARBO	70 100		MOGANO SIPO	
ORDINATE ZONA ALBERO (12,13,14)	LATO GARBO	110 100		MOGANO SIPO	
BAGLI COPERTA	LATO GARBO	70 60		MOGANO SIPO	
BAGLI TUGA	LATO GARBO	70 60		MOGANO SIPO	
BAGLI CUPOLINO	LATO GARBO	70 50		MOGANO SIPO	
FASCIAME CUPOLINO			10+4+4	MOGANO KAHIA	
DORMIENTI	LATO GARBO	40 120		MOGANO SIPO	
PARATIE B,C,D,F,G,H,I PARATIE A,E			20 25	COMP. M. KAHIA	
FASCIAME SCAFO			11+ 5+5+5+5	MOGANO SIPO MOGANO KAHIA	VEDERE SCHEMA
FASCIAME COPERTA			6+8	COMP. M.KAHIA	
FASCIAME TUGA			8+8	COMP. M.KAHIA	
FASCIAME FIANCHI TUGA			8+8	COMP. M.KAHIA	
FASCIAME FONDO E FIANCHI POZZETTI			10+10	COMP. M.KAHIA	
SPECCHIO DI POPPA			5+5+5+5	COMP. M.KAHIA	

PER DETTAGLIO CUPOLINO VEDERE DISEGNO "FORME TUGA E POZZETTI"
DETTAGLIO MASTRA VEDERE DISEGNO PARTICOLARE
DETTAGLIO LANDA STRALLO TRINCHETTA VEDERE DISEGNO PARTICOLARE
DETTAGLIO LANDA STRALLO VEDERE DISEGNO PARTICOLARE
DETTAGLIO LANDA PATERAZZO VEDERE DISEGNO PARTICOLARE
RINFORZO VERRICELLO SALPA–ANCORA
TONDO DI PRUA IN IROKO LAMELLARE
BOCCOLE ASSE TIMONE VEDERE DISEGNO "ASSE TIMONE" RINFORZO BOCCOLE MOGANO KAHIA
GENERATORE STAMEGNA
PER ESATTA POSIZIONE ASSE TIMONE, VEDERE DISEGNO PARTICOLARE
ELICA A TRE PALE
ASTUCCIO ASSE ELICA TIPO SEMI–FLOTTANTE RADICE O EQUIVALENTE ATTENZIONE!! VERIFICARE L'ACCESSIBILITA' AL PRESSATRECCE A PRUAVIA DEL GENERATORE
MOTORE AUSILIARIO YANMAR 4LH HTE
MADIERI IN ACCIAIO
PARAMEZZALE IN ACCIAIO INOX AISI 316 VEDERE DISEGNO PARTICOLARE
FINE TONDO DI PRUA
PERPENDICOLARE AVANTI (PPAV)
S10,5 S10 S9,5 S9 S8,5 S8 S7,5 S7 S6,5 S6 S5,5 S5 S4,5 S4 S3,5 S3 S2,5 S2 S1,5 S1 S0,5
S11
FILO ESTERNO SEZ.
LAMELLARE AGGIUNTIVO sp.18 mm
BAGLIO TUGA 70X60
LAMELLARE 18 mm
SEZ. C–C PARTICOLARE GIUNZIONE CUPOLINO–TUGA SCALA 1:5
BAGLIO
DORMIENTE 120X40
NOTA BENE
1) GLI INCOLLAGGI DEI FASCIAMI SARANNO ESEGUITI CON RESINA EPOSSIDICA O COLLA TIPO WEST–SYSTEM GLI INCOLLAGGI DI CHIGLIA E ORDINATE SARANNO ESEGUITI CON COLLA AERODUX 185 (COLLA ROSSA)
2) PER DETTAGLI RINFORZI COPERTA VEDERE DISEGNO "RINFORZI COPERTA"
3) LA COPERTA SARA' FINITA CON UNO STRATO DI TEAK COSTITUITO DA DOGHE DI SPESSORE 10 mm
4) IL FASCIAME DELLO SCAFO SARA' FINITO ALL'ESTERNO CON N° 2 STRATI DI TESSUTO DI VETRO "E" DA 200 gr/mq IMPREGNATI CON RESINA EPOSSIDICA
5) IL FASCIAME DELLO SCAFO SARA' IRRIGIDITO CON FASCE DI CARBONIO "UDR" DA 175 g/mq DISPOSTE A +/– 45° SECONDO IL DISEGNO "SCHEMA RETICOLATO CARBONIO"
N.B.: PER DETTAGLI FORME POZZETTI VEDERE DISEGNO "FORME TUGA E POZZETTI"
DORMIENTE 120X40
Alloggiamento per zolfiera tipo Artimar 8 P.
DORMIENTE 120X40
DORMIENTE 120X40

Atlantis

Atlantis was a project for which the owner made some very specific requests: an elegant style tending towards the classic, an optimized hull for the IMS rule, and a general layout of the boat allowing her to be sailed shorthanded. The boat was built by the Crosato yard in Treviso (Nautica C4) in wood and epoxy resin; the planking was made up of an internal layer of spruce to which were added four diagonal layers of kahia mahogany, while the deck was made of marine plywood covered in high-quality teak. The bonding of the various layers was done using a vacuum technique. A carbon-fiber mesh was inserted between the third and fourth layers. The internal design was based on the principle of giving the hull and coachroof an aesthetic function too; in fact, where possible, the hull frames and the deck beams were left exposed without resorting to furnishings that concealed them, except where these were indispensable. *Atlantis* was rigged as a cutter with a furling jib, staysail, and full-batten mainsail.

Year 1996
Builder Nautica C4 Treviso
Construction Cold molded wood
LOA 15.685 m
B MAX 4.58 m
D 2.8 m
DSPL 15 t
Mainsail 69 mq
Headsail 99 mq
Spi 199 mq

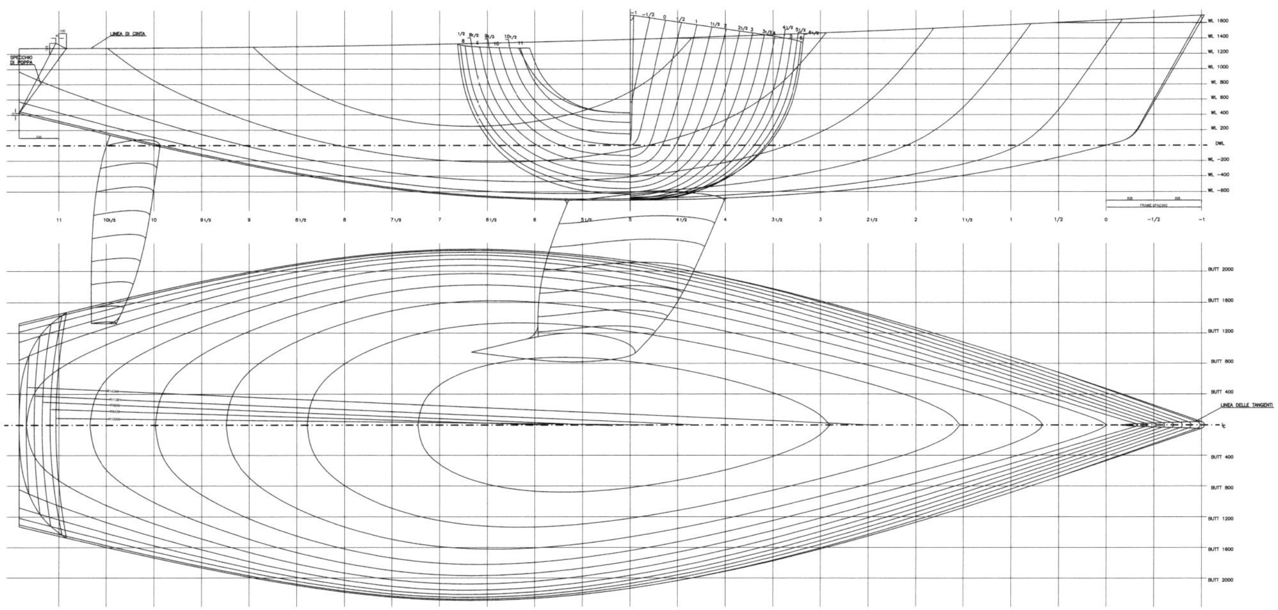

Lines plan (stations, waterlines, buttocks).

General arrangement.

Guest and owner's cabins. *Atlantis* from astern.

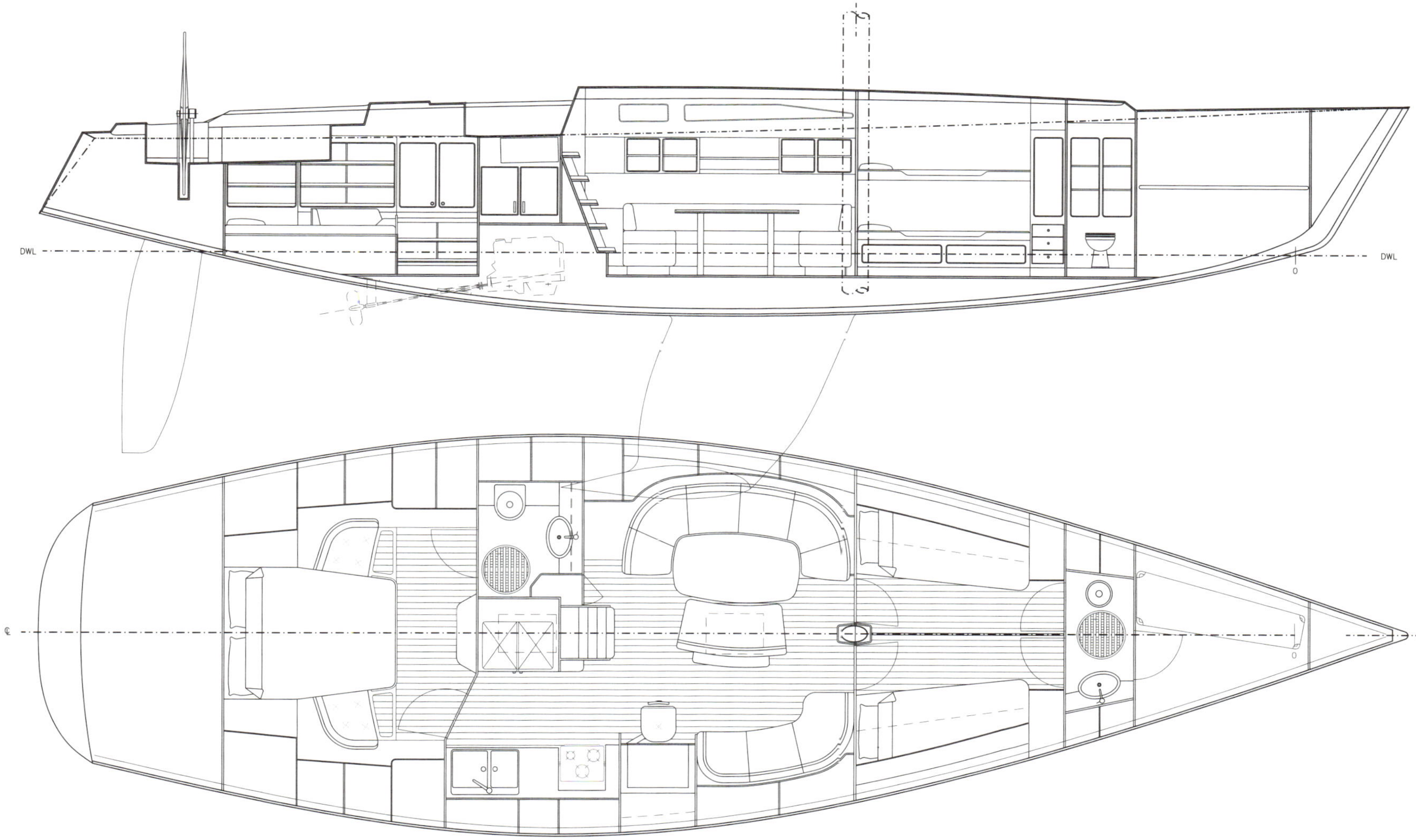
DWL
DWL

Atlantis
Douglas

Virtuelle

She was the first new-generation fast-cruiser that we designed, making use of construction techniques and methodologies from racing boats and delegating a large part of her sail handling to hydraulic and electrically assisted systems. Compared to the fast cruisers of her generation, *Virtuelle* had a large and rounded stern: a choice that made it possible to combine a greater wetted area with greater form stability. She also had a vertical stem to maximize waterline length. The underwater plane included a deep keel fin with bulb, and twin rudder blades inclined transversely by 15°. This configuration, then in use in the Open 60s, represented a novelty for fast cruisers. *Virtuelle* was rigged as a sloop with five sets of carbon spreaders, an unusual solution for this type of boat. Subsequently, to improve performance in light winds, the keel was replaced, lightening, and lengthening it. The flush deck was organized with all sheets led aft of the steering wheels, to leave the central cockpit, dedicated to guests, safe and free of equipment. The design of *Virtuelle* saw the important contribution of Philippe Starck in the design of the superstructures and interior furnishings. He chose an unusual and characteristic solution with the use of cotton laminates impregnated with epoxy resin, colored in such a way as to simulate the color of bakelite and thus what bakelite itself simulated, namely mahogany. This deliberately gave an uncertain perception of the material, in a sort of provocation played on semantic ambiguities. The boat was built by the Tencara yard with an Airex sandwich core and pre-impregnated epoxy carbon, Kevlar and glass fiber. The lamination was done using the vacuum technique, while the curing was carried out in a special oven. In 2001 *Virtuelle* won the Maxi Rolex Cup in her category and the Tre Golfi Sailing Week in 2003.

Year 1997
Builder Tencara Marghera (VE)
Construction Composite
LOA 23.935 m
B MAX 6.2 m
D 3.92 m
DSPL 33 t
Mainsail 155 mq
Headsail 194 mq
Spi 410 mq

Virtuelle racing.

Sketches for the profile, deck plan and general arrangement. Pencil sketches by Andrea Vallicelli, except N°5 by Philippe Starck.

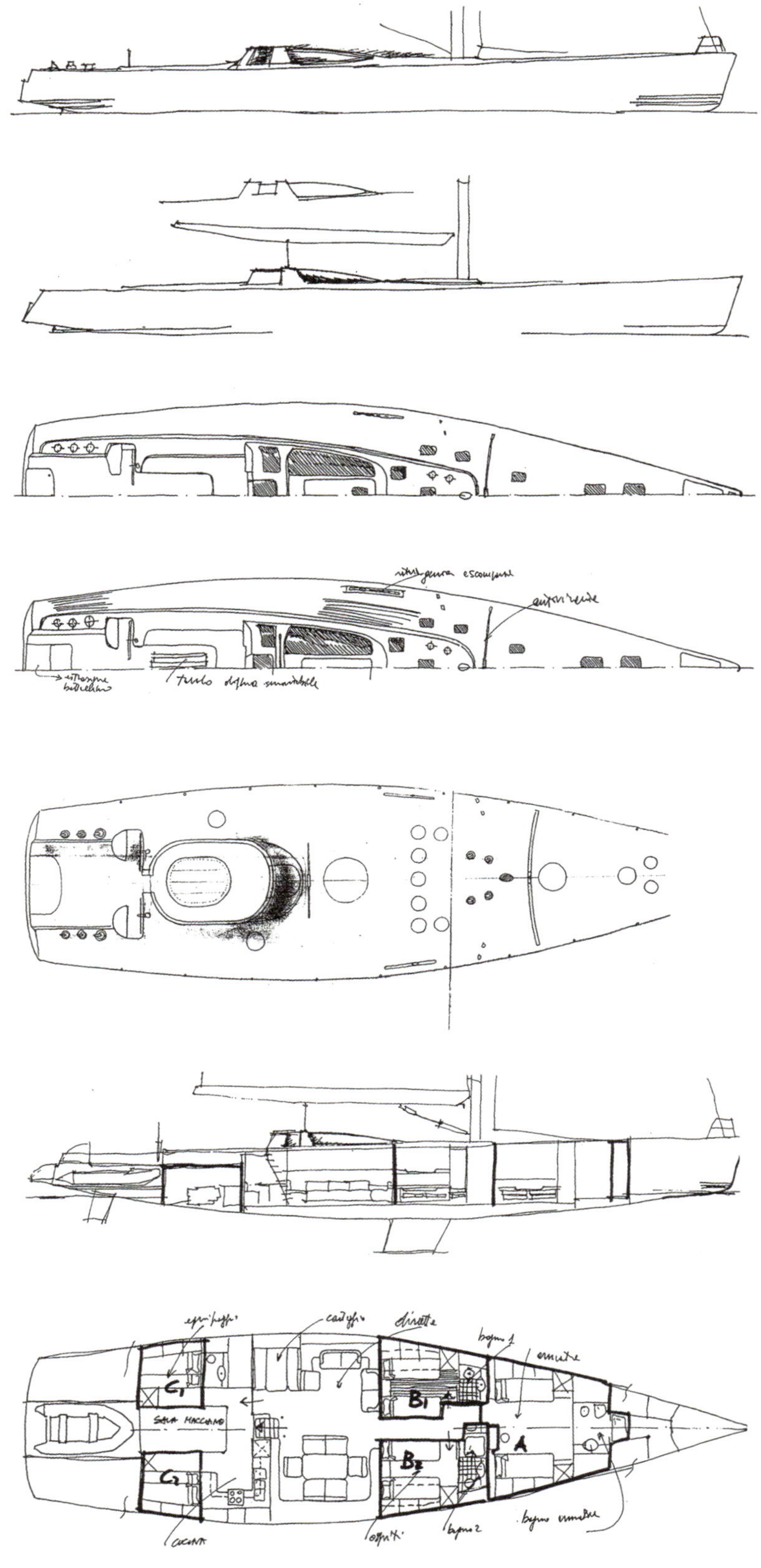

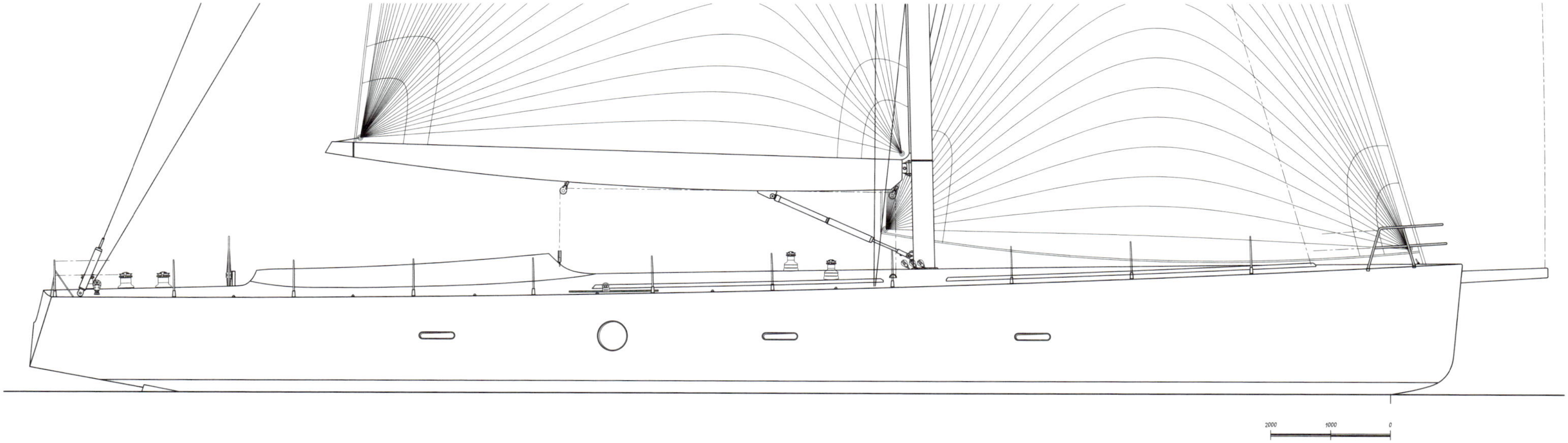

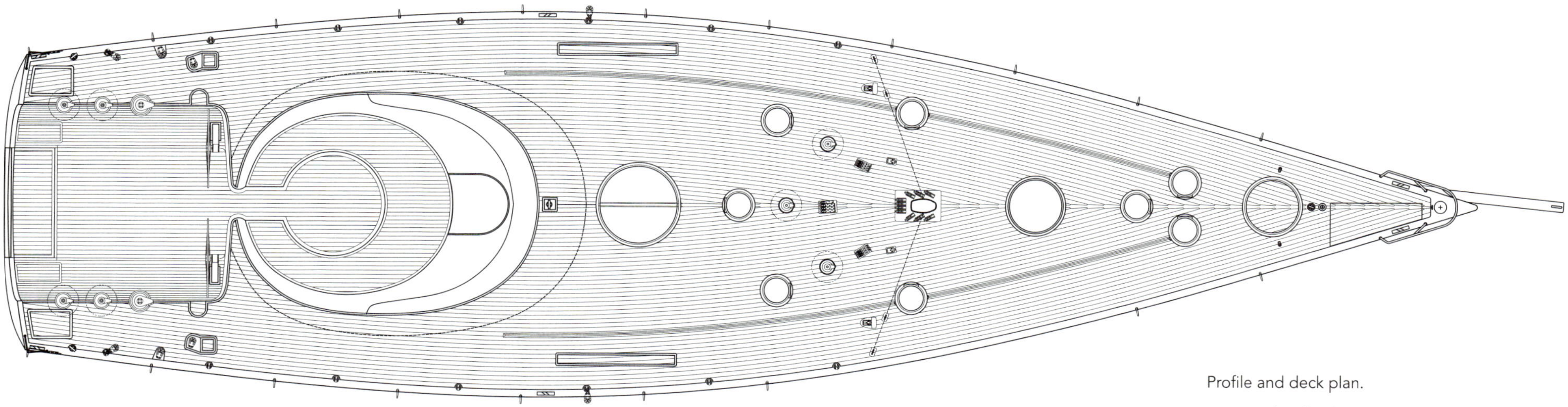

Profile and deck plan.

Exterior details – bowsprit, circular hatch on deck, helm, companionway.

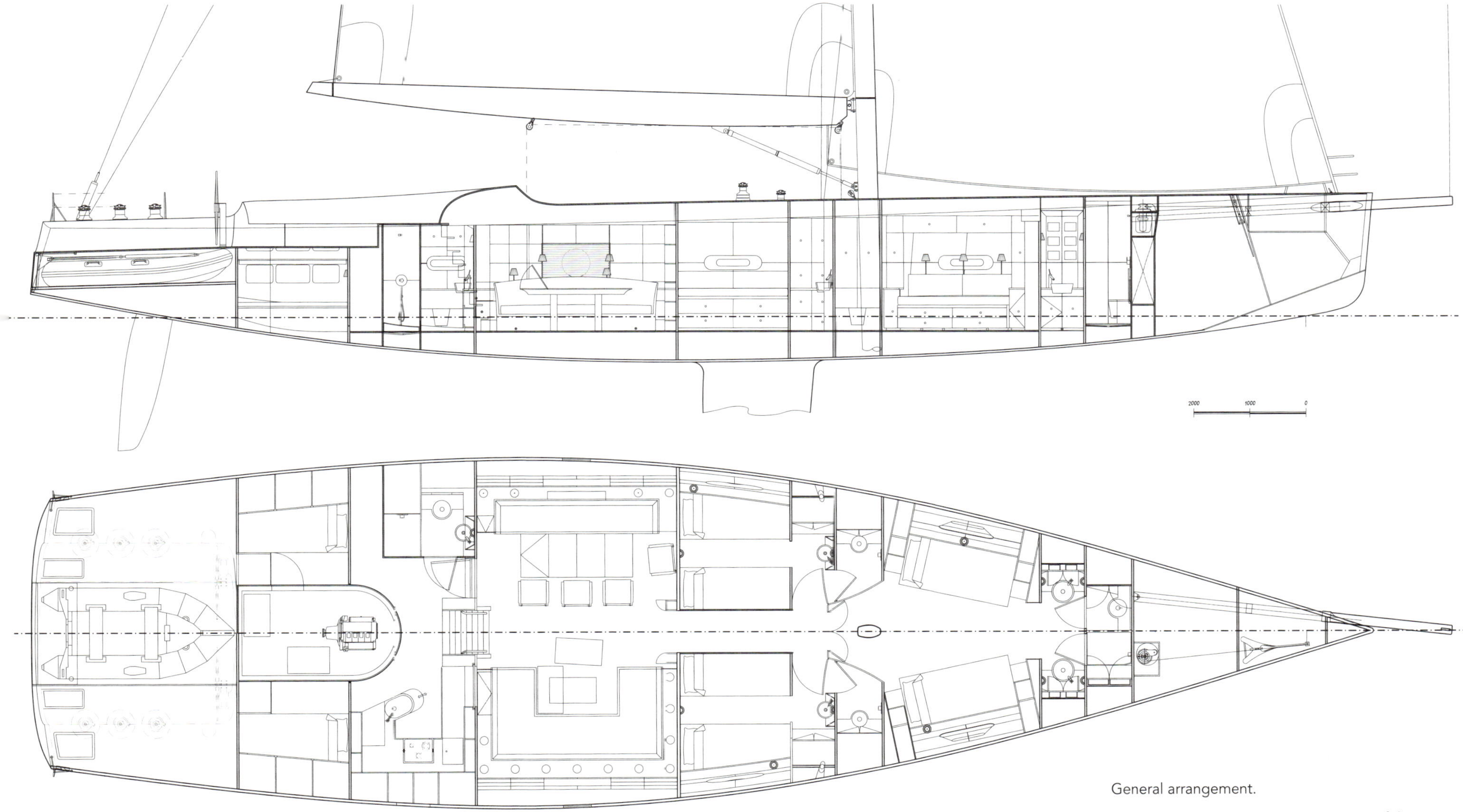

General arrangement.

Interiors – navigation station, two views of the saloon, circular deck hatch seen from below.

Construction shots in the Tencara yard in Maghera.

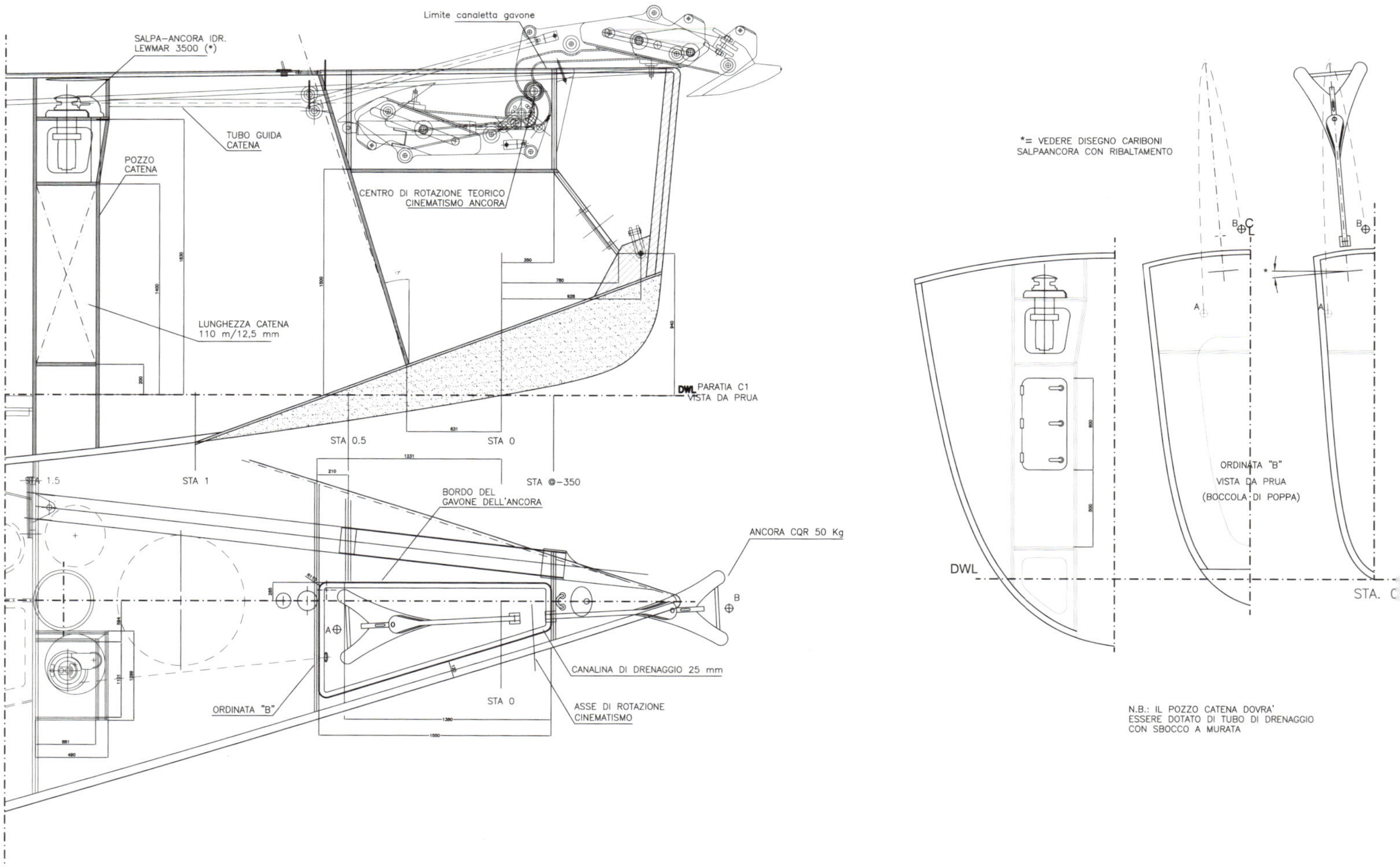

Anchor mechanism detail.

Left to right: Michelangelo Valllicelli, Andrea Vallicelli, Philippe Starck.

Amor

Amor was an Open 55 class; her hull was built in composite materials by Fabio Soleri and completed by the Se.Ri.Gi yard. The boat was less extreme than *Riviera di Rimini*, but had a maximum beam located well aft and was fitted with a stainless steel lifting keel with a lead bulb. The rudder featured twin blades, a solution that we had adopted for the first time with *Virtuelle*. She was also equipped with water ballast and had a 9/10 rig with a retractable bowsprit and the gooseneck located on deck. Compared to our other open classes, she had a greater maximum beam and a flatter bottom. The geometry of the hull was characterized by a bow with a very slight angle of attack and lines that progressively widened towards the stern, forming a broad transom. The deckhouse design was built with very rounded shapes, in which the large lenticular portholes stand out, a solution that we adopted in several successive boats. As for the interiors, the special features were a chart table just below the entrance companionway and a U-shaped dinette positioned in a rather unusual, but functional way in the bow area.

Year 2000
Builder Serigi Aquileia (UD)
Construction Composite
LOA 16.345 m
B MAX 5.1 m
D 4 m
DSPL 9 t
Mainsail 105 mq
Headsail 63 mq
Spi 185 mq

Cockpit and deck saw from the mast.

Deck plan and general arrangement.

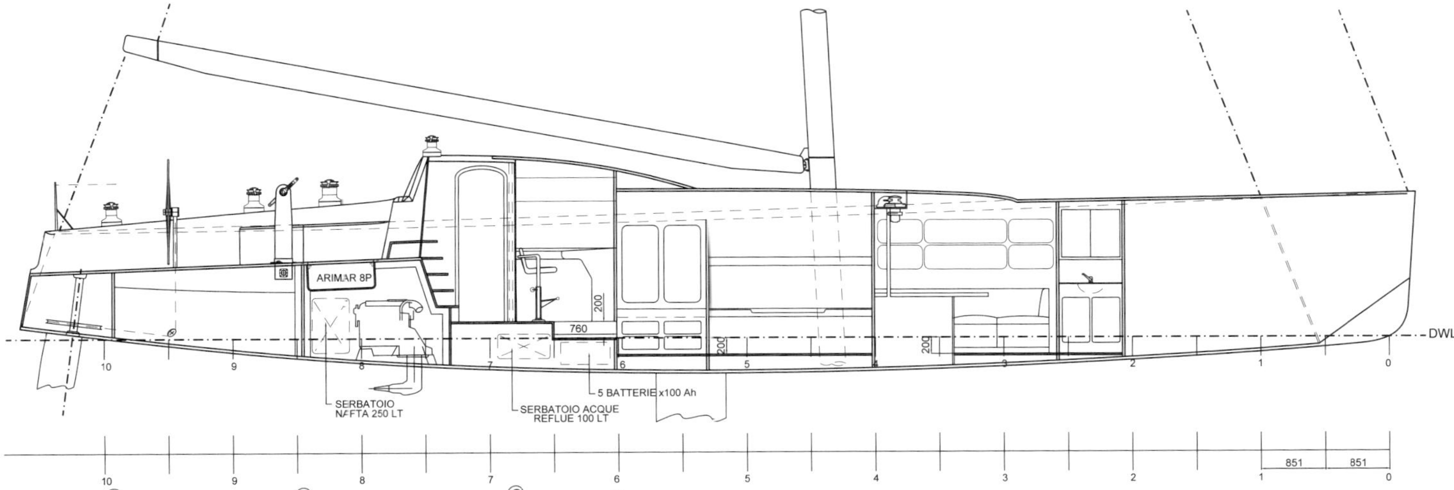
ARIMAR 8P
DWL
SERBATOIO NAFTA 250 LT
SERBATOIO ACQUE REFLUE 100 LT
5 BATTERIE x100 Ah
851
851

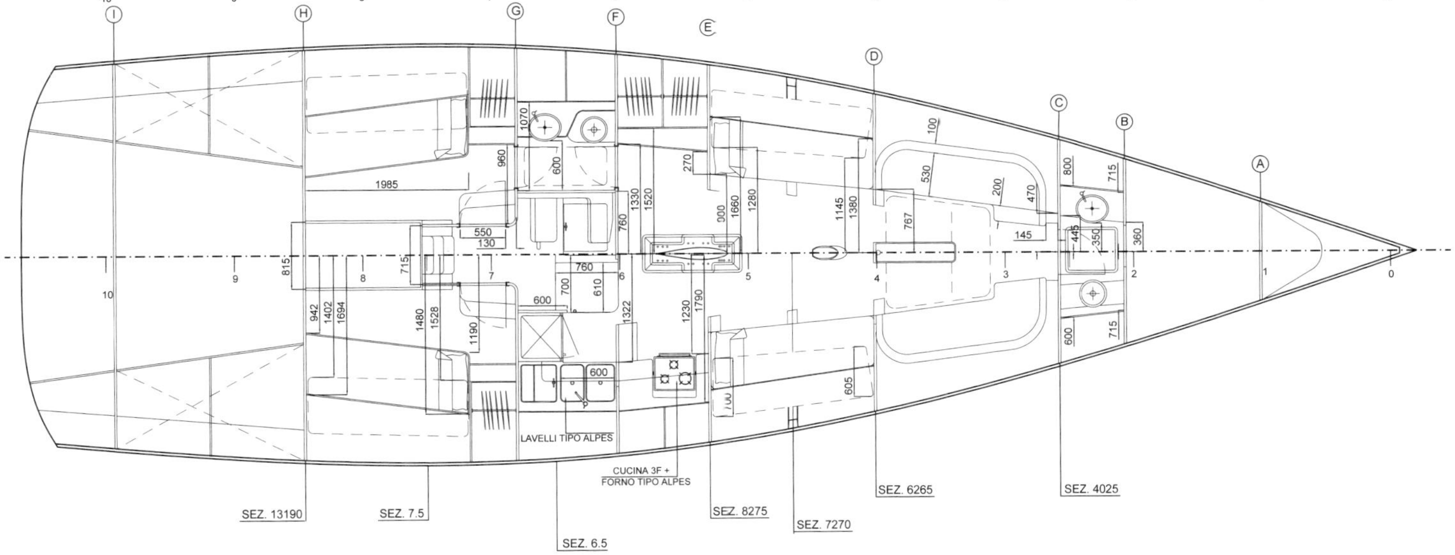
LAVELLI TIPO ALPES
CUCINA 3F + FORNO TIPO ALPES
SEZ. 13190
SEZ. 7.5
SEZ. 6.5
SEZ. 8275
SEZ. 7270
SEZ. 6265
SEZ. 4025

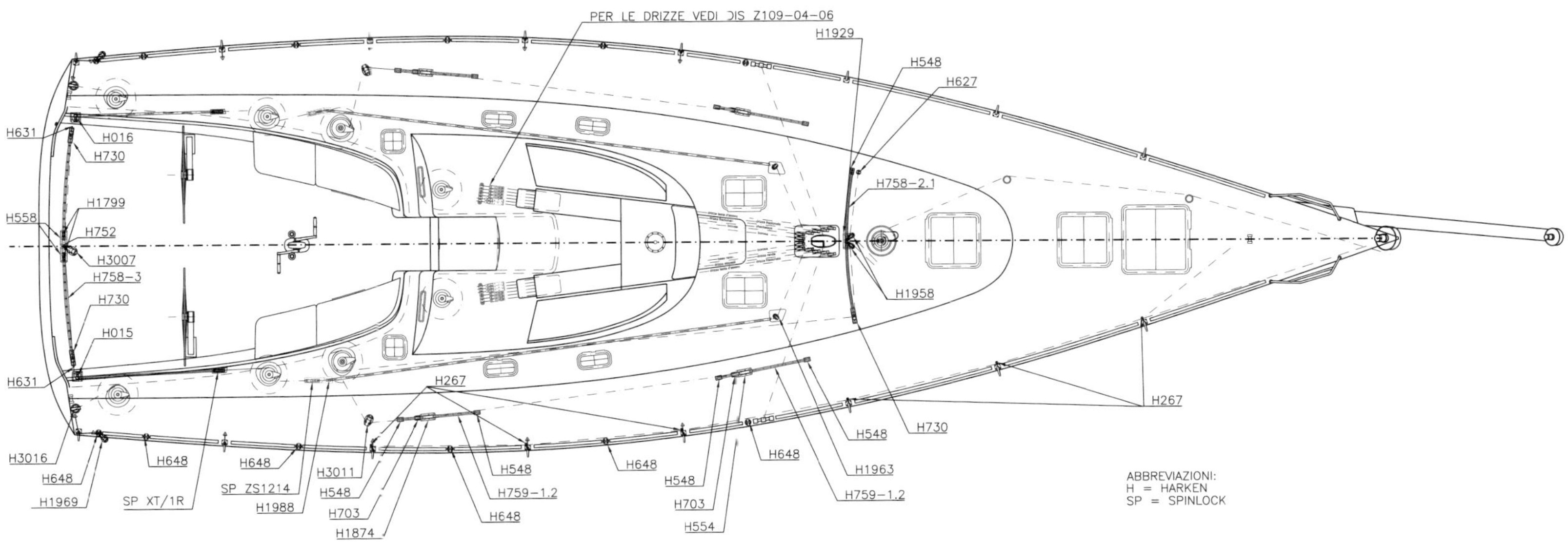
PER LE DRIZZE VEDI DIS Z109-04-06
H1929
H548
H627
H758-2.1
H1958
H631
H016
H730
H1799
H558
H752
H3007
H758-3
H730
H015
H631
H267
H267
H3016
H648
H1969
H648
SP XT/1R
H648
SP ZS1214
H3011
H548
H1988
H703
H1874
H548
H759-1.2
H648
H648
H548
H703
H554
H648
H548
H1963
H759-1.2
H730
ABBREVIAZIONI:
H = HARKEN
SP = SPINLOCK

Amor under sail in Trieste.

Line plan (stations, waterlines, buttocks).

Launch and main cabin were looking astern.

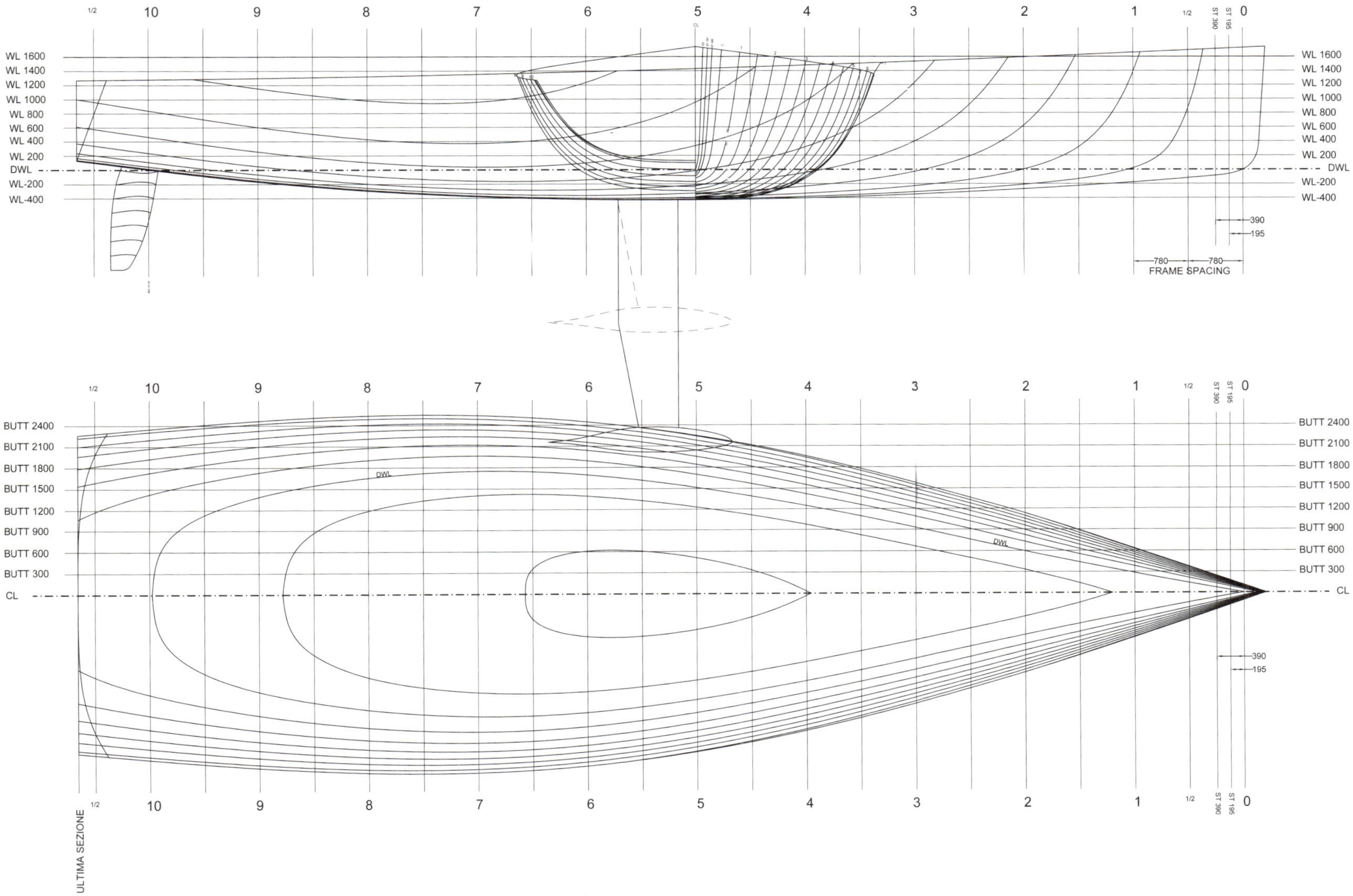
WL 1600
WL 1400
WL 1200
WL 1000
WL 800
WL 600
WL 400
WL 200
DWL
WL-200
WL-400
FRAME SPACING
780
390
195
ST 390
ST 195
BUTT 2400
BUTT 2100
BUTT 1800
BUTT 1500
BUTT 1200
BUTT 900
BUTT 600
BUTT 300
CL
ULTIMA SEZIONE
1/2 10 9 8 7 6 5 4 3 2 1 1/2 0

Penelope

She was a modern 60-foot cutter intended for cruising but with features that made her suitable for racing. Two yards participated in her construction: Cantieri Brazzoni for the composite part and the Alto Adriatico yard for the interiors. The sandwich hull skin and the carbon mast made it possible to create a very light and agile boat. For the Cutter 60', we again used the configuration of a twin rudder, a solution that makes it possible to maintain a balanced hull when heeling and, at the same time, reduce the strain on the wheels. The keel had a very elongated fin with bulb, to lower the center of gravity and reduce the displacement. The deck plan was designed to separate between the technical areas from those dedicated to the comfort of guests. For the sail plan, the choice was that of a 9/10 rig, with either a 110% genoa or a yankee coupled with a staysail set on a removable inner forestay. The mast is made of carbon with spreaders swept back 20° to ensure good longitudinal stability without resorting to running backstays.

Year 2001
Builder Alto Adriatico San Giorgio di Nogaro (UD)
Construction Composite
LOA 18.31 m
B MAX 5.25 m
D 3 m
DSPL 18 t
Mainsail 95 mq
Headsail 87 mq
Spi 240 mq

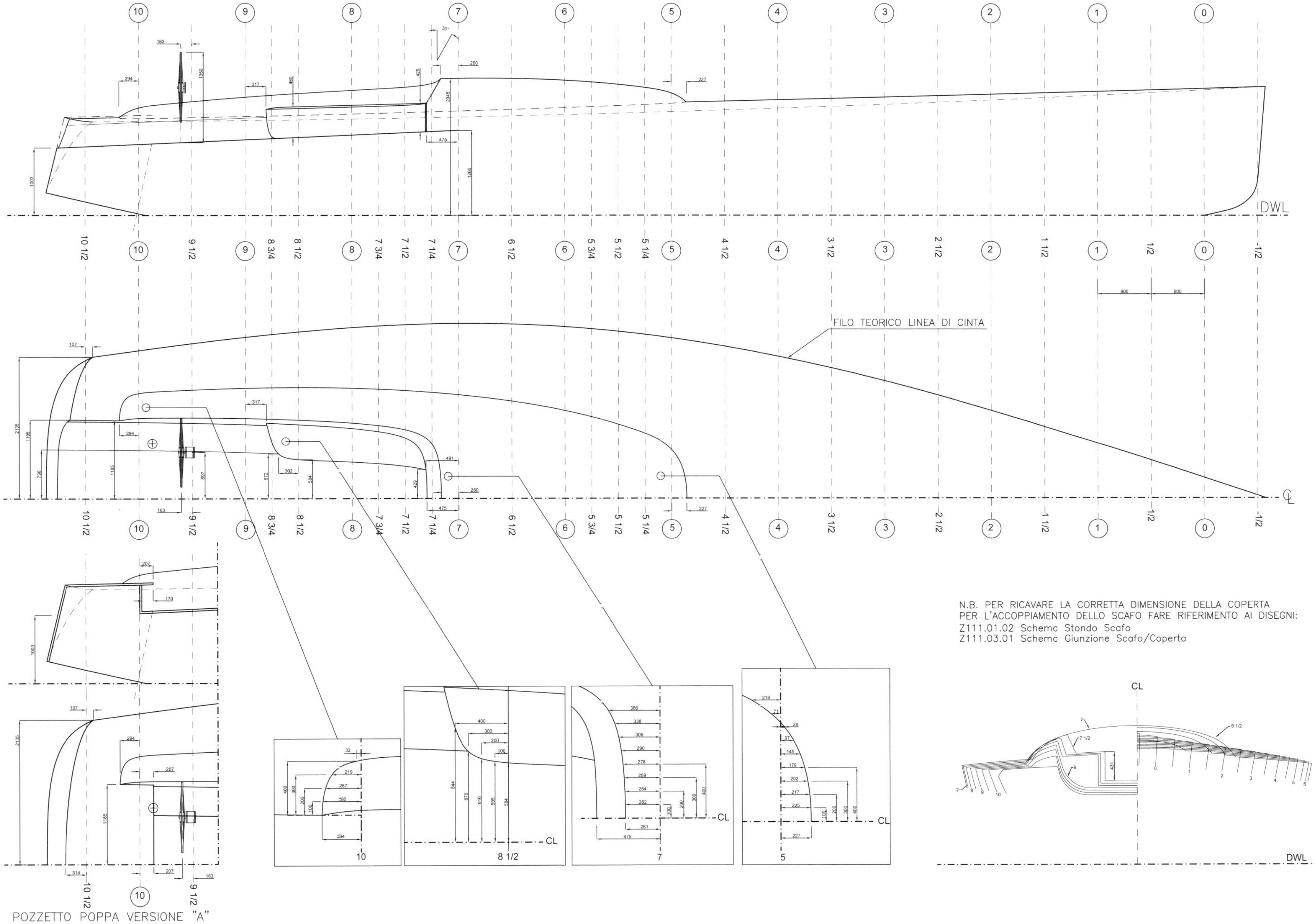

The deck saw from the mast.

Lines for the deckhouse and cockpit.

H_2O

H_2O was a fast cruiser of slightly over 78 feet, built on a mold in composite materials by the C.N. Yacht 2000 shipyard. Her architecture was characterized by a deck saloon (or raised saloon), a streamlined hull with very soft lines and with a little overhang. The deck, covered in teak, was arranged around a central cockpit for guests, where the outdoor seating and dining tables were arranged. Adjacent to this area were the twin steering wheels with their instrument consoles and seating for the helmsman. Immediately behind the wheels was the battery of electric winches, dedicated to the mainsail and genoa sheets. Continuing towards the stern, there was a large sun deck, beneath which there was a large locker suitable for stowing a medium-sized tender, which could be launched using the boom. The interior layout revolved around the large deck saloon from which an almost 360 degree external view can be enjoyed while comfortably seated in the two large sofas, thanks to the large lenticular glazing of the deckhouse and the portholes in the boat's hull. Forward of the saloon, beneath which the engine room was located, there were the guest cabins and the owner's cabin. In the after area, there was a fourth guest cabin, the galley, and the crew's quarters. Through an elaborate study for the design of *H_2O*, we succeeded in harmonizing the volumes of the deck saloon and those of the hull, giving the boat a remarkable aesthetic balance thanks to the perfect ratios between freeboard, sheer line and deckhouse profile. Despite being a 2004 project, *H_2O* won the Rolex Cup in her category in 2015, 2017, 2018 and 2019. From the same mold, adopting other architectural solutions for the deck and the interiors, other similar boats were produced by Yacht 2000, such as *Wizard* and *Ladisea*.

Year 2006
Builder C.N. Yacht 2000 Salvirola (CR)
Construction Composite
LOA 23.81 m
B MAX 6.015 m
D 2.9 m
DSPL 35 t
Mainsail 160 mq
Headsail 100 mq
Spi 410 mq

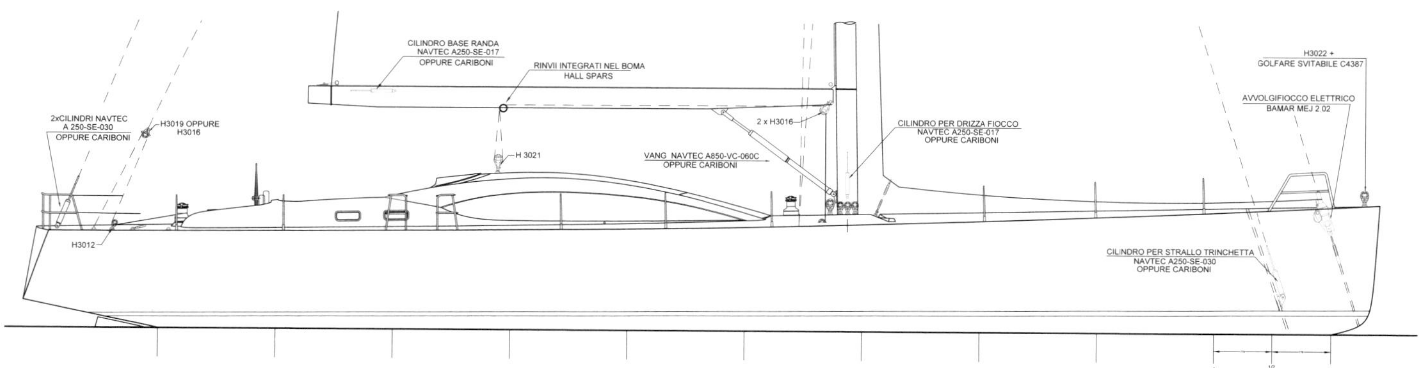

Deck plan profile.

H_2O racing in Porto Cervo.

Deck plan.

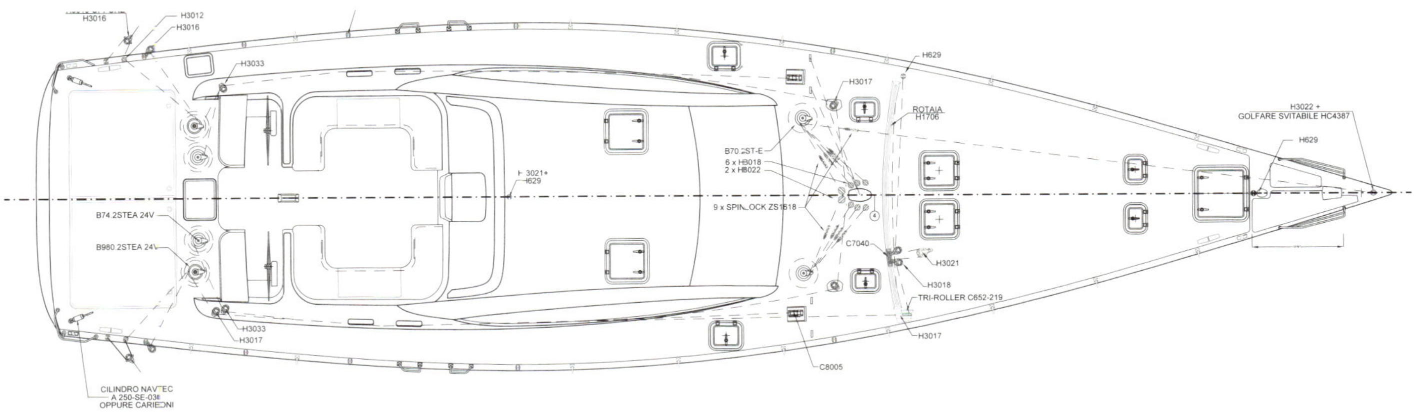
H3016
H3012
H3016
H3033
B74.2STEA 24V
B980.2STEA 24V
H3033
H3017
CILINDRO NAVTEC
A 250-SE-03
OPPURE CARIBONI
3021+
H629
B70.2ST-E
6 x H3018
2 x H3022
9 x SPINLOCK ZS1618
H3017
H629
ROTAIA
H1706
C7040
H3021
H3018
TRI-ROLLER C652-219
H3017
C8005
H3022 +
GOLFARE SVITABILE HC4387
H629

H_2O shorthanded.

Sail plan.

H_2O during the Maxi Rolex Cup in Sardegna.

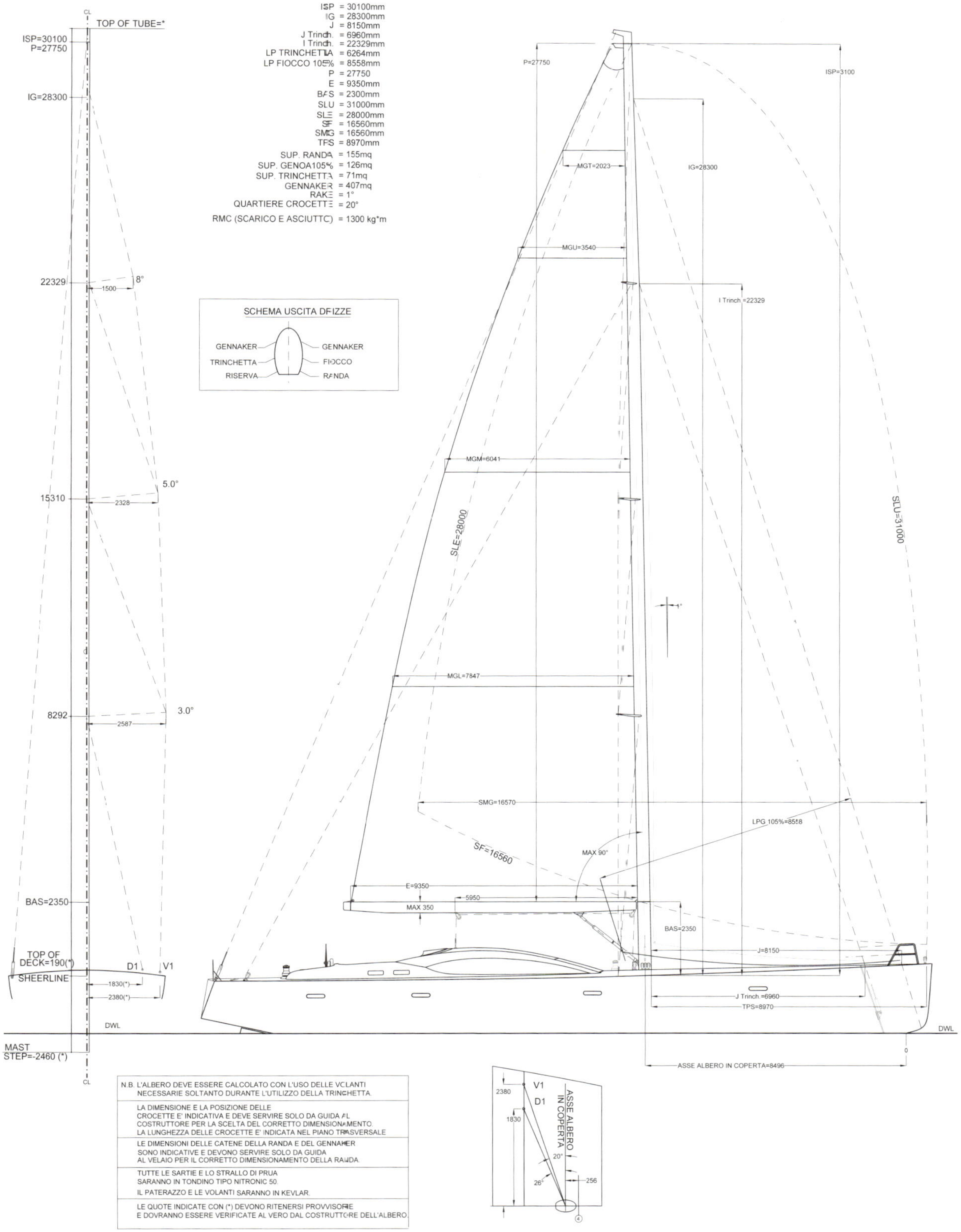

ISP = 30100mm
IG = 28300mm
J = 8150mm
J Trinch. = 6960mm
I Trinch. = 22329mm
LP TRINCHETTA = 6264mm
LP FIOCCO 105% = 8558mm
P = 27750
E = 9350mm
BAS = 2300mm
SLU = 31000mm
SLE = 28000mm
SF = 16560mm
SMG = 16560mm
TFS = 8970mm
SUP. RANDA = 155mq
SUP. GENOA105% = 126mq
SUP. TRINCHETTA = 71mq
GENNAKER = 407mq
RAKE = 1°
QUARTIERE CROCETTE = 20°
RMC (SCARICO E ASCIUTTO) = 1300 kg*m
TOP OF TUBE=*
ISP=30100
P=27750
IG=28300
22329
15310
8292
BAS=2350
TOP OF DECK=190(*)
SHEERLINE
MAST STEP=-2460 (*)
SCHEMA USCITA DRIZZE
GENNAKER
TRINCHETTA
RISERVA
GENNAKER
FIOCCO
RANDA
SLE=28000
SLU=31000
SF=16560
SMG=16570
LPG 105%=8558
E=9350
MAX 350
BAS=2350
J=8150
J Trinch.=6960
TPS=8970
DWL
ASSE ALBERO IN COPERTA=8496
ASSE ALBERO IN COPERTA
N.B. L'ALBERO DEVE ESSERE CALCOLATO CON L'USO DELLE VOLANTI NECESSARIE SOLTANTO DURANTE L'UTILIZZO DELLA TRINCHETTA.
LA DIMENSIONE E LA POSIZIONE DELLE CROCETTE E' INDICATIVA E DEVE SERVIRE SOLO DA GUIDA AL COSTRUTTORE PER LA SCELTA DEL CORRETTO DIMENSIONAMENTO. LA LUNGHEZZA DELLE CROCETTE E' INDICATA NEL PIANO TRASVERSALE
LE DIMENSIONI DELLE CATENE DELLA RANDA E DEL GENNAKER SONO INDICATIVE E DEVONO SERVIRE SOLO DA GUIDA AL VELAIO PER IL CORRETTO DIMENSIONAMENTO DELLA RANDA.
TUTTE LE SARTIE E LO STRALLO DI PRUA SARANNO IN TONDINO TIPO NITRONIC 50.
IL PATERAZZO E LE VOLANTI SARANNO IN KEVLAR.
LE QUOTE INDICATE CON (*) DEVONO RITENERSI PROVVISORIE E DOVRANNO ESSERE VERIFICATE AL VERO DAL COSTRUTTORE DELL'ALBERO.

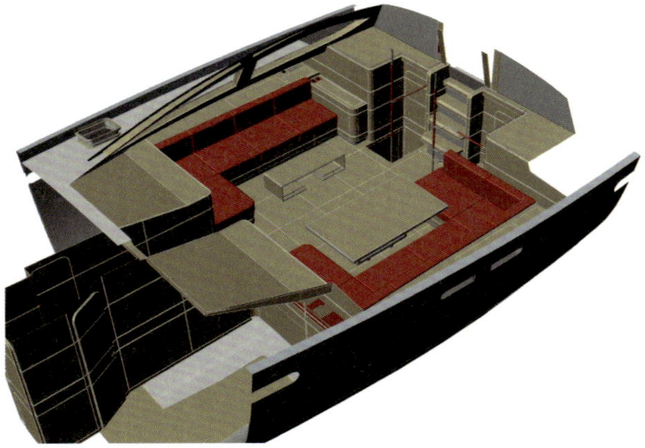

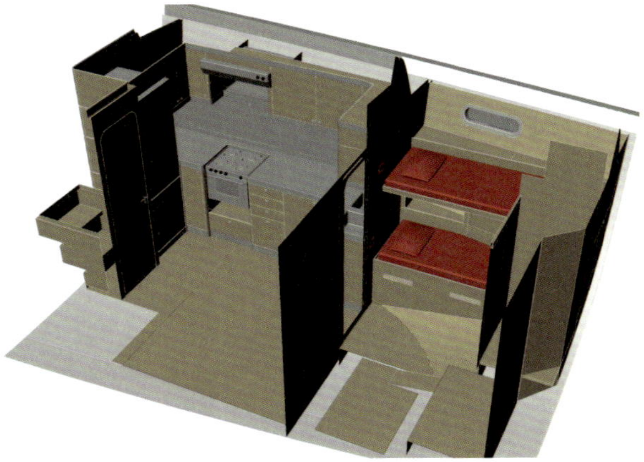

Rendering of the design for the main saloon, galley and crew's quarters.

General arrangement.

Views of the main saloon.

Detail of the opening transom and stowage for the tender.

Rudder shape and hardware, steering controls.

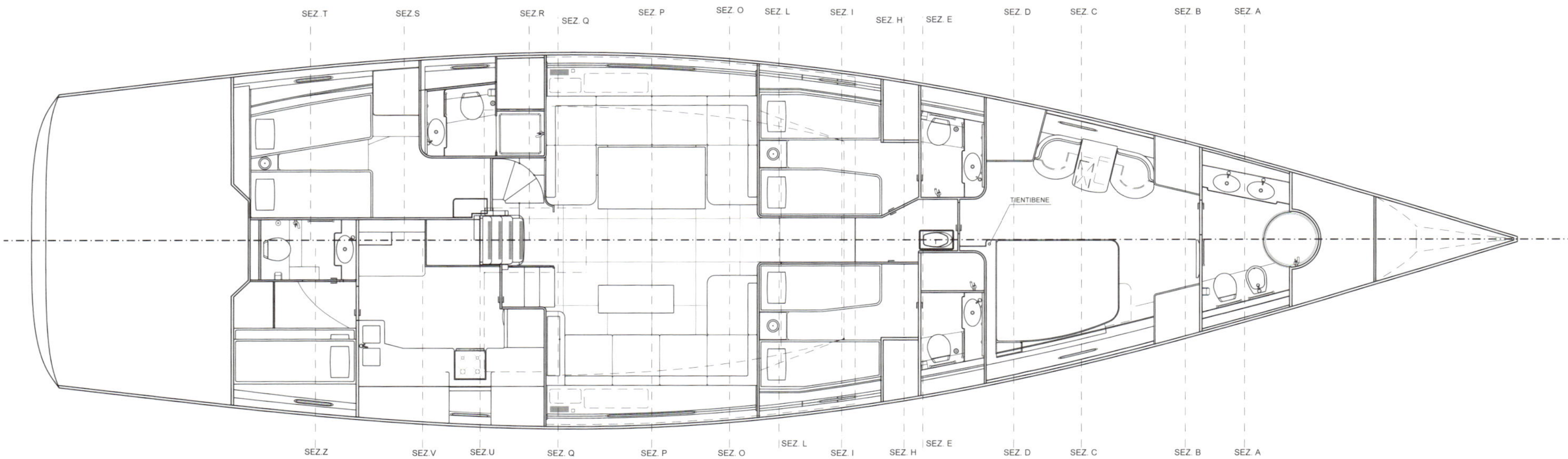

H2O

MERCURY

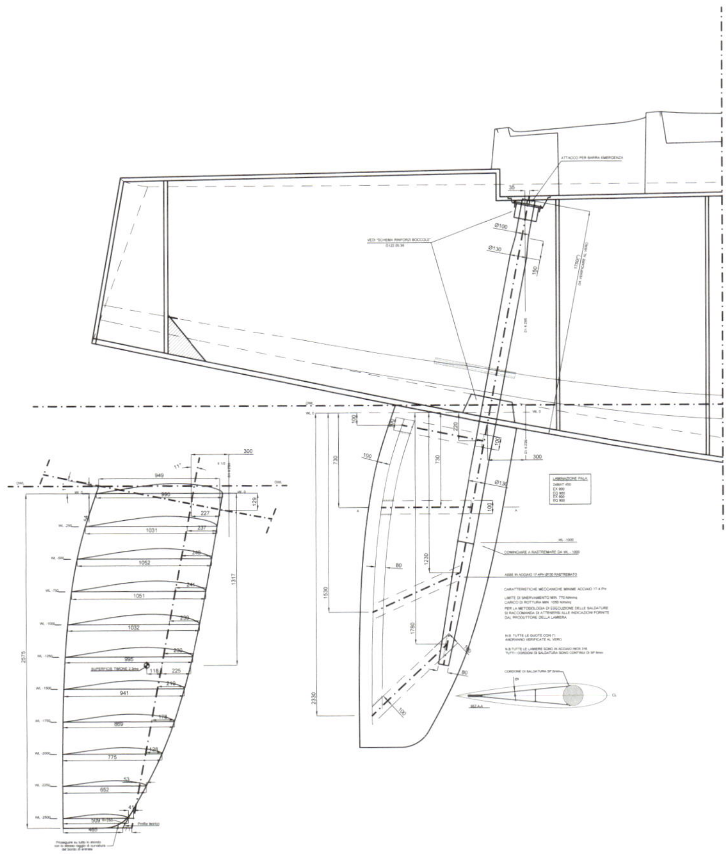

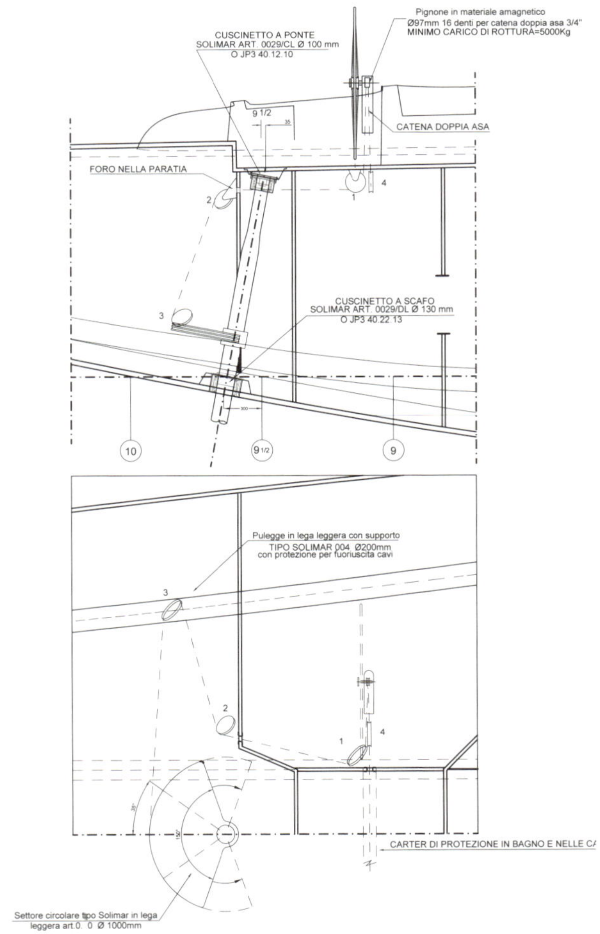
Pignone in materiale amagnetico
Ø97mm 16 denti per catena doppia asa 3/4"
MINIMO CARICO DI ROTTURA=5000Kg
CUSCINETTO A PONTE
SOLIMAR ART. 0029/CL Ø 100 mm
O JP3 40.12.10
CATENA DOPPIA ASA
FORO NELLA PARATIA
CUSCINETTO A SCAFO
SOLIMAR ART. 0029/DL Ø 130 mm
O JP3 40.22.13
Pulegge in lega leggera con supporto
TIPO SOLIMAR 004 Ø200mm
con protezione per fuoriuscita cavi
CARTER DI PROTEZIONE IN BAGNO E NELLE C
Settore circolare tipo Solimar in lega
leggera art.0 0 Ø 1000mm

Gala

Gala was created to be built in series production as a small one-design, halfway between an Olympic class and a sporty mini-day cruiser. She was designed and equipped to be sailed single-handed when rigged with just a mainsail and jib, and by two crew members when the rig included a gennaker. The hull, built of polyester resin, had a maximum length of 7.5 meters and a displacement of 650 kg, including the bulb. Her sail plan, with a total area of 24 square meters for the mainsail and jib, was set on a free standing (i.e., without shrouds or stays) rotating wing mast. The boat was designed for racing, especially in match races, and adopted the criterion of "compulsory sitting," which prohibits the use of the trapeze or any other hiking system, to allow anyone (and at any age) to compete in all weather conditions.

Year 2007
Builder Fanatek S.p.a. Roma
Construction GRP
LOA 7.574 m
B MAX 1.77 m
D 1.5 m
DSPL 0.75 t
Mainsail 15 mq
Headsail 7 mq
Spi 33 mq

Gala upwind and downwind.

Deck plan and profile.

GALA

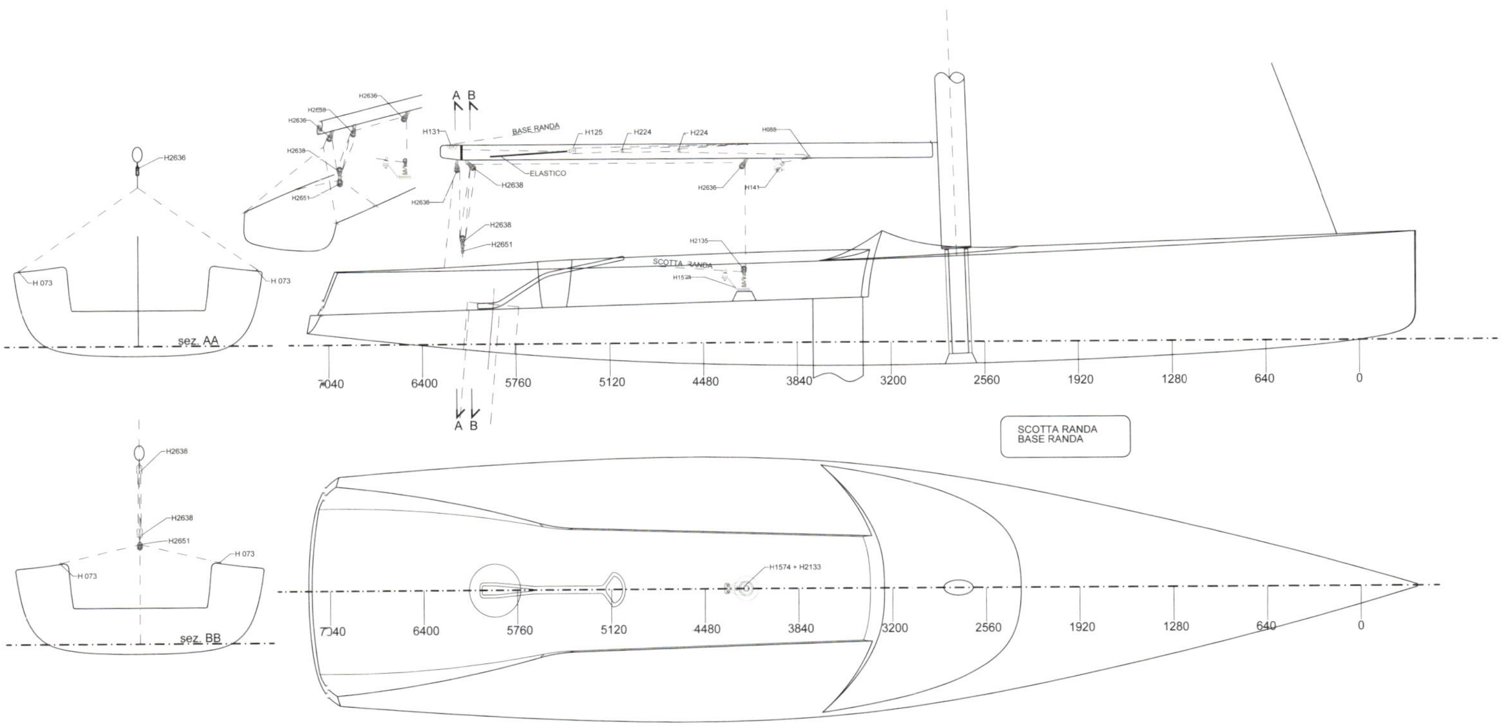
A B
BASE RANDA
ELASTICO
SCOTTA RANDA
sez. AA
7040
6400
5760
5120
4480
3840
3200
2560
1920
1280
640
0
A B
SCOTTA RANDA
BASE RANDA
sez. BB
H1574 + H2133
7040
6400
5760
5120
4480
3840
3200
2560
1920
1280
640
0

Shadow

Launched in 2011, *Shadow* was the first maxi yacht that we designed for Comar of Fiumicino. She was a 31 meter fast cruising design with a displacement of 68 tons and could be considered a stylistic evolution, as well as a dimensional one, of other series models such as the Comet 62 RS and Comet 52 RS, or fast cruisers such as *H_2O* and *Wizard*. She almost represents a compendium of two design themes that characterized our studio at the time. The boat had an extremely uncluttered deck plan, equipped with a continuous cockpit divided by the two wheels of the helm. The crew's work area was located in the forward part, while a large sun deck was available towards the stern, beneath which the tender was stowed. During the design phase, our attention was focused on obtaining maximum equilibrium between the principle architectural profiles. With this objective, we sought to maintain a continuous sheer from the bow to the transom. The latter had an open design with lateral connections so that the section of the hull harmonized with the cockpit. For *Shadow's* deckhouse, we adopted a profile with enveloping framing that outlined the long side windows, a choice that characterized many of our projects of that period. Below, the boat had six cabins with 12 berths and six bathrooms.

Year 2011
Builder Comar Fiumicino (RM)
Construction Composite
LOA 30.862 m
B MAX 6.7 m
D 4.8 m
DSPL 65 t
Mainsail 283 mq
Headsail 216 mq
Spi 703 mq

Shadow underway.

Profile, general arrangement and deck plan.

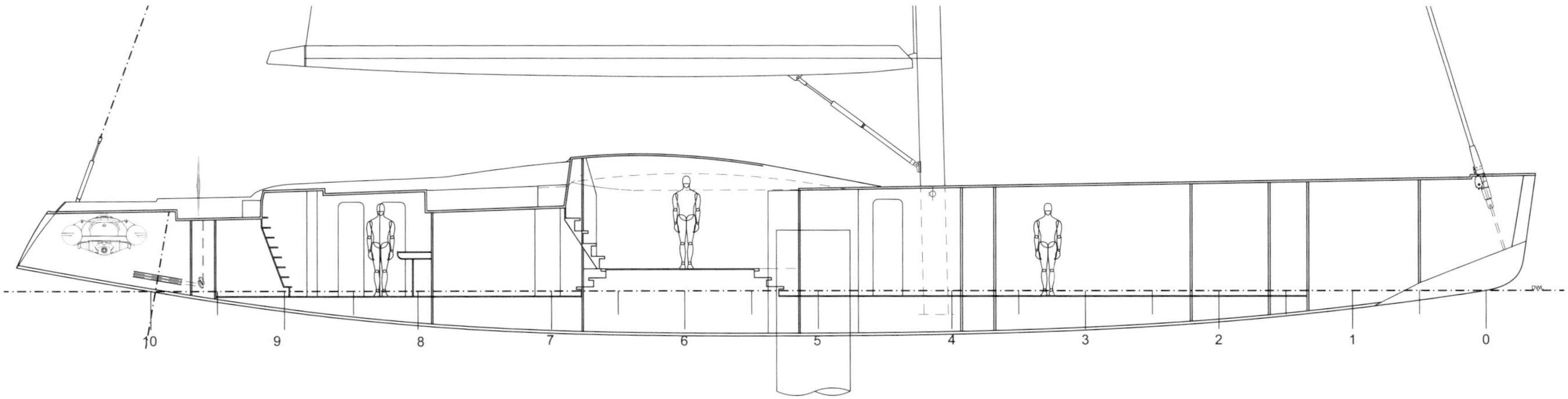
10
9
8
7
6
5
4
3
2
1
0

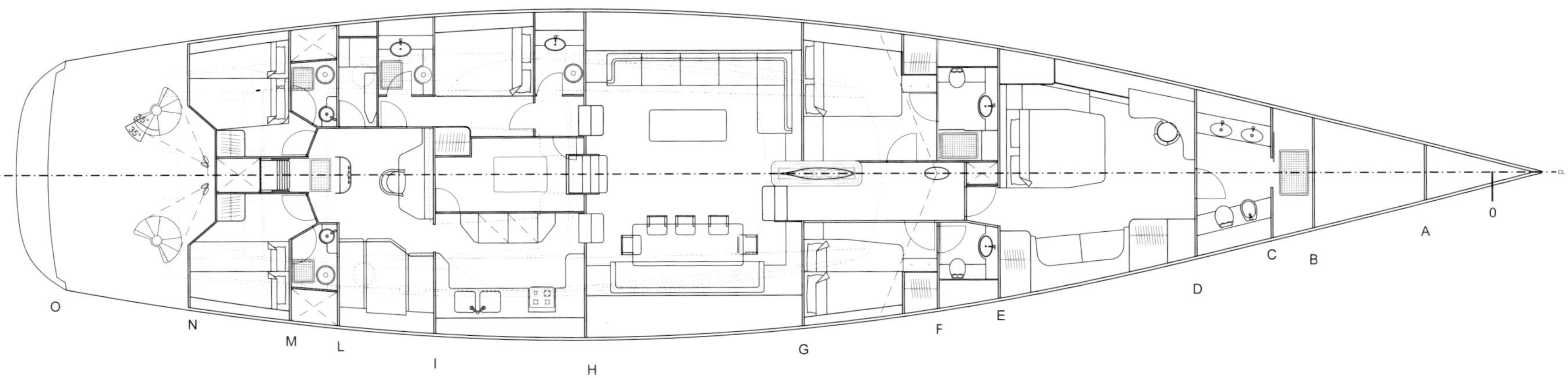
0
A
B
C
D
E
F
G
H
I
L
M
N
O

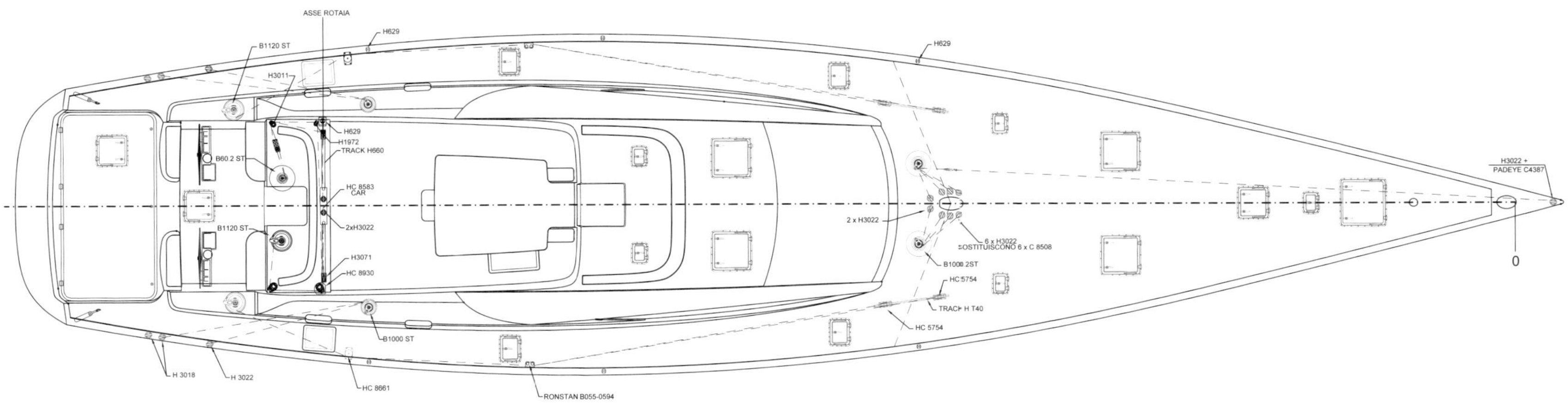
ASSE ROTAIA
H629
B1120 ST
H3011
H629
H1972
TRACK H660
B60.2 ST
HC 8583
CAR
2xH3022
B1120 ST
H3071
HC 8930
B1000 ST
H 3018
H 3022
HC 8661
RONSTAN B055-0594
H629
2 x H3022
6 x H3022
SOSTITUISCONO 6 x C 8508
B1000.2ST
HC 5754
TRACK H T40
HC 5754
H3022 +
PADEYE C4387
0

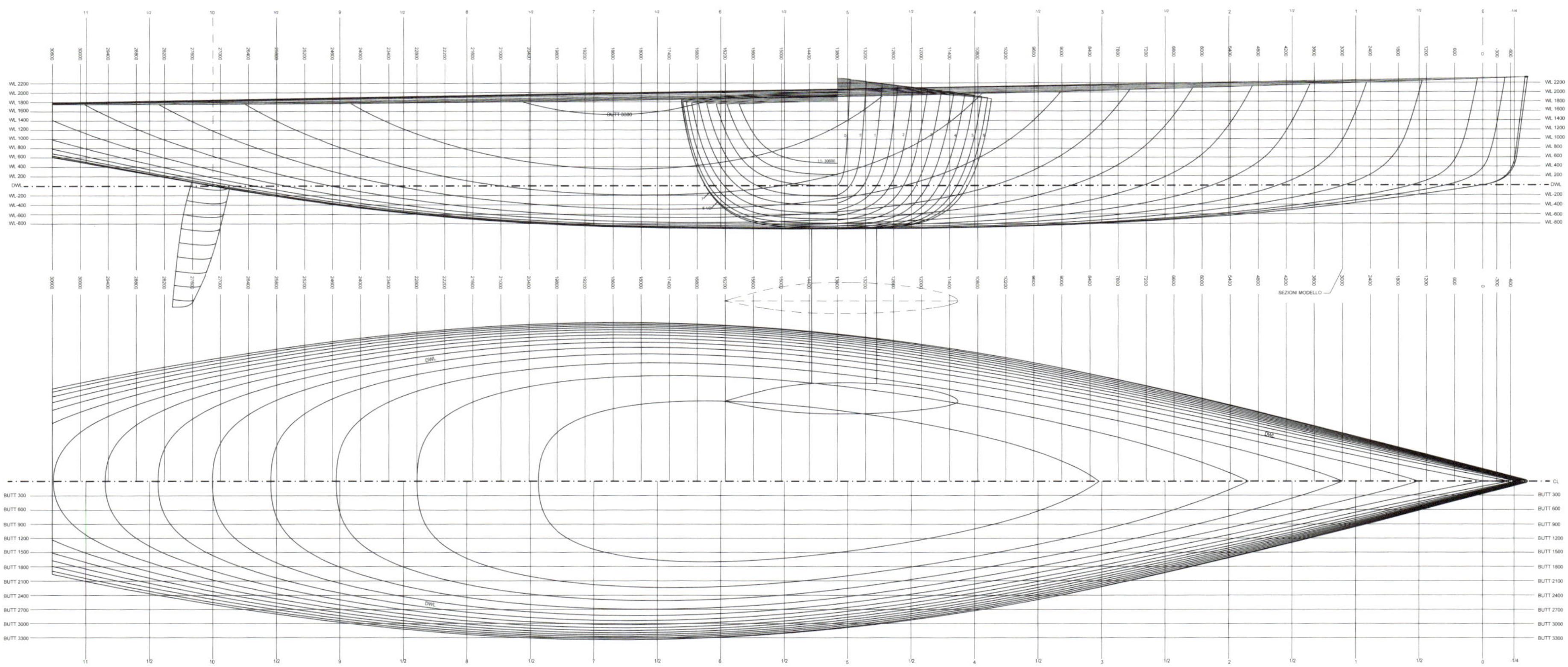

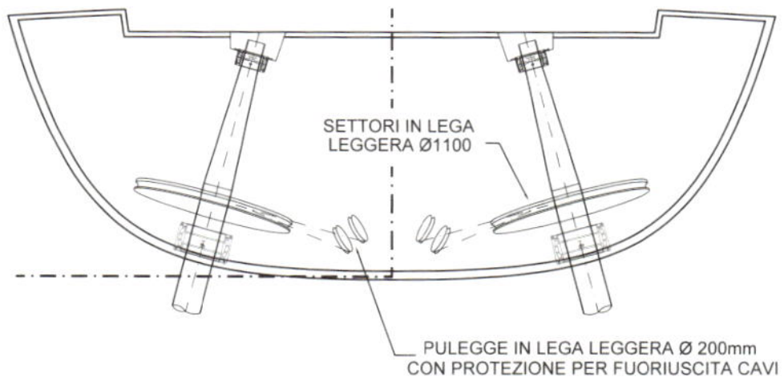

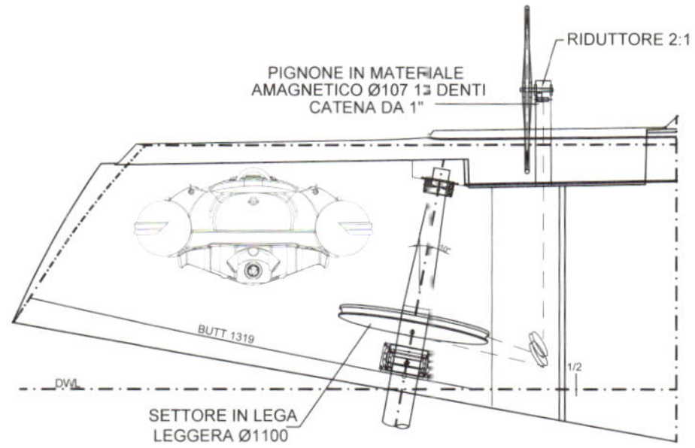

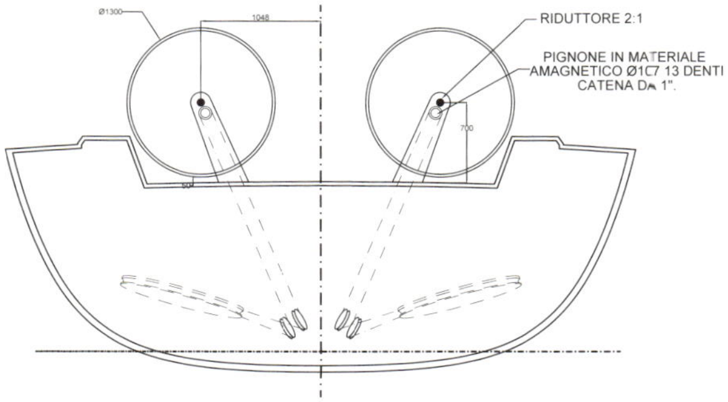

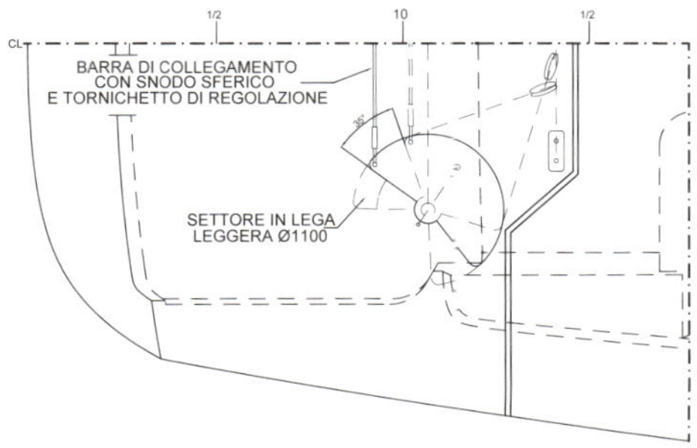

In construction at the Comar yard in Fiumicino – extracting the hull from the mold.

Lines and appendages plan (stations, waterlines, buttocks).

Details of the steering mechanisms. Plans for the rudder blade and post.

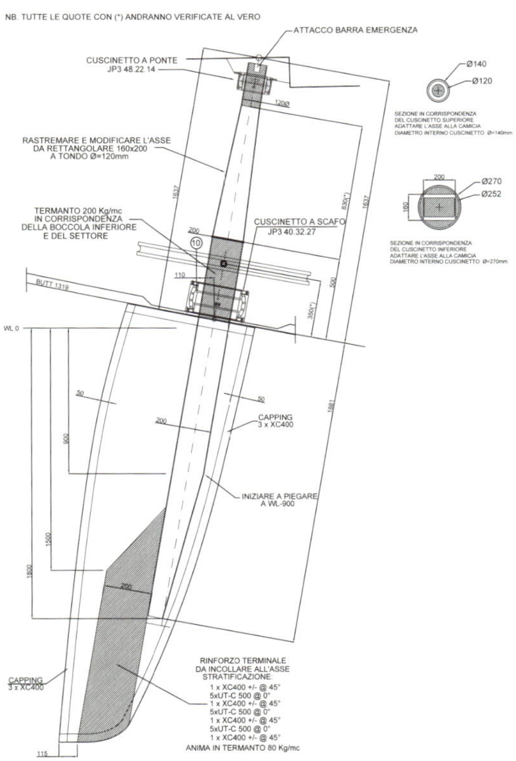

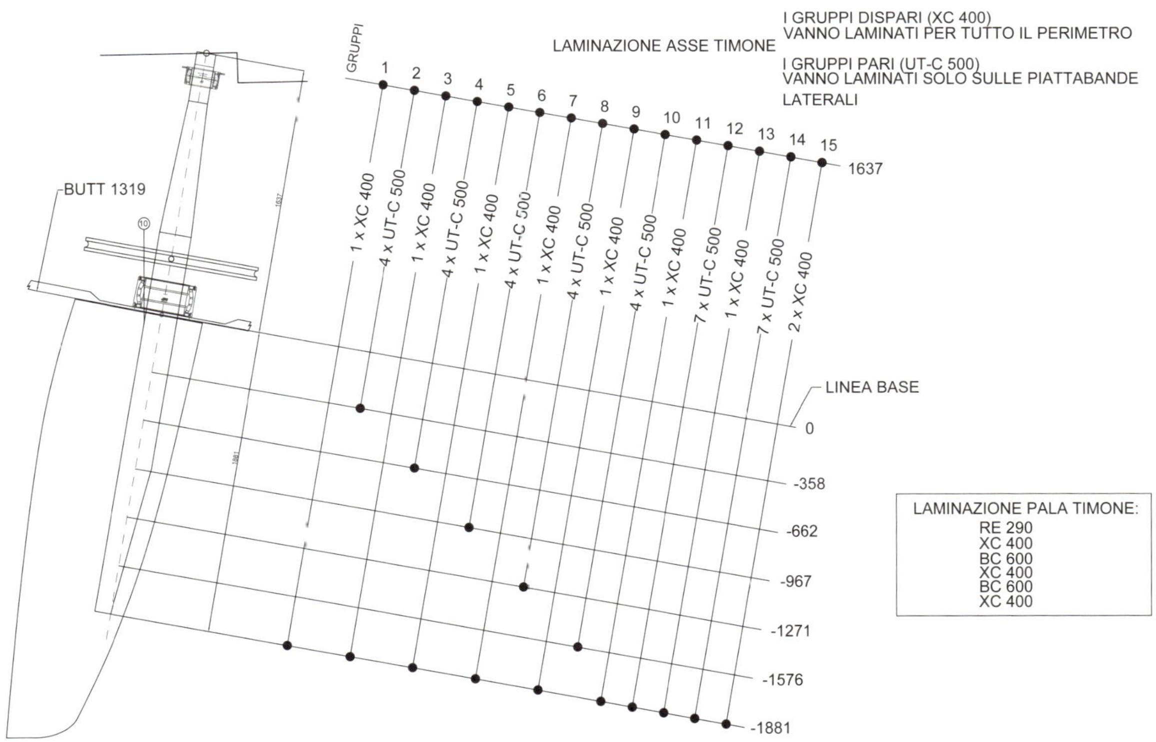

PRODUCTION YACHTS

One of the characteristics of Italian offshore sailing is that it has used hulls originally designed for racing and then adapted them for cruising. This approach may be ascribed to a typically Italian attraction to the *gran turismo* class. As a consequence, *Ziggurat*, *Gattone* and many other hulls Studio Vallicelli designed to win races went on to become commercially successful as mass-production boats. At the end of the ’70s, the Studio produced its first designs expressly for mass production, the Avventura 703 and the Canados 33, followed by the Marchi 39 and the Canados 37, all boats whose derivation from racing hulls is evident. The following decade, the Comet 12 marked a change of course. This model showed that mass-production yachts had diverged from racing yachts, in part because racing yachts had become more and more extreme and, therefore, less and less adaptable to the needs of cruising. The Studio went on to design a whole series of models for Comar based on new aesthetic and functional criteria, including the Genesi and Phoenix, which combined classic lines with innovative solutions such as panoramic windows, self-tacking jibs and, inside, a double cabin aft with a central bathroom. During those years, we also designed several ULDBs (Ultra-Light Displacement Boats) for VR Nautica, plus the Este 35 and Este 39 models for the D’Este yards. In the 2000s, we designed a new line of very successful boats for Comar, first and foremost the Comet 51, which in addition to being a commercial success won the Italian IMS Championship in 2006, and the Comet 45, which won the ORC World Championship in Athens in 2008.

Avventura 703

She was one of the smallest boats intended for series production that we designed at the end of the '70s. She was conceived as a small cruiser, but also to be attractive looking for competent sailors. Indeed, the aim was for her to look more like a large racer than other vessels of similar size, which at the time were being produced primarily in France, and were characterized by a huge volume in relation to their length. Avventura 703 was therefore designed with a more slender hull than the earlier *Ziggurat*, with a shift from a masthead rig to a 7/8 rig. The deck layout offered a large cockpit and interiors where the bow cabin, set up with a classic triangular bed, was followed by a saloon with two bunks, a chart table, and a unit containing the sink and stove. The boat, equipped with a chemical toilet placed beneath the bow berths, was designed with both inboard and outboard propulsion. With the Avventura 703, built as a production model by the Nauta yard, the graphic and design choices of *Ziggurat* and *Argento vivo* were retained.

Year 1976
Builder Nauta S.n.c. Milano
Construction GRP
LOA 7.03 m
B MAX 2.48 m
D 1.45 m
DSPL 0.85 t
Mainsail 12 mq
Headsail 18 mq
Spi 45 mq

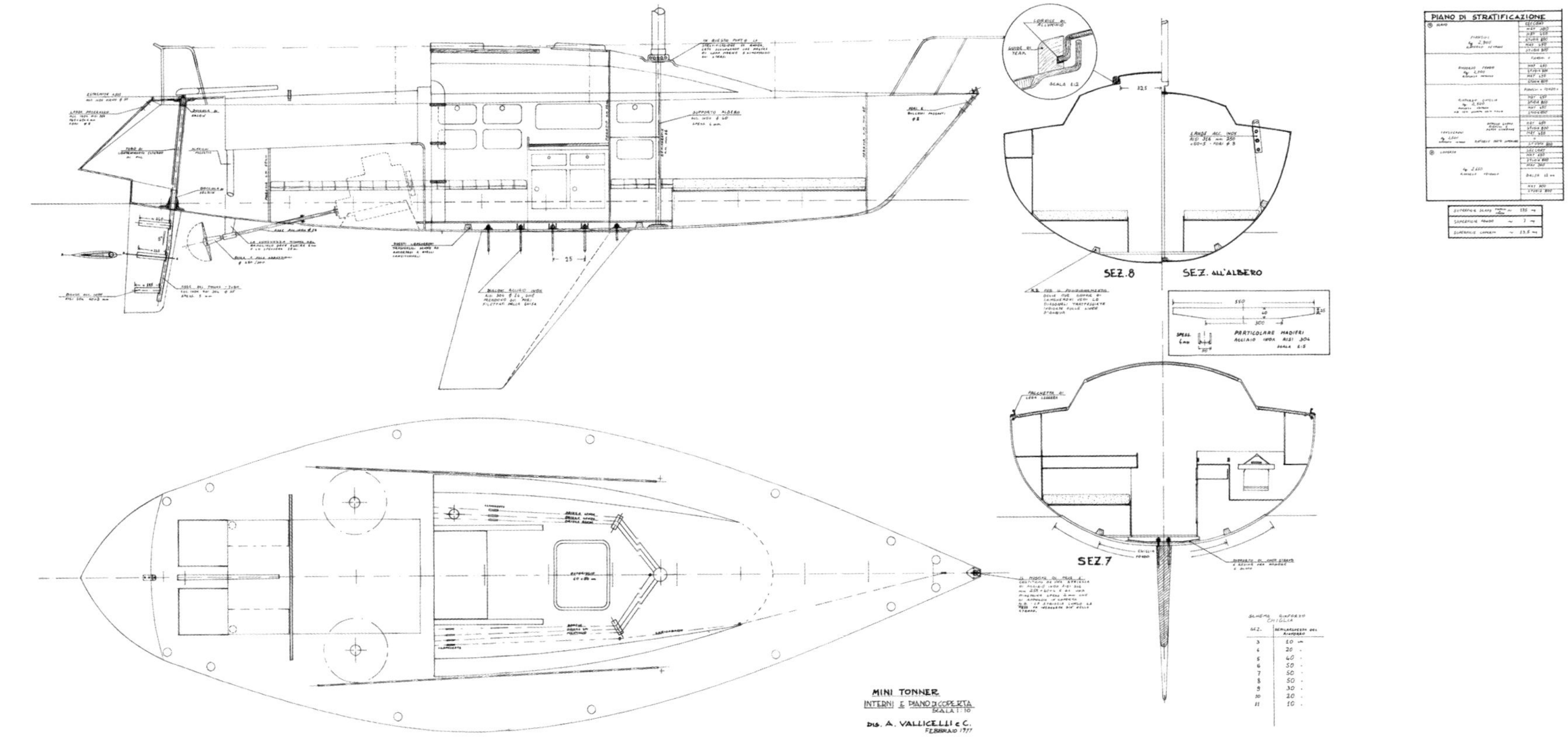

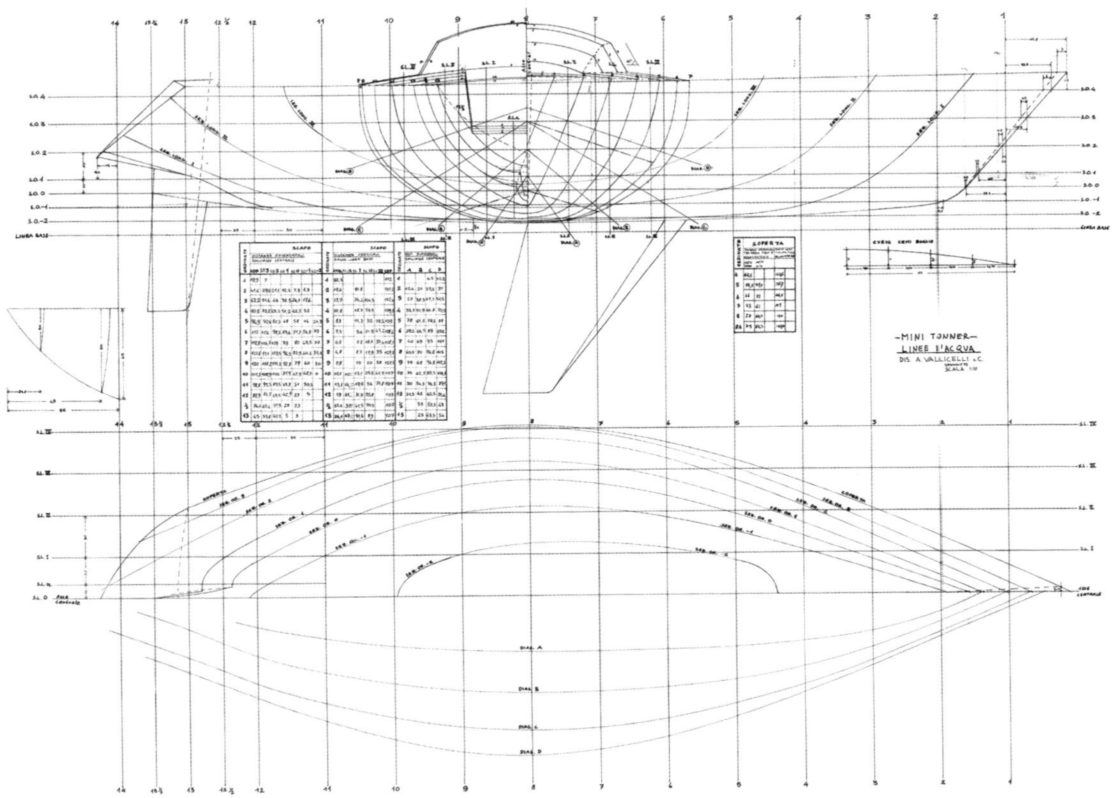

One of the first Avventura 703's under sail.

Lines plan (stations, waterlines, buttocks, diagonal sections).

General arrangement, deck plan.
Pencil and ink drawings on tracing paper.

Canados 37

The Canados 37 was designed for Cantieri Navali di Ostia after the success of the Canados 33; she was a one tonner of the *Filo da torcere* generation and represents one of those projects in which we sought to combine the canons of the American school of medium-heavy displacement hulls with the New Zealand one of light displacements. The boat, however, was intended for recreational use and designed to offer great flexibility in the arrangement of the interiors, which included two alternatives: a traditional solution and another, unusual one for a boat of that size, with an aft owner's cabin. To obtain this production variant, we designed a mold where it was possible to modify the after part by adding a section. Despite the boat being born for cruising, the hull turned out to be very fast, so much so that at the 1980 One Ton Cup in Naples, *Excelsior* (a standard Canados 37) helmed by Paul Elvestrom, was in contention for the final victory until the last race.

Year 1979
Builder Canados Ostia (RM)
Construction GRP
LOA 11.3 m
B MAX 3.63 m
D 1.98 m
DSPL 6.2 t
Mainsail 27 mq
Headsail 55 mq
Spi 110 mq

A Canados 37 sailing off Fiumicino.

Sail plan, deck plan, details of cockpit variation and transom.

Interiors.

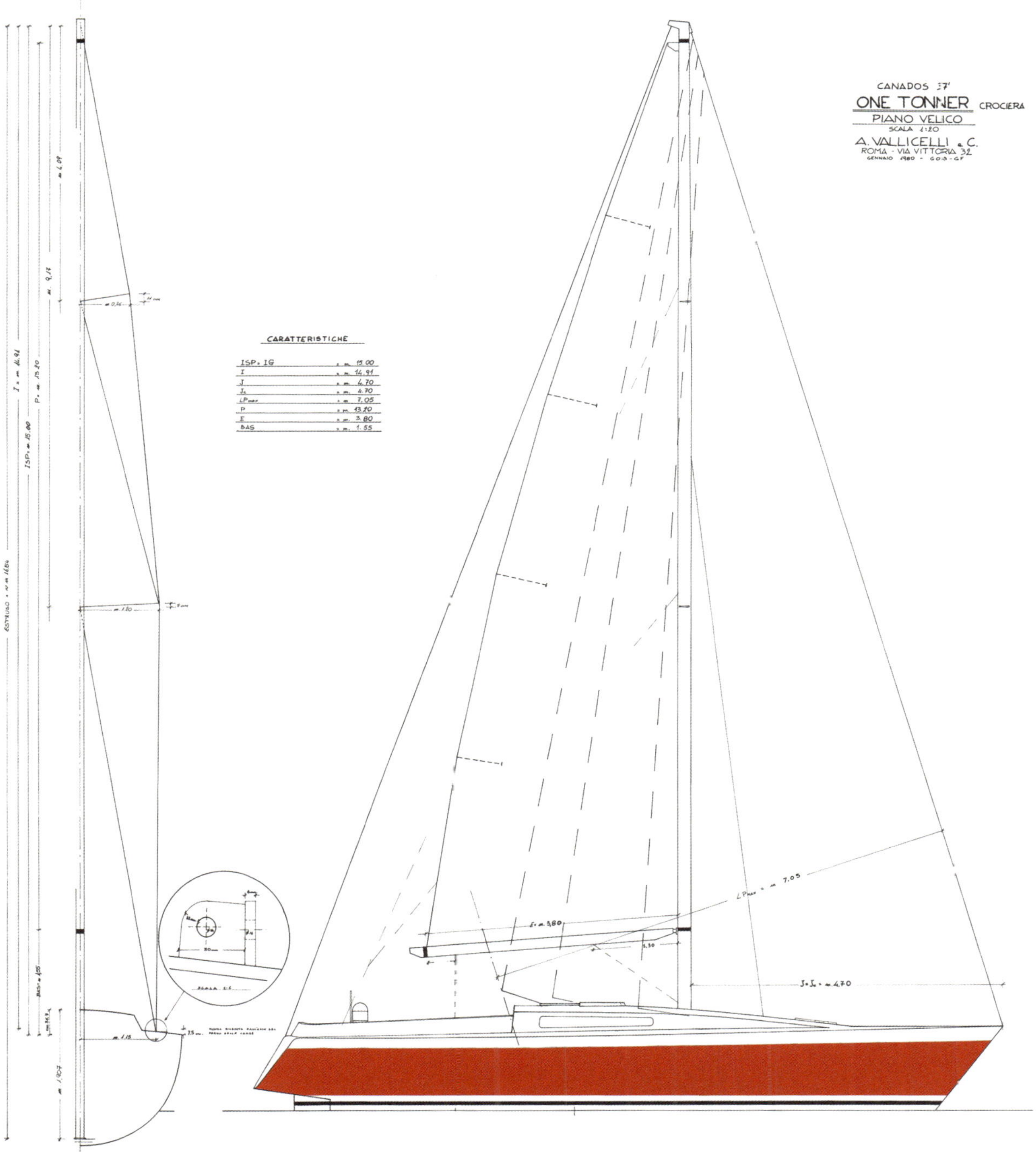
CANADOS 37'
ONE TONNER CROCIERA
PIANO VELICO
SCALA 1:20
A. VALLICELLI & C.
ROMA - VIA VITTORIA 32
GENNAIO 1980
CARATTERISTICHE
ISP. IG = m. 15.00
I = m. 14.91
J = m. 4.70
J_1 = m. 4.70
LP_{max} = m. 7.05
P = m. 13.10
E = m. 3.80
BAS = m. 1.55

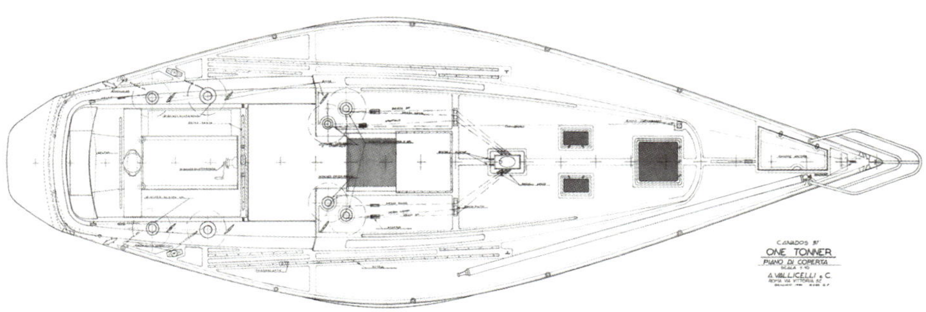
CANADOS 37'
ONE TONNER
PIANO DI COPERTA
A. VALLICELLI & C.

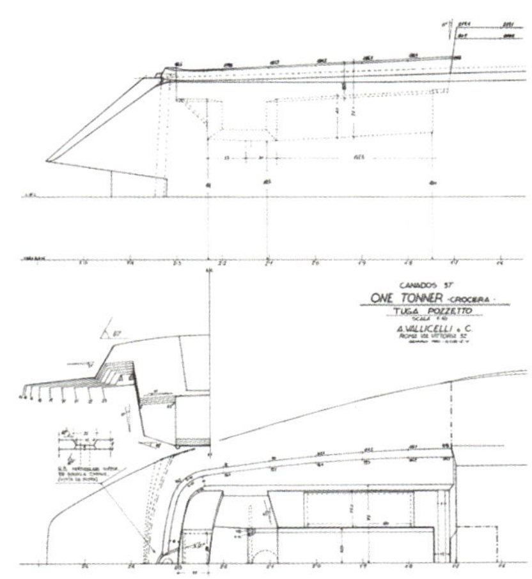
CANADOS 37'
ONE TONNER -CROCIERA-
TUGA POZZETTO
A. VALLICELLI & C.

Comet 12

This is the first boat we designed, in 1983, for Comar of Forlì. She enjoyed great commercial success, and in less than ten years, more than 200 boats were produced in the various versions (12m and 420). The hull, fairly closely related to the IORs of the time, had a generous maximum beam, a rather vertical bow and rounded stern sections leading gently to the transom. The layout of the deck included a classic cockpit aft of the coachroof, with the sheet winches placed on the coamings and halyards led back to the sides of the entrance hatch. Inside there were two double berths with heads aft, then came the chart table and galley and, further forward, a large dinette and a spacious cabin with en-suite heads. The boat was marketed with the possibility of choosing between cruising or racing sail plans and appendages. In 1987 a re-styling was done, marketed as the Comet 420, which involved a large part of the superstructure, adapting it to new trends. In the following years, following the direction inaugurated by the Comet 12, and again for the Forlì yard, we designed the Comet 333 and 303, the latter with molded interiors; this feature made the boat more expensive and heavier but provided it with an unusual refinement for the time. The long collaboration with Comar in those years enabled us to acquire a great deal of experience and skill in dealing with the problems concerning series production.

Year 1983
Builder Comar Forlì
Construction GRP
LOA 12.621 m
B MAX 3.9 m
D 2.5 m
DSPL 8.5 t
Mainsail 31 mq
Headsail 60 mq
Spi 135 mq

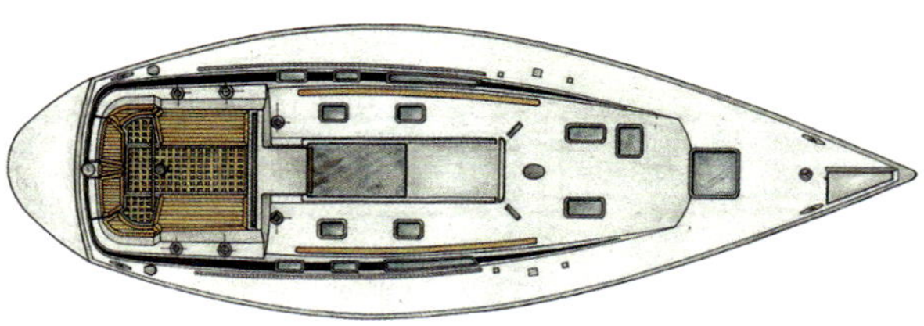

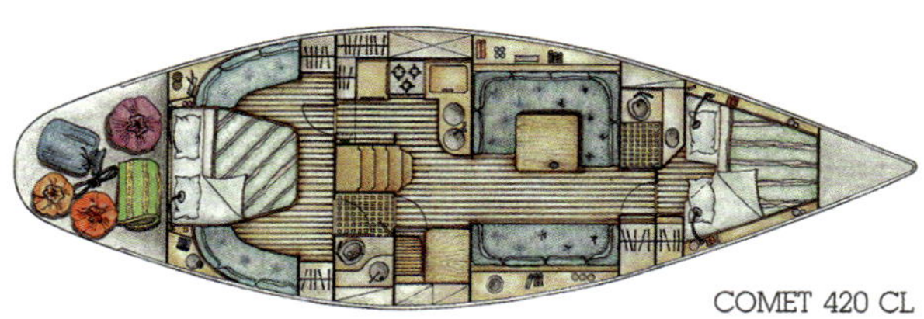

COMET 420 CL

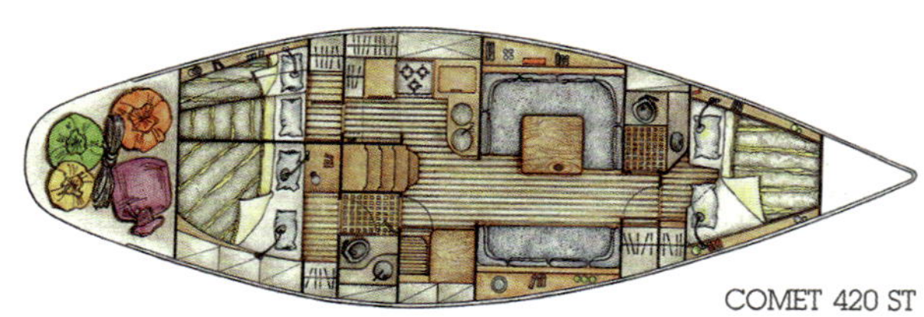

COMET 420 ST

Deck plan and two versions of the interior layout.
Ink and crayon illustrations.

Sail plan, general arrangement and deck plan.

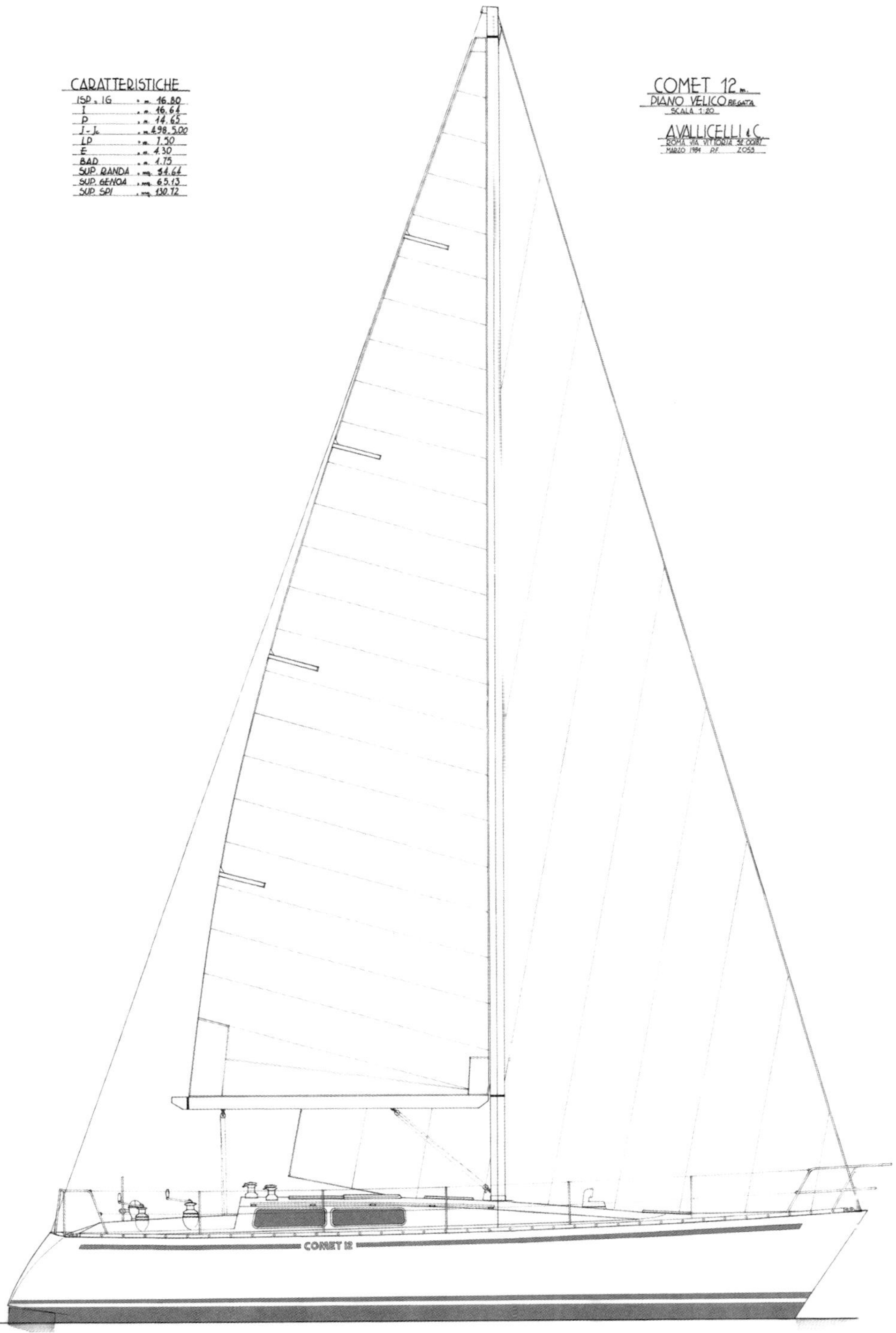

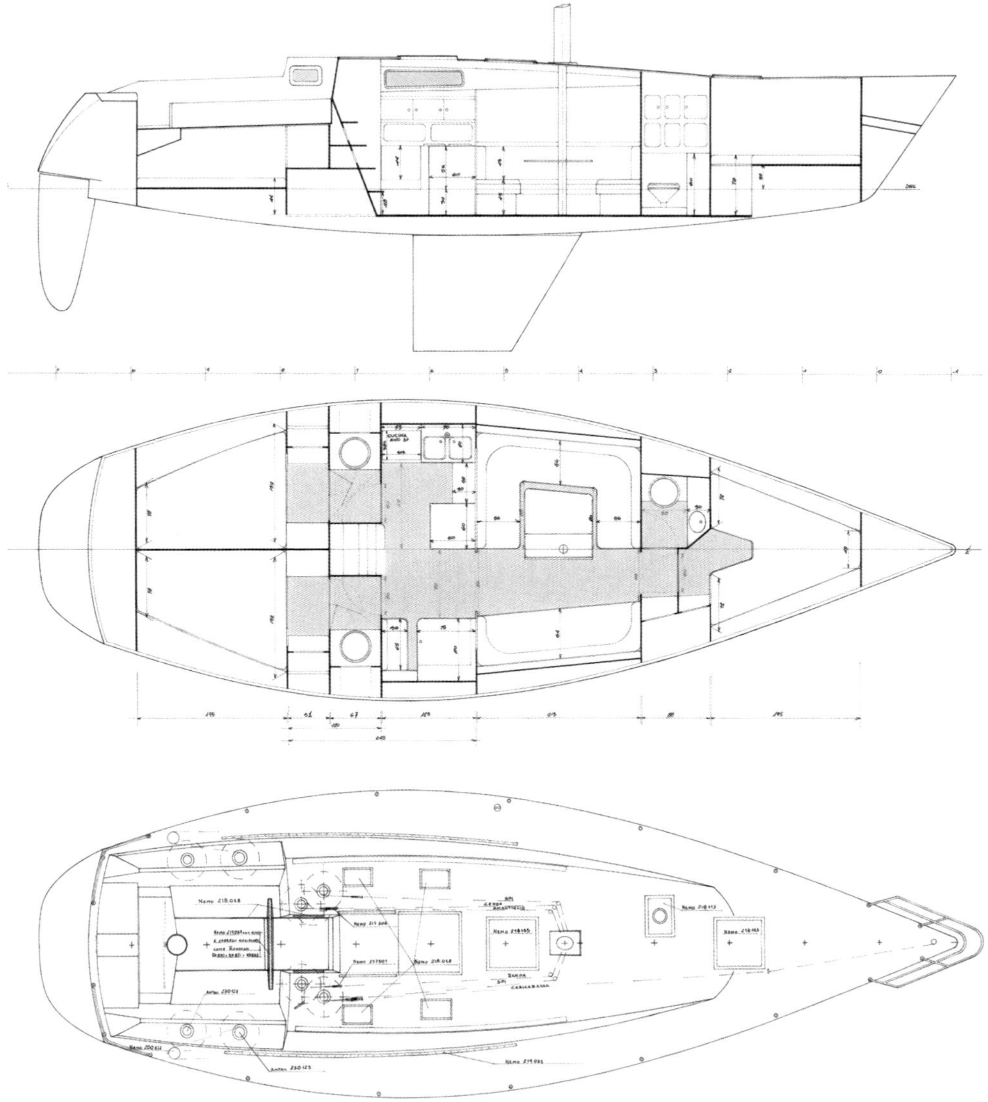

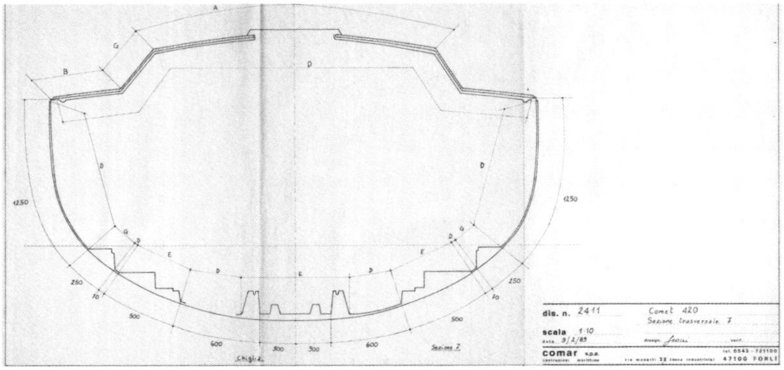

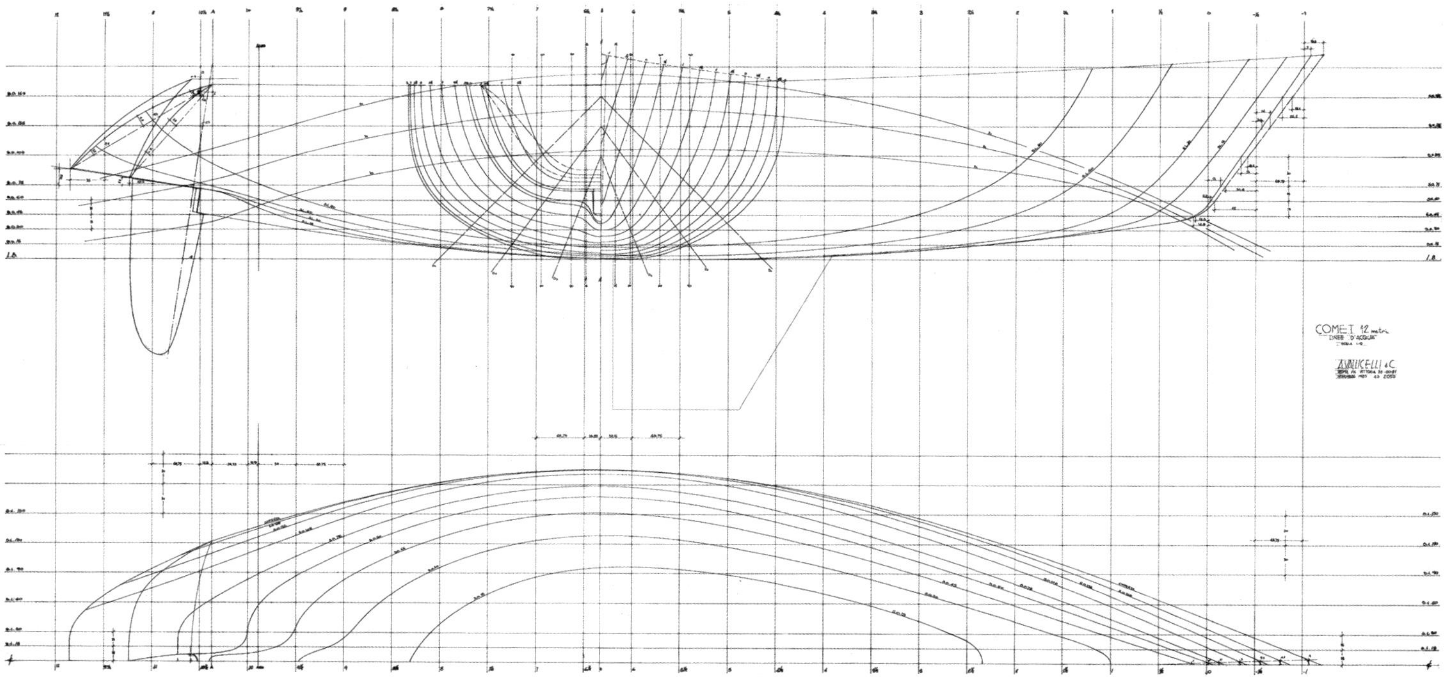

Copy of yard drawing showing structural reinforcements at station N°7.

Lines plan (stations, waterlines, buttocks, diagonal sections).

Wooden scale model of the deck and hull to make the fiberglass mold.

Comet 12 running downwind.

Hull and deck assemblies in the yard.

comet 12 metri

ULDB 65

Building on the experiences gained with light displacement boats such as *Caccia alla volpe*, *Asabranca* and *Nonno Gigi*, at the end of the 1980s we were commissioned by VR Nautica of Bergamo to produce yet another light displacement yacht, able to exploit the canons of what at the time was already identifiable as the Californian school of ULDB (Ultra Light Displacement Boat) design with long, narrow, light boats with small sail plans: but in a Mediterranean key. The result was a 65-footer that differs from Californian standards in that it had a larger maximum beam, broader stern, more sail area and a more rounded turn of the bilge. The boat was designed to be built in series, in different versions. Among these was *Nastro Azzurro*, the racing version of the standard ULDB 65', with a longer keel and larger sail plan than the standard version. In 1992, *Nastro Azzurro* finished second overall in the ocean race from Genoa to Puerto Rico, a competition held on the occasion of the Colombiadi (one of the celebrations for the 500th anniversary of the discovery of America) and in the same year she won the San Pellegrino Trophy, held in the waters of the Hudson River in front of Manhattan.

Year 1987
Builder VR Yachts Orio al Serio (BG)
Construction GRP
LOA 19.7 m
B MAX 4.9 m
D 2.8 m
DSPL 15 t
Mainsail 70 mq
Headsail 114 mq
Spi 224 mq

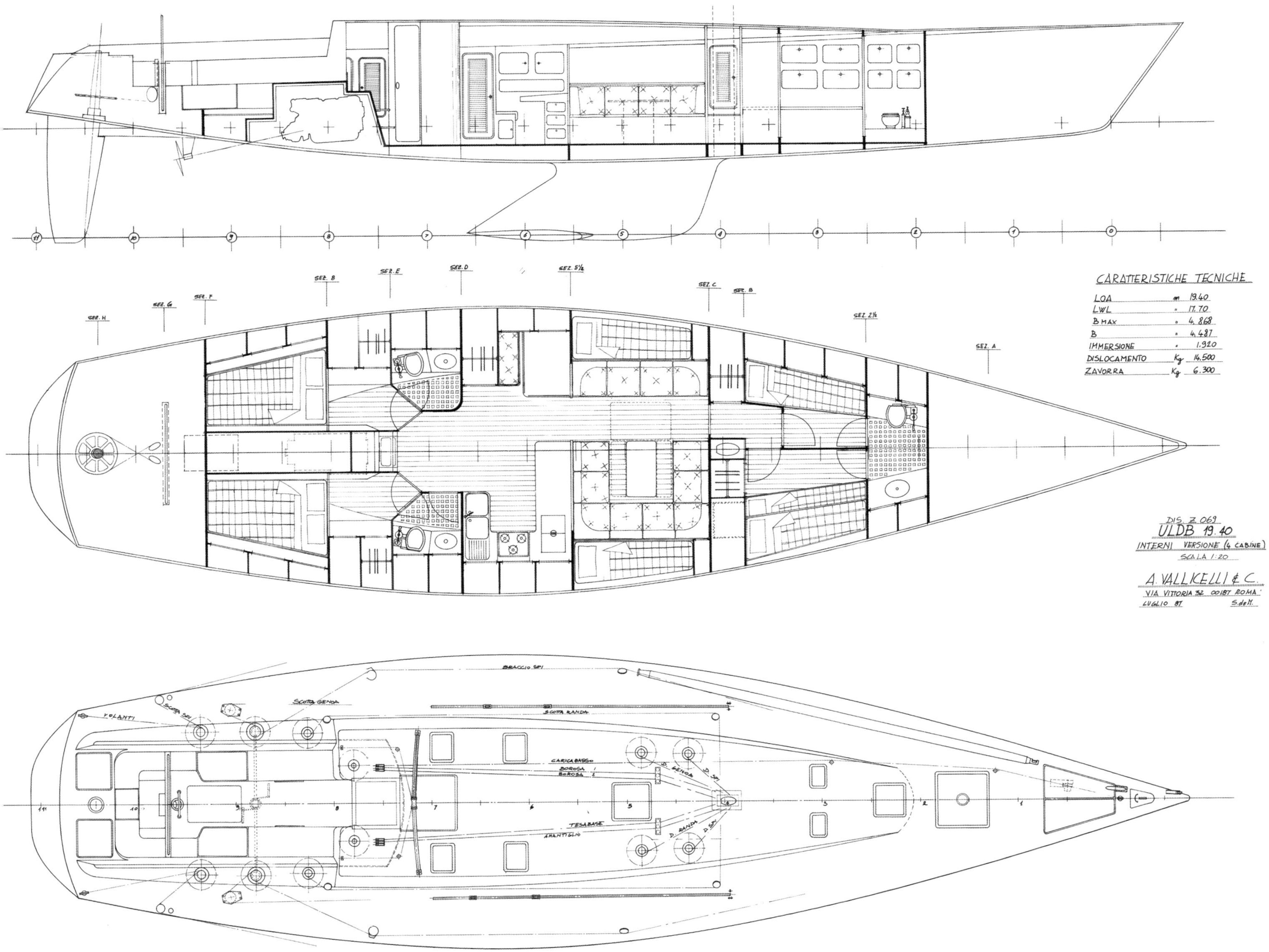

Nastro Azzurro and *Nitissima*.

Four cabin cruising version general arrangement, deck plan.

Genesi

The Genesi marked a fundamental moment in our project development dedicated to production boats for racing and cruising, indeed almost a generational watershed. We designed her for Comar of Forlì together with two other boats that should have opened a new phase in the yard's commercial policy: *Phoenix*, 45 feet long, and *Diva*, a ketch that unfortunately remained on paper. Genesi represented a line of products designed in close harmony with Paolo Villani (commercial and marketing director of the time) who wanted to offer design-oriented boats, including their sailing equipment and accessories. Masts, gangways, portholes, and other accessories were specially designed and supplied by companies, such as Nemo, working for Comar. The boat had a fractional rig with swept-back spreaders and a self-tacking jib, which was innovative for cruise boats of the time. The deck plan reflected that lightening of components that was emerging in those years. The coachroof, which gently blends into the deck, was characterized by a wraparound window providing the connotative element of the external design. The internal layout offered original solutions, such as the position of the heads for the two after cabins, located beneath the entrance steps in such a way as to be accessible from both cabins. For this reason, the engine was moved forward and housed under the central sofa bench. In the saloon, in front of the oval-shaped dinette, there was the chart table and a linear galley: a solution that was becoming increasingly popular in those years. In designing the Genesi, to harmonize the volume of the coachroof with the hull, we carefully studied how to make the sheer as perceptible as possible. To this end, we drew the section of the gunwale to taper towards the ends, generating a kind of cone (typical of racing boats), which was very elegant but also quite expensive to make on a mold. The construction of the boat was in fiberglass sandwich with a PVC core, making use of many counter-molded elements. The lamination was done in a vacuum, a process not much used at the time because of its high cost.

Year 1989
Builder Comar Forlì
Construction GRP
LOA 13.1 m
B MAX 4.02 m
D 2.4 m
DSPL 9.1 t
Mainsail 43 mq
Headsail 57 mq
Spi 115 mq

Lines plan (stations, waterlines, buttocks).

Left: vacuum bag lamination technique.
Right: Interior assembly.

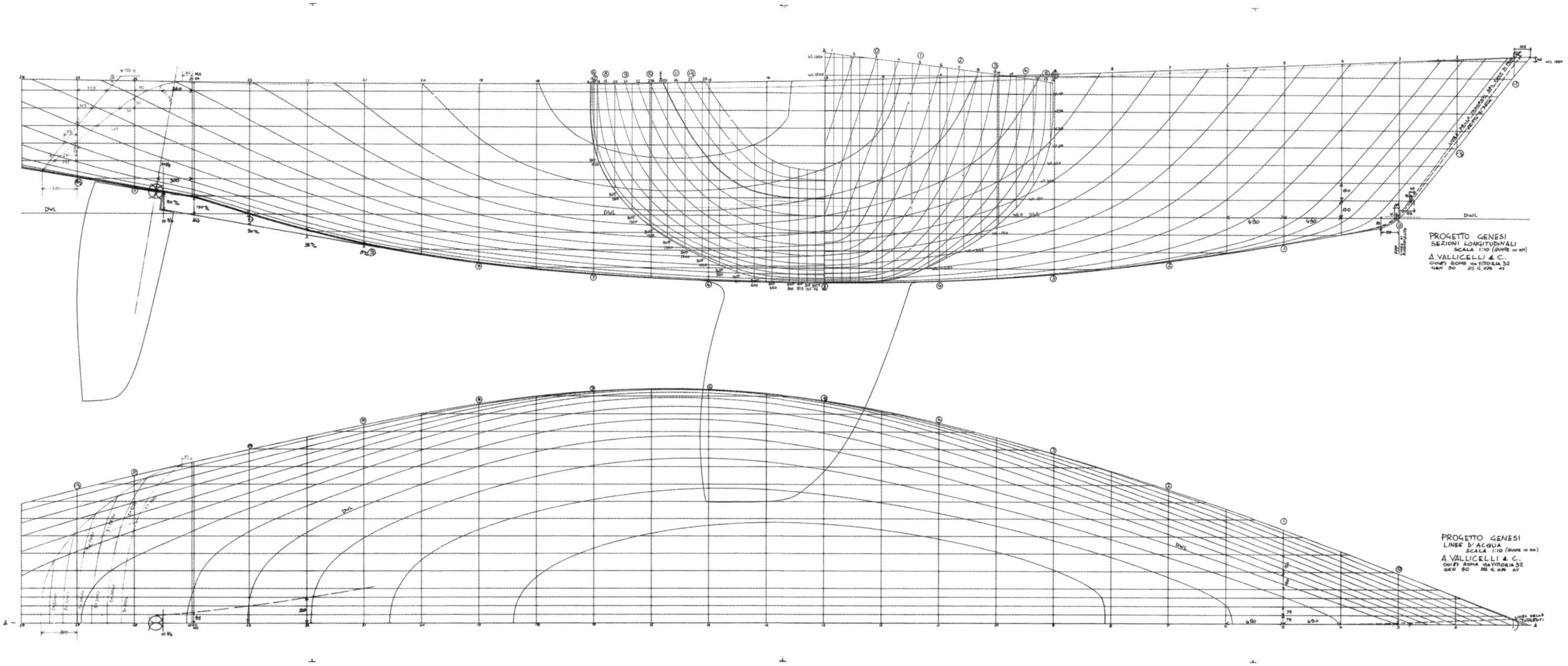
PROGETTO GENESI
SEZIONI LONGITUDINALI
SCALA 1:10
A. VALLICELLI & C.
ROMA VIA VITTORIA 32
PROGETTO GENESI
LINEE D'ACQUA
SCALA 1:10
A. VALLICELLI & C.
ROMA VIA VITTORIA 32
GEN 90

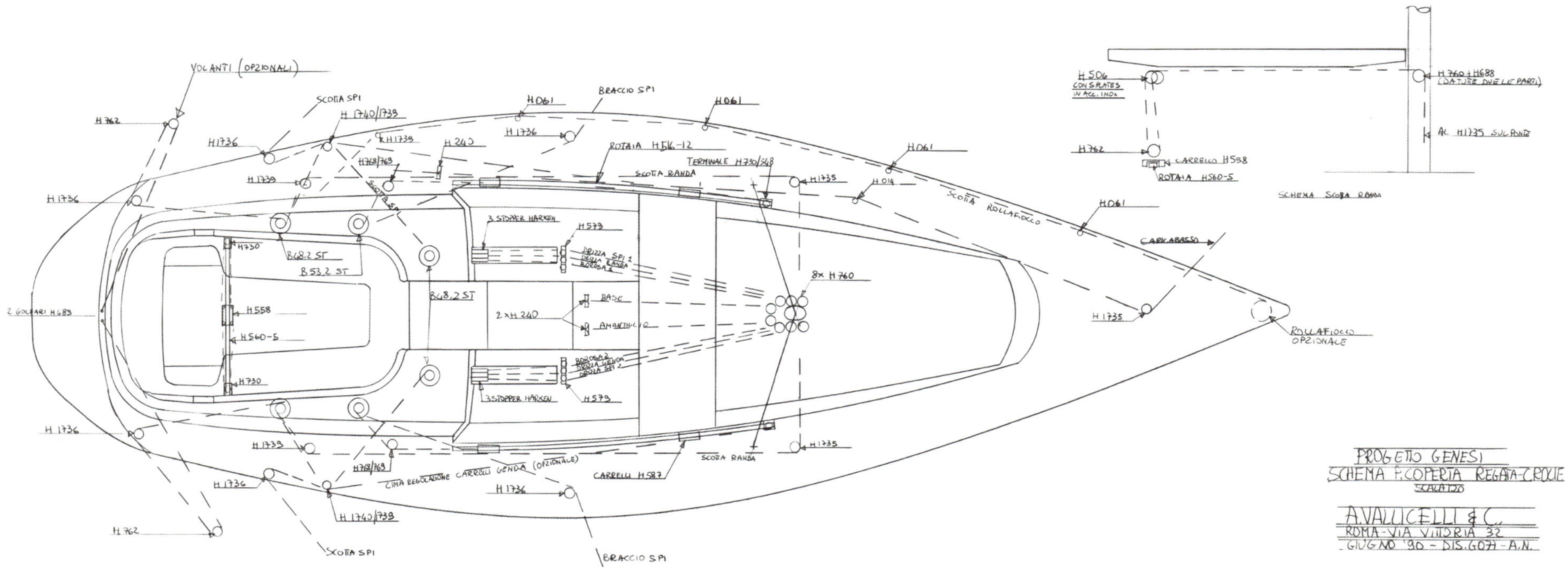

Deck plan draft.

Plan of the deck.

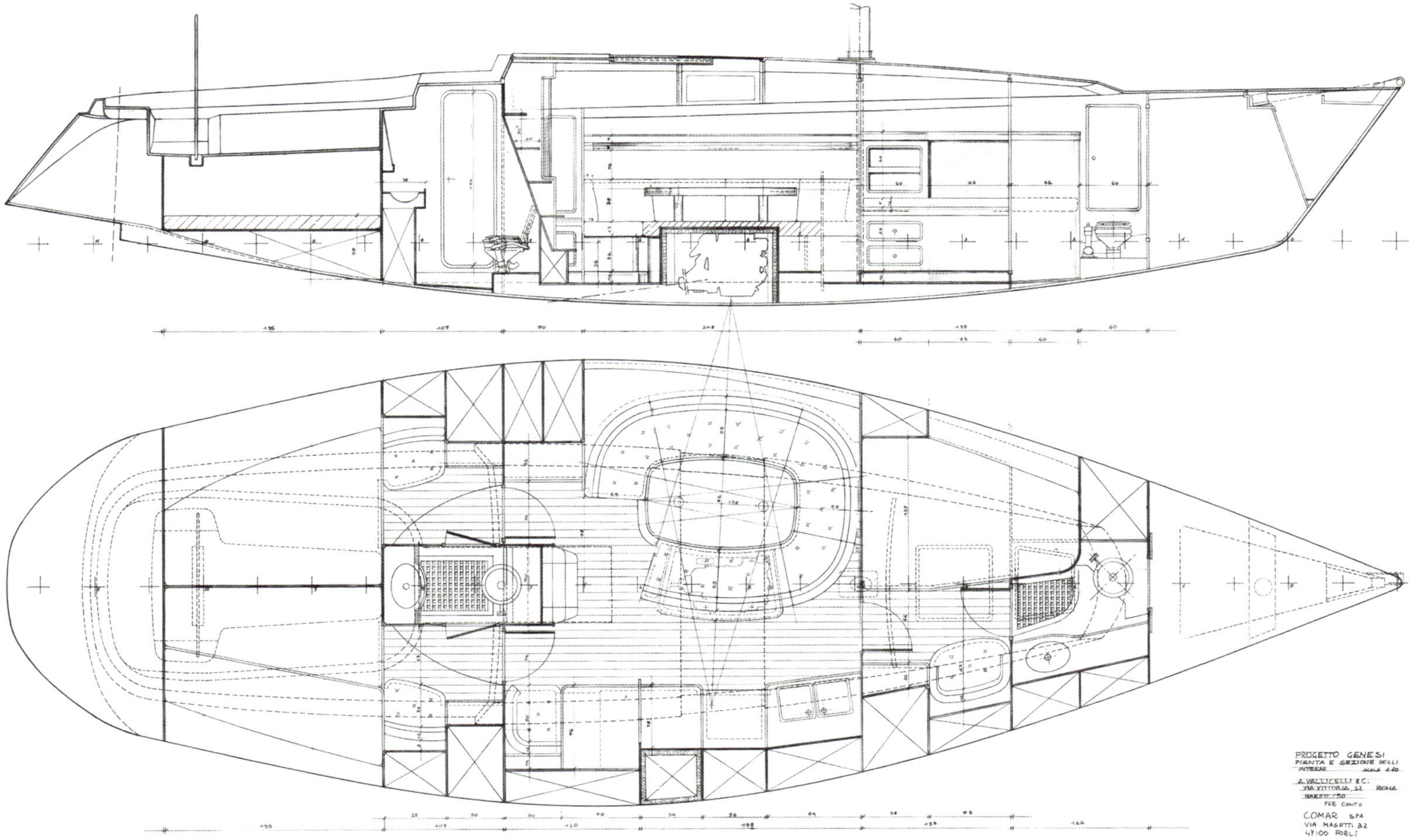

General arrangement.

Genesi underway.

Este 35

She was one of the first boats designed by us, taking into consideration the IMS rules and was made of molded fiberglass by the D'Este yard in Fiumicino. The shallow hull is well rounded, with minimal overhangs. The deck plan included a generously sized cockpit, set quite far forward to bring the weight of the crew as close to the center of the boat as possible. This configuration also allowed us to create a large sundeck at the stern, holding two lockers and a micro-cockpit, which in a regatta could be used by the mainsail trimmer. The interiors of the boat are minimal and include a galley and chart table, set immediately aft from the entrance steps, along with a large symmetrical saloon and the bow cabin with the V-berth. In her first competitions, the Este 35 won Alassio Week, the regattas of Portofino and Santo Stefano, the Italian IMS Championships in Punta Ala and the Copa del Rey in Palma de Mallorca. Following the same all-round design philosophy, with lines as soft and streamlined as possible, soon after, we designed her larger sister, the Este 39, which won the Copa del Rey in 1995 and finished second in the IMS European Championship.

Year 1991
Builder Builder D'Este Fiumicino (RM)
Construction Composite
LOA 10.6 m
B MAX 3.27 m
D 1.9 m
DSPL 3.5 t
Mainsail 34 mq
Headsail 20 mq
Spi 90 mq

Whisky Bravo racing.

Hull and deck lamination plan.

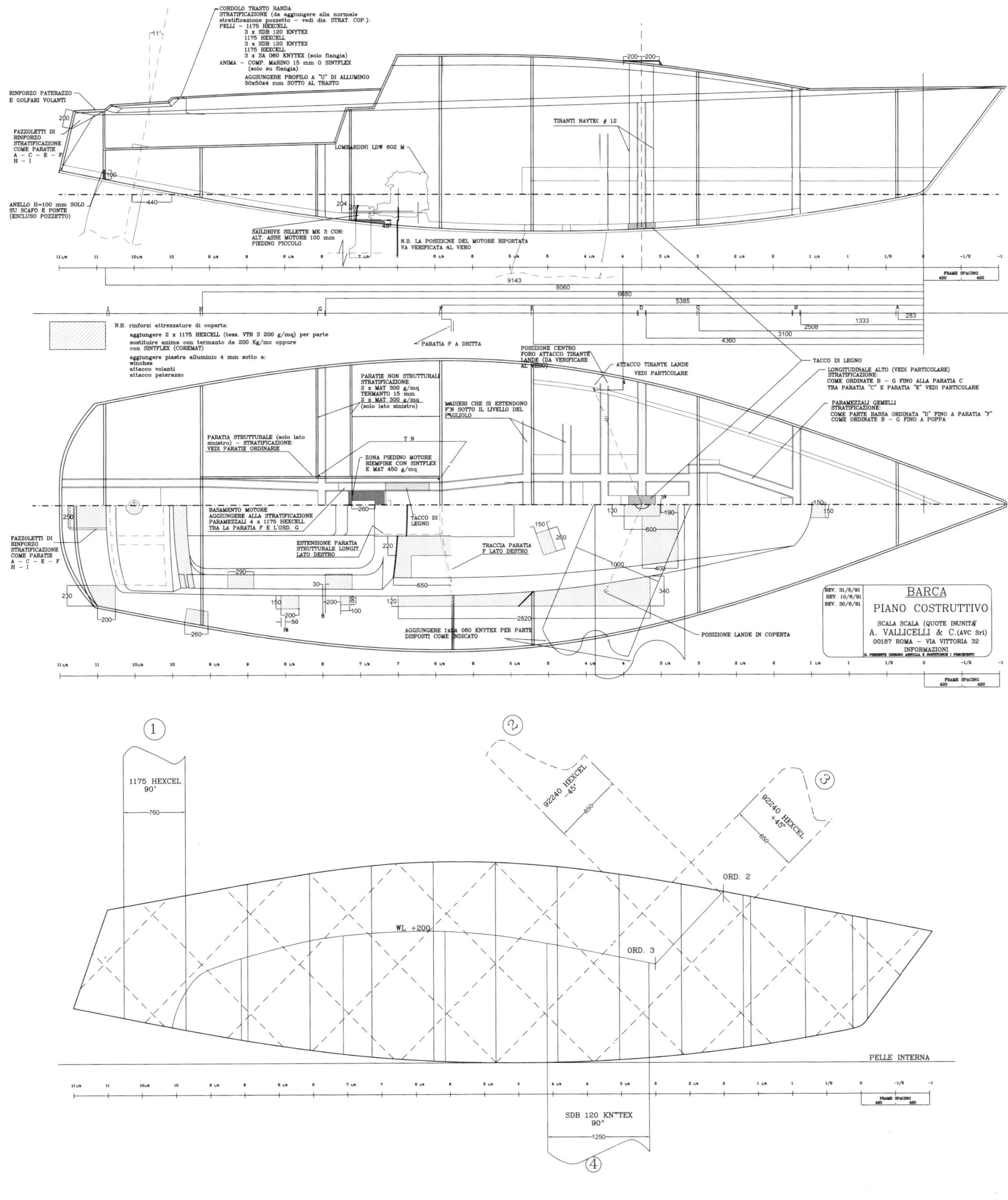

CORDOLO TRASTO RANDA
STRATIFICAZIONE (da aggiungere alla normale stratificazione pozzetto – vedi dis. STRAT. COP.):
PELLI – 1175 HEXCELL
3 x SDB 120 KNYTEX
1175 HEXCELL
3 x SDB 120 KNYTEX
1175 HEXCELL
3 x SA 060 KNYTEX (solo flangia)
ANIMA – COMP. MARINO 15 mm O SINTFLEX (solo su flangia)
AGGIUNGERE PROFILO A "U" DI ALLUMINIO 50x50x4 mm SOTTO AL TRASTO
RINFORZO PATERAZZO E GOLFARI VOLANTI
FAZZOLETTI DI RINFORZO STRATIFICAZIONE COME PARATIE A – C – E – F H – I
TIRANTI NAVTEC # 12
LOMBARDINI LDW 602 M
ANELLO H=100 mm SOLO SU SCAFO E PONTE (ESCLUSO POZZETTO)
SAILDRIVE SILLETTE MK 3 CON: ALT. ASSE MOTORE 100 mm PIEDINO PICCOLO
N.B. LA POSIZIONE DEL MOTORE RIPORTATA VA VERIFICATA AL VERO
N.B. rinforzi attrezzature di coperta:
aggiungere 2 x 1175 HEXCELL (tess. VTR S 200 g/mq) per parte
sostituire anima con termanto da 200 Kg/mc oppure con SINTFLEX (COREMAT)
aggiungere piastra alluminio 4 mm sotto a:
winches
attacco volanti
attacco paterazzo
PARATIA F A DRITTA
POSIZIONE CENTRO FORO ATTACCO TIRANTE LANDE (DA VERIFICARE AL VERO)
ATTACCO TIRANTE LANDE VEDI PARTICOLARE
TACCO DI LEGNO
LONGITUDINALE ALTO (VEDI PARTICOLARE) STRATIFICAZIONE: COME ORDINATE B – G FINO ALLA PARATIA C TRA PARATIA "C" E PARATIA "E" VEDI PARTICOLARE
PARAMEZZALI GEMELLI STRATIFICAZIONE: COME PARTE BASSA ORDINATA "D" FINO A PARATIA "F" COME ORDINATE B – G FINO A POPPA
PARATIE NON STRUTTURALI STRATIFICAZIONE: 2 x MAT 300 g/mq TERMANTO 15 mm 2 x MAT 300 g/mq (solo lato sinistro)
MADIERI CHE SI ESTENDONO FIN SOTTO IL LIVELLO DEL PAGLIOLO
PARATIA STRUTTURALE (solo lato sinistro) – STRATIFICAZIONE: VEDI PARATIE ORDINARIE
ZONA PIEDINO MOTORE RIEMPIRE CON SINTFLEX E MAT 450 g/mq
BASAMENTO MOTORE AGGIUNGERE ALLA STRATIFICAZIONE PARAMEZZALI 4 x 1175 HEXCELL TRA LA PARATIA F E L'ORD. G
TACCO DI LEGNO
ESTENSIONE PARATIA STRUTTURALE LONGIT. LATO DESTRO
TRACCIA PARATIA F LATO DESTRO
AGGIUNGERE 1x SA 060 KNYTEX PER PARTE DISPOSTI COME INDICATO
POSIZIONE LANDE IN COPERTA
BARCA
PIANO COSTRUTTIVO
A. VALLICELLI & C. (AVC Srl)
00187 ROMA – VIA VITTORIA 32
INFORMAZIONI
1175 HEXCEL 90°
92240 HEXCEL -45°
92240 HEXCEL +45°
SDB 120 KNYTEX 90°
WL +200
ORD. 2
ORD. 3
PELLE INTERNA

Oceano 54

The Oceano 54 was our first new-generation series boat, incorporating the market demands that were emerging in the early 1990s. The boat was designed for long cruises, including oceangoing ones, but could be sailed shorthanded. For the hull design, we took advantage of the studies conducted on offshore racing boats to optimize the symmetry of the hull shapes at all angles of heel. This was above all to create a hull capable of maintaining a constant trim on the longitudinal plane as much as possible. The productive collaboration with the VR Nautica yard allowed us to experiment with some innovative solutions: a cockpit with a large table positioned aft of the helm and a raised chart area between the cockpit and the saloon. This arrangement recalled the deck plans of the 60-footers for solo round-the-world races but was unusual for a relatively small cruiser. This area, equipped with the chart table and a small sofa, provided shelter in high seas and winds while maintaining visibility. In the design, it was challenging to find a solution that was morphological as well as functional, to find a way to harmonize the dome aesthetically with the other coachroof volumes. The boat was rigged as a cutter with swept-back spreaders and the sail plan made use of a full-batten mainsail, a furling genoa, and a self-tacking staysail, which could also be furled.

Year 1992
Builder VR Yacht Orio al Serio (BG)
Construction Composite
LOA 16.4 m
B MAX 4.92 m
D 2.6 m
DSPL 16.8 t
Mainsail 66 mq
Headsail 104 mq
Spi 160 mq

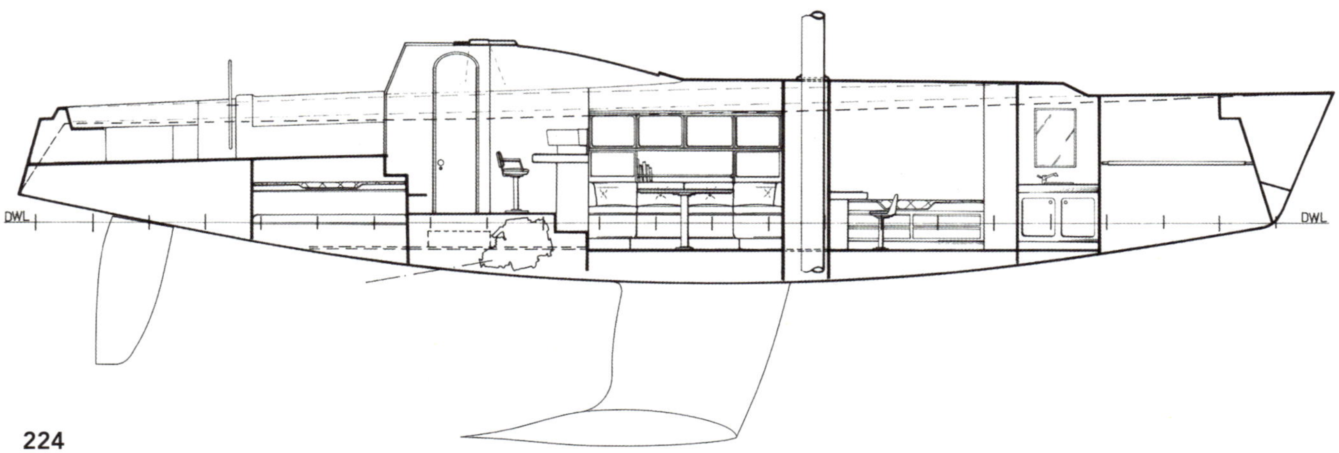

Longitudinal section.

Ocean 54 portrait.

Deck and cabin line plans.

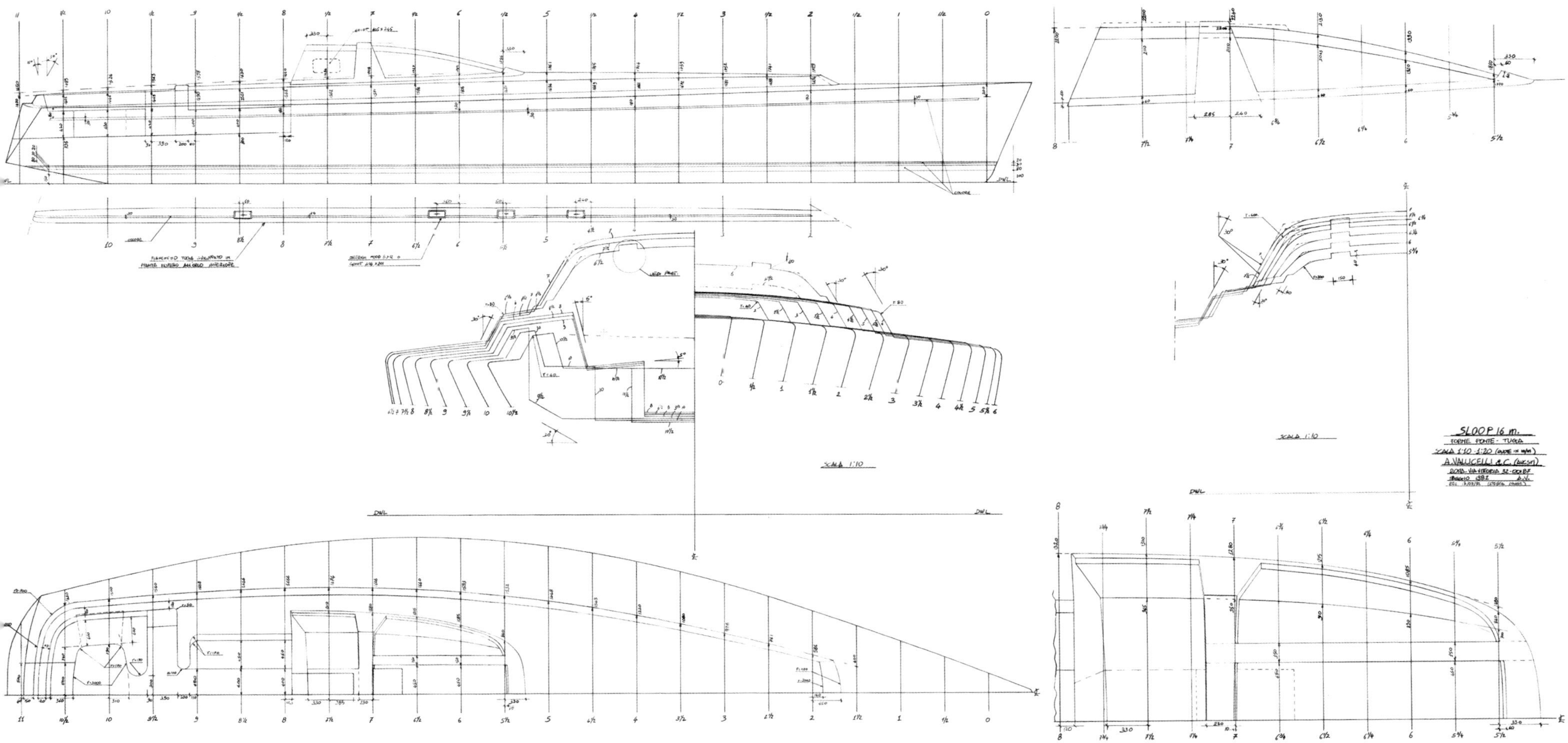

Comet 51

This was a cruiser/racer boat of about fifteen meters; she was designed to perform well when racing but also offered very comfortable interiors: under the cockpit, there were two double cabins and, after that, the heads, galley and a spacious dinette. Forward of the mast, there was a master cabin with a double bed and its heads. The boat came with a fractional rig, with a mast with three pairs of swept-back spreaders. The deck was sober and uncluttered, with a large cockpit equipped with twin steering, and the deck itself was covered with elegant teak strips. Great attention was paid to the external design, so much so that this particularly successful project gave life to a new family line characterizing Comar's production for many years to come. The coachroof, geometrically essential in form, was built on rounded, continuous sections, very flattened and without corners, marked only by the windows in which the polycarbonate was recessed from the external surface to give an effect of sculptural depth. The coaming was very low and integrated into the coachroof with a connecting element where the halyard winches were positioned. To give the boat an elegant and aggressive style, we decided to leave the transom open, harmonizing its internal curvature with that of the hull. The hull shapes of the Comet 51, developed without particular attention to the IMS formula, were streamlined longitudinally and rounded transversely in the stern sections, a feature that we had successfully tested in *Virtuelle*. Although it was not conceived to exploit the measurement rules as much as possible, the Comet 51 proved to be a very competitive boat, winning the Italian IMS overall Championship in 2006, eight years after being designed.

Year 2000
Builder Comar Fiumicino (RM)
Construction Composite
LOA 15.025 m
B MAX 4.69 m
D 3.04 m
DSPL 12 t
Mainsail 77 mq
Headsail 62 mq
Spi 168 mq

Comet 51 with cruising sails.

Steering scheme.

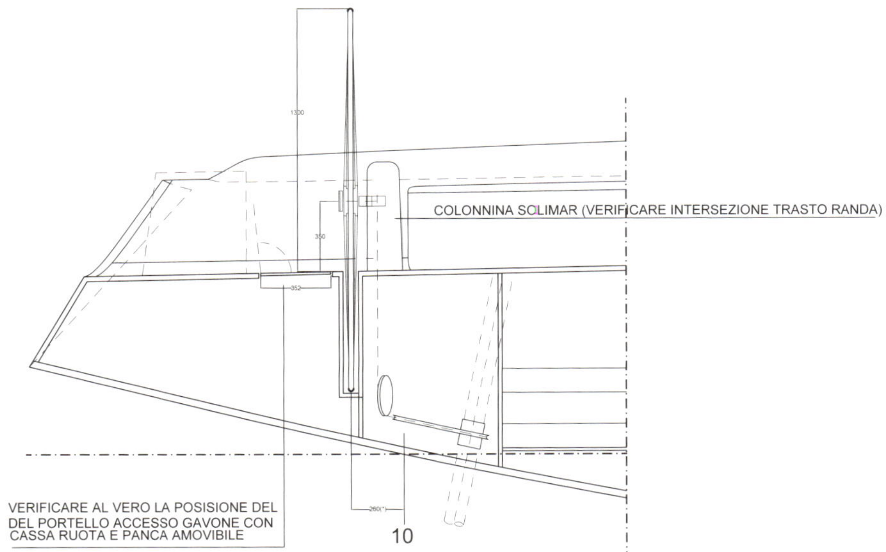
COLONNINA SCLIMAR (VERIFICARE INTERSEZIONE TRASTO RANDA)
VERIFICARE AL VERO LA POSISIONE DEL
DEL PORTELLO ACCESSO GAVONE CON
CASSA RUOTA E PANCA AMOVIBILE
10

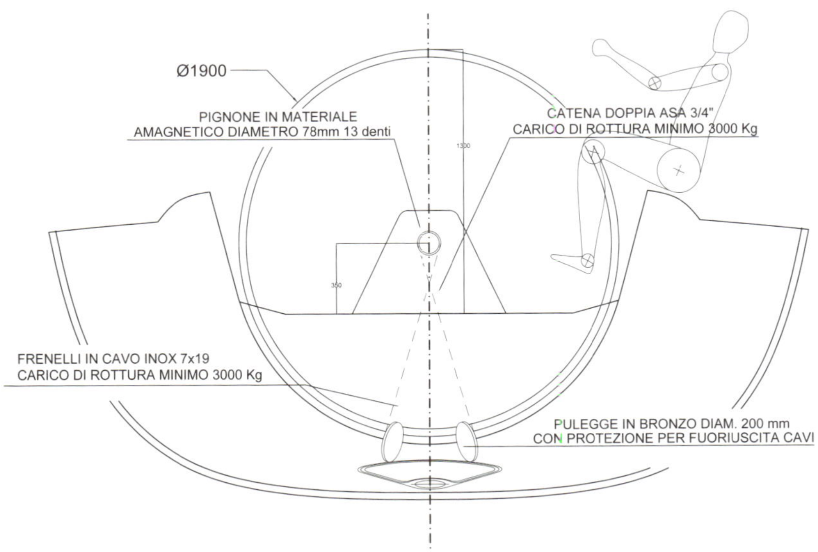
Ø1900
PIGNONE IN MATERIALE
AMAGNETICO DIAMETRO 78mm 13 denti
CATENA DOPPIA ASA 3/4"
CARICO DI ROTTURA MINIMO 3000 Kg
FRENELLI IN CAVO INOX 7x19
CARICO DI ROTTURA MINIMO 3000 Kg
PULEGGE IN BRONZO DIAM. 200 mm
CON PROTEZIONE PER FUORIUSCITA CAVI

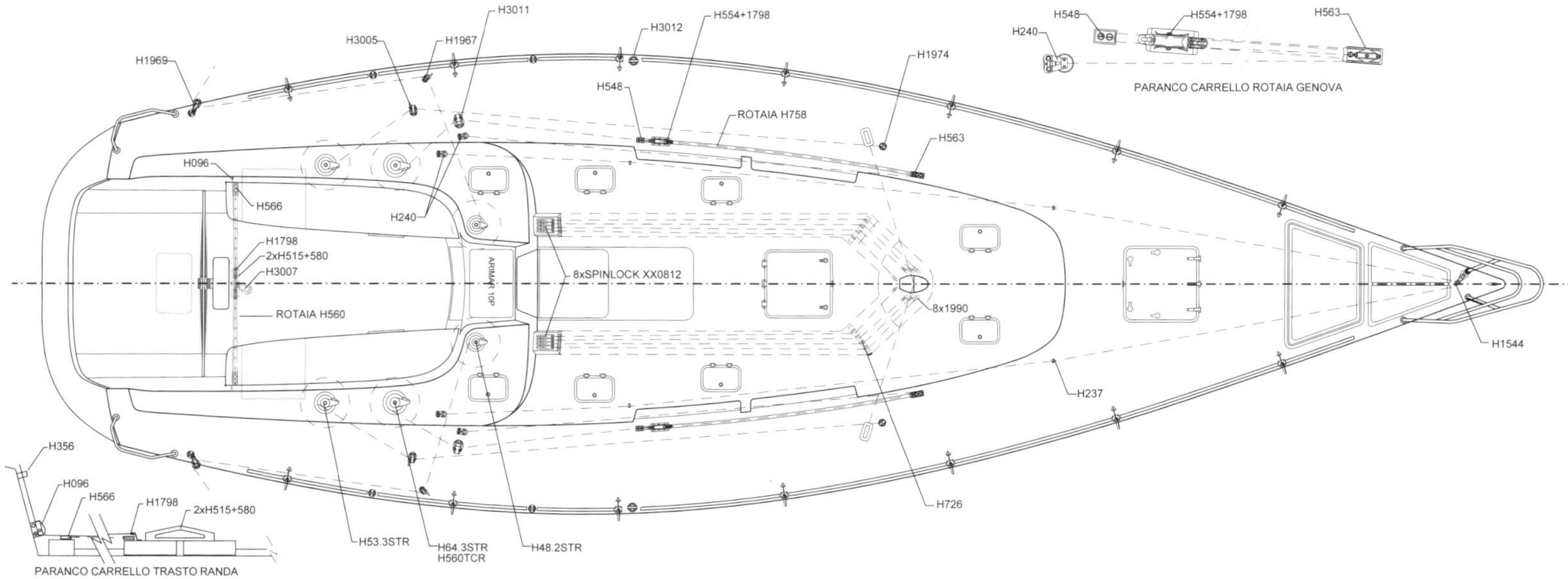
H3011
H3005
H1967
H3012
H554+1798
H1969
H1974
H548
ROTAIA H758
H563
H096
H566
H240
H1798
2xH515+580
H3007
8xSPINLOCK XX0812
ROTAIA H560
8x1990
H1544
H237
H726
H356
H53.3STR
H64.3STR
H560TCR
H48.2STR
PARANCO CARRELLO TRASTO RANDA
PARANCO CARRELLO ROTAIA GENOVA

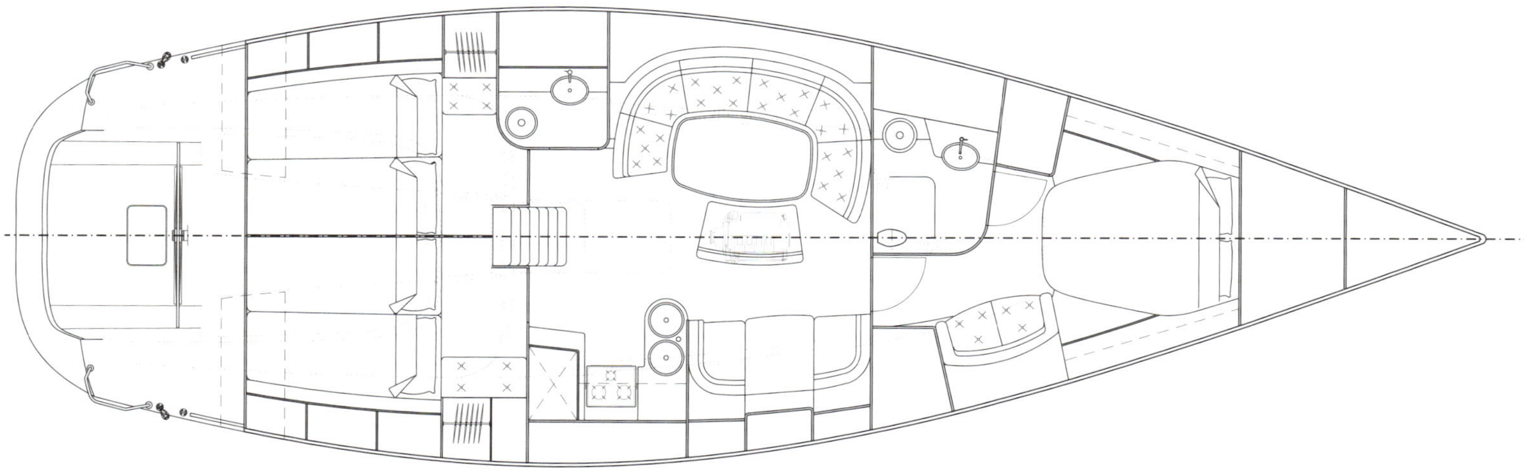

Deck plan and general arrangement.

Interiors.

Sail plan.

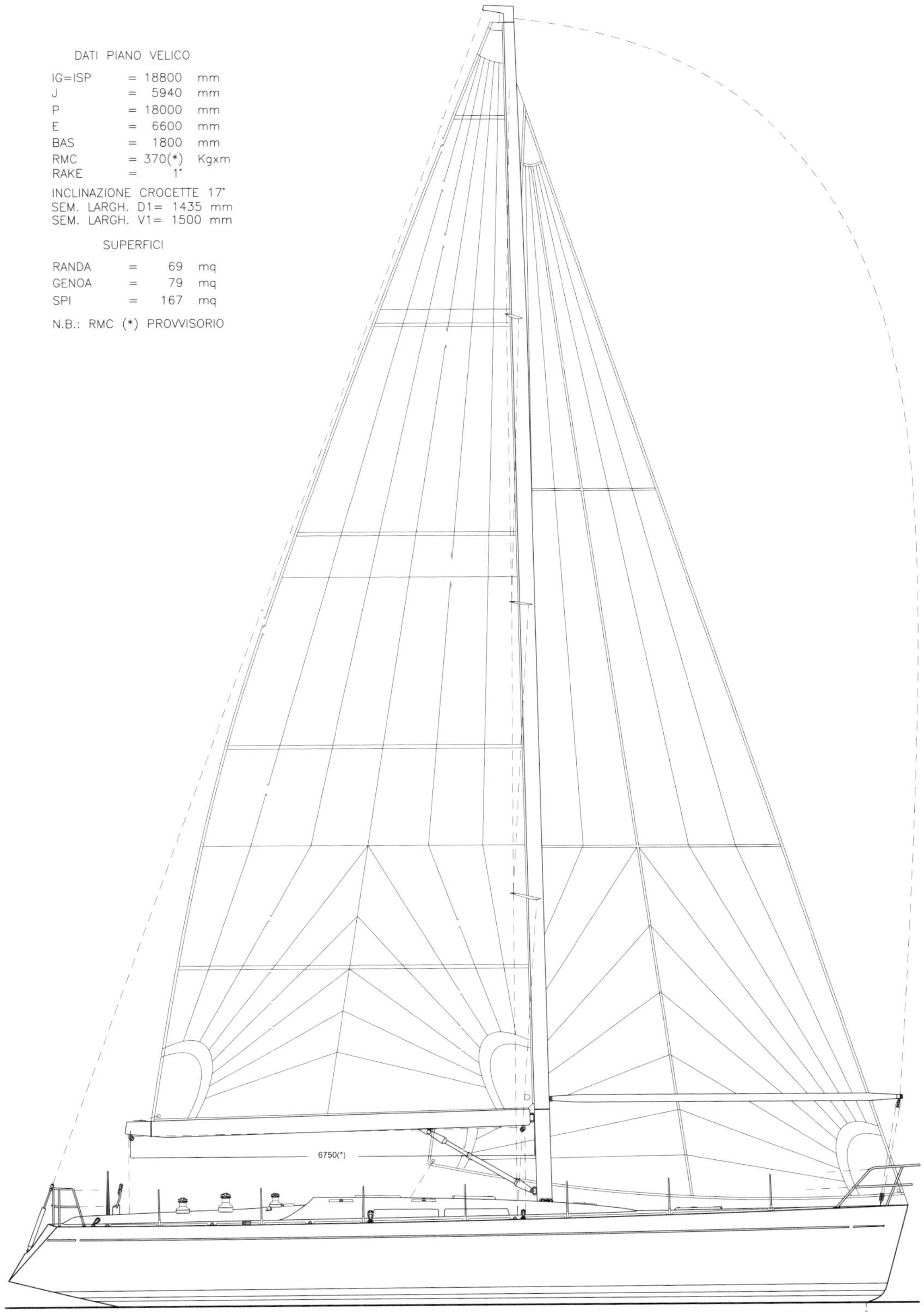

Comet 41

In the wake of the 51 and 45 footers, in 2003, Comar commissioned a 40 footer from us: a size in which there was fierce competition in the yachting market. The Comet 41 S was born, a typical cruiser racer of the time, designed in such a way as to combine a fast hull, able to perform well in regattas, with comfortable interiors. The external architecture of the boat was an evolution of the family line of our previous models, while the hull was developed with the aim of obtaining good form stability but without exaggerating, and modulating the volumes with fine bow sections and wide and rounded stern sections. In this model, too, as in the previous ones, the transom was left open. The coachroof, built on sections that give the impression of being almost semi-circular, extended with the same form along its entire length and connected to the cockpit coamings in a gentle line. The interiors consisted of three cabins, two heads and a large saloon. The Comet 41 S also performed well on the racecourse, among other things winning the 2006 European IMS Championship.

Year 2004
Builder Comar Fiumicino (RM)
Construction Composite
LOA 12.45 m
B MAX 3.92 m
D 2.4 m
DSPL 7 t
Mainsail 55 mq
Headsail 54 mq
Spi 125 mq

The Comet 41 *Galatea* during sea trails for a magazine.

Sail plan.

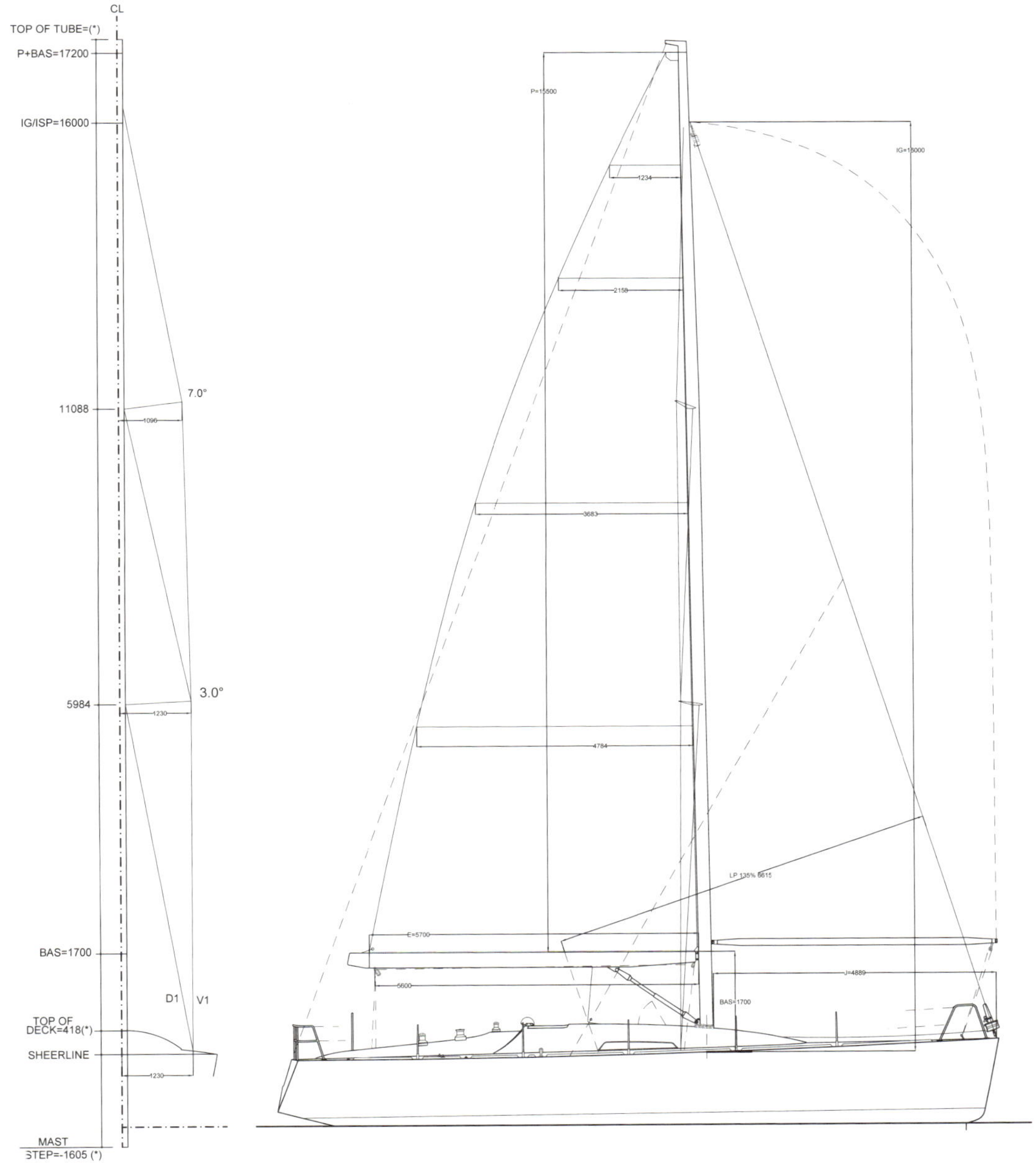

ISP = 16000mm
IG = 16000mm
J = 4900mm
SPL = 4900mm
LP FIOCCO 135% = 6615mm
P = 15500mm
E = 5700mm
BAS = 1700mm
SUP. RANDA = 52mq
SUP. GENOA135% = 55mq
SPI = 116mq
RAKE = 1°
QUARTIERE CROCETTE = 20°
RMC (SCARICO E ASCIUTTO) = 170kg*m

RMC (PIENO CARICO) RM 30° (PIENO CARICO) RM 40° (PIENO CARICO) RM MAX (PIENO CARICO)	= I DATI FINALI VERRANNO COMUNICATI AL COSTRUTTORE DELL'ALBERO

N.B. L'ALBERO DEVE ESSERE CALCOLATO SENZA L'USO DELLE VOLANTI

LA DIMENSIONE E LA POSIZIONE DELLE CROCETTE E' INDICATIVA E DEVE SERVIRE SOLO DA GUIDA AL COSTRUTTORE PER LA SCELTA DEL CORRETTO DIMENSIONAMENTO. LA LUNGHEZZA DELLE CROCETTE E' INDICATA NEL PIANO TRASVERSALE

LE DIMENSIONI DELLE CATENE DELLA RANDA SONO INDICATIVE E DEVONO SERVIRE SOLO DA GUIDA AL VELAIO PER IL CORRETTO DIMENSIONAMENTO DELLA RANDA.

LE QUOTE INDICATE CON (*) DEVONO RITENERSI PROVVISORIE E DOVRANNO ESSERE VERIFICATE AL VERO DAL COSTRUTTORE DELL'ALBERO.

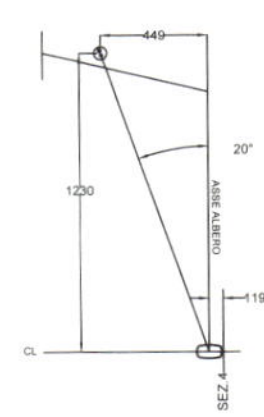

General arrangement with sections.
Keel construction details and sections.

Lines plan (stations, waterlines, buttocks).

Construction at the Comar yard in Fiumicino.

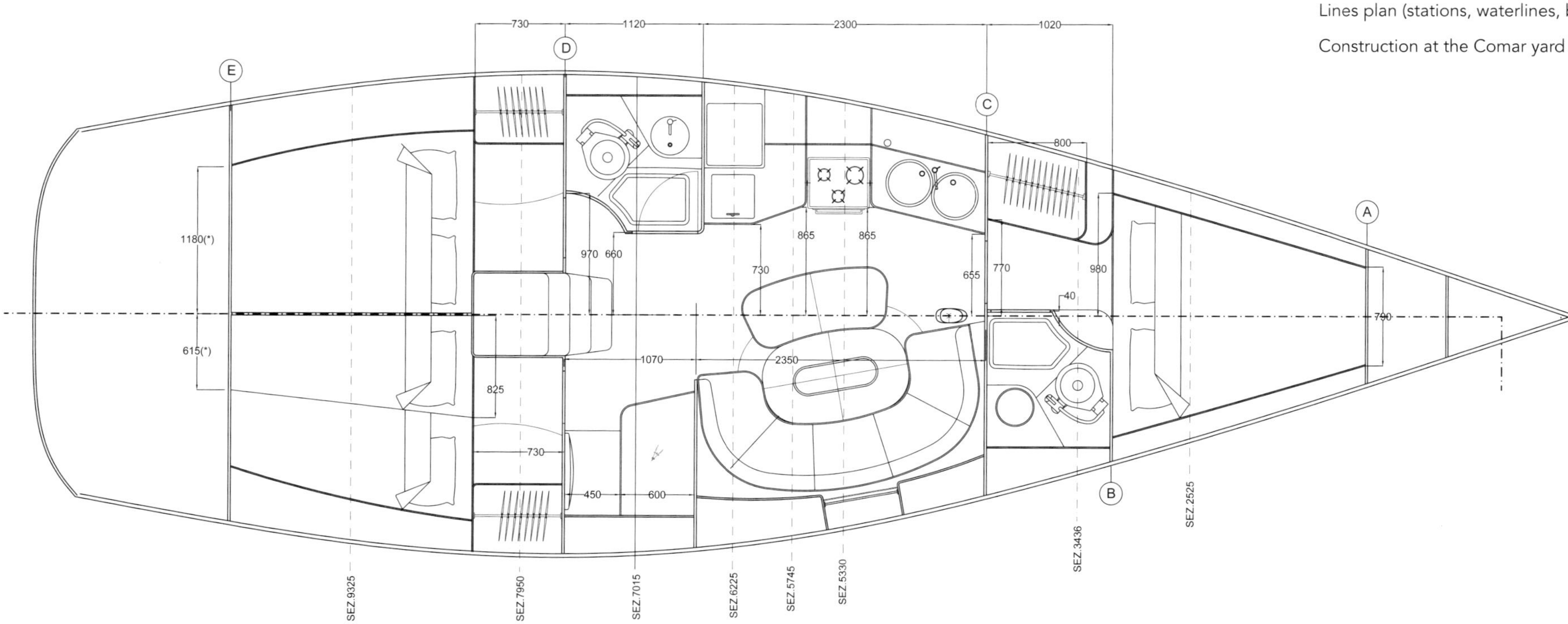

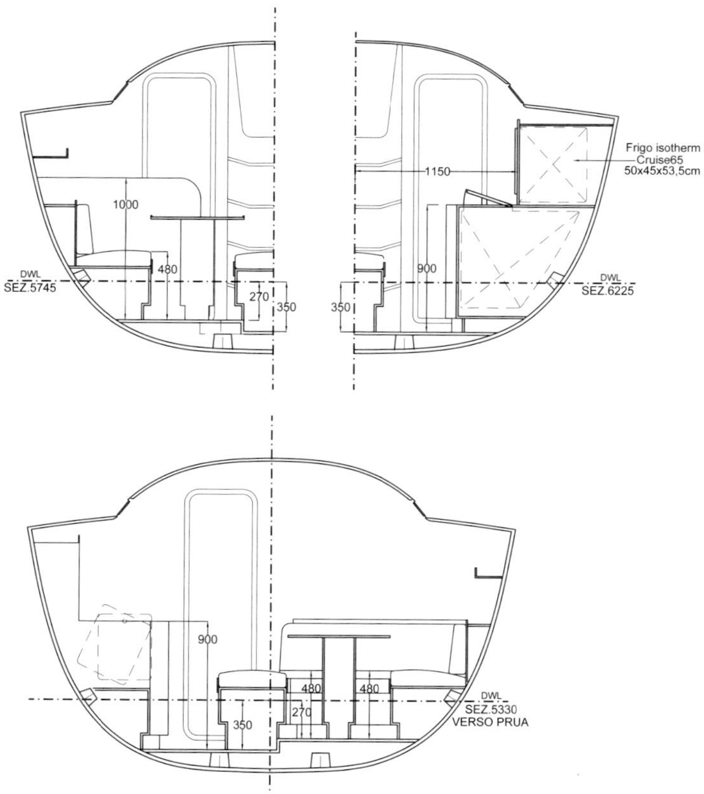

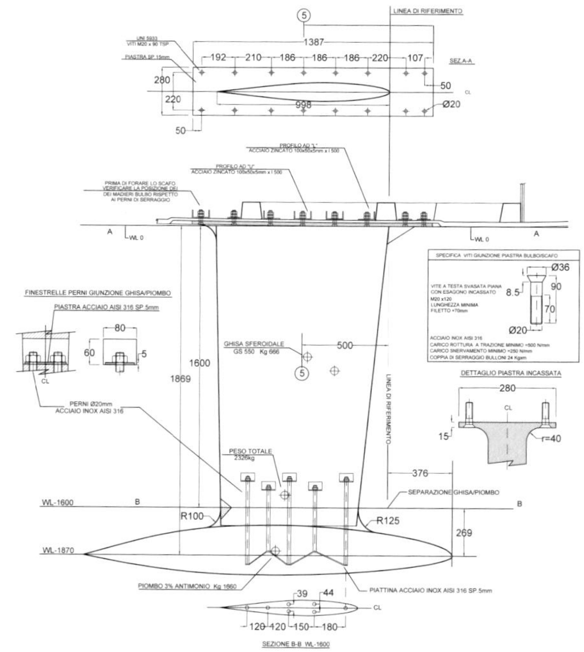

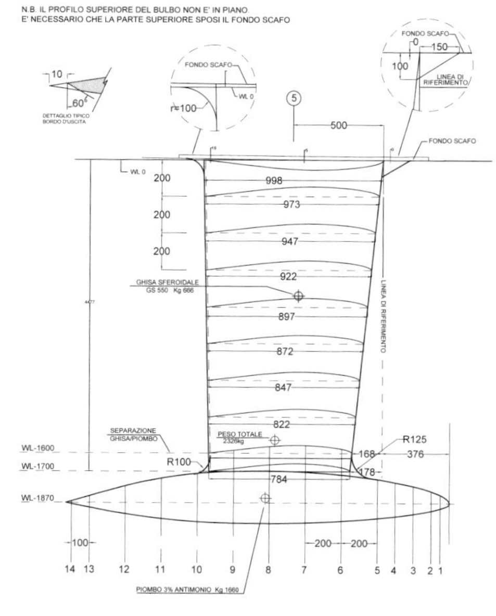

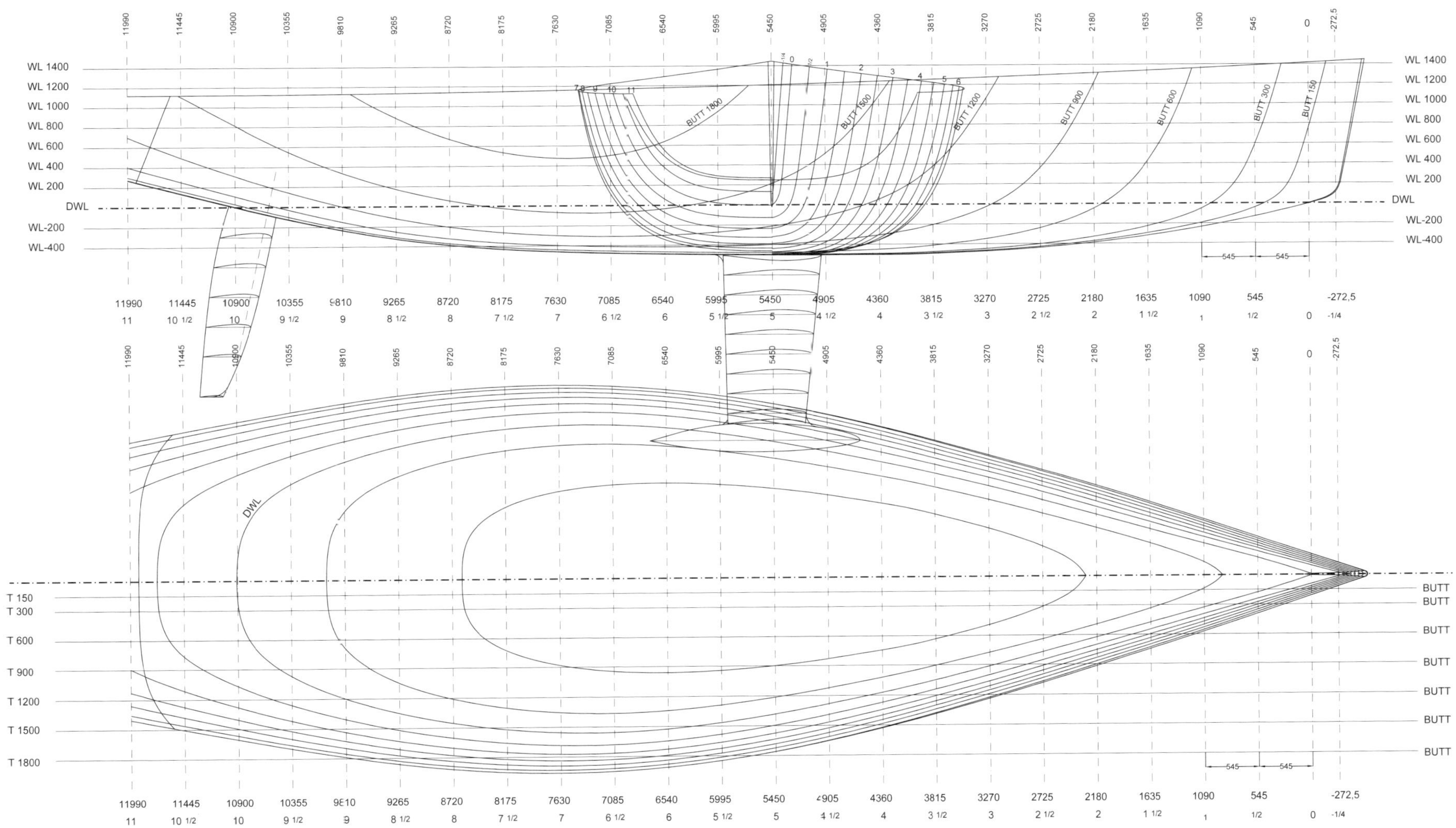
WL 1400
WL 1200
WL 1000
WL 800
WL 600
WL 400
WL 200
DWL
WL-200
WL-400
BUTT 1800
BUTT 1500
BUTT 1200
BUTT 900
BUTT 600
BUTT 300
BUTT 150
11990 11445 10900 10355 9810 9265 8720 8175 7630 7085 6540 5995 5450 4905 4360 3815 3270 2725 2180 1635 1090 545 0 -272,5
11 10 1/2 10 9 1/2 9 8 1/2 8 7 1/2 7 6 1/2 6 5 1/2 5 4 1/2 4 3 1/2 3 2 1/2 2 1 1/2 1 1/2 0 -1/4
DWL
T 150
T 300
T 600
T 900
T 1200
T 1500
T 1800
BUTT
545
545

Comet 52 RS

This was the first of the series production models set up to establish a family line of comfortable, elegant, but also fast cruisers. With this in mind, a raised saloon deckhouse was designed for the Comet 52 RS, drawn to assure both a large and bright interior and good looks. In the general architectural layout, a peculiarity of the boat takes the form of a special space located between the cockpit and the actual interior, from which it was possible to enjoy the sailing while staying sheltered. The hull design was studied using a comparative analysis process of numerous configurations, with the help of computerized simulation techniques similar to those used by us for racing boats. Among the various hypotheses examined, the one that was chosen assured good speed associated with low pitch inertia: a solution that aimed at greater comfort under sail. The stylistic research was oriented instead towards creating strongly expressive features, such as the shape of the framing outlining the large porthole and the large forward-facing window of the deckhouse. The boat received the European Boat of the Year 2008 award at the Düsseldorf boat show.

Year 2005
Builder Comar Fiumicino (RM)
Construction Composite
LOA 15.3 m
B MAX 4.6 m
D 3.1 m
DSPL 13 t
Mainsail 75 mq
Headsail 55 mq
Spi 170 mq

Comet 52 close hauled.

Sail plan.

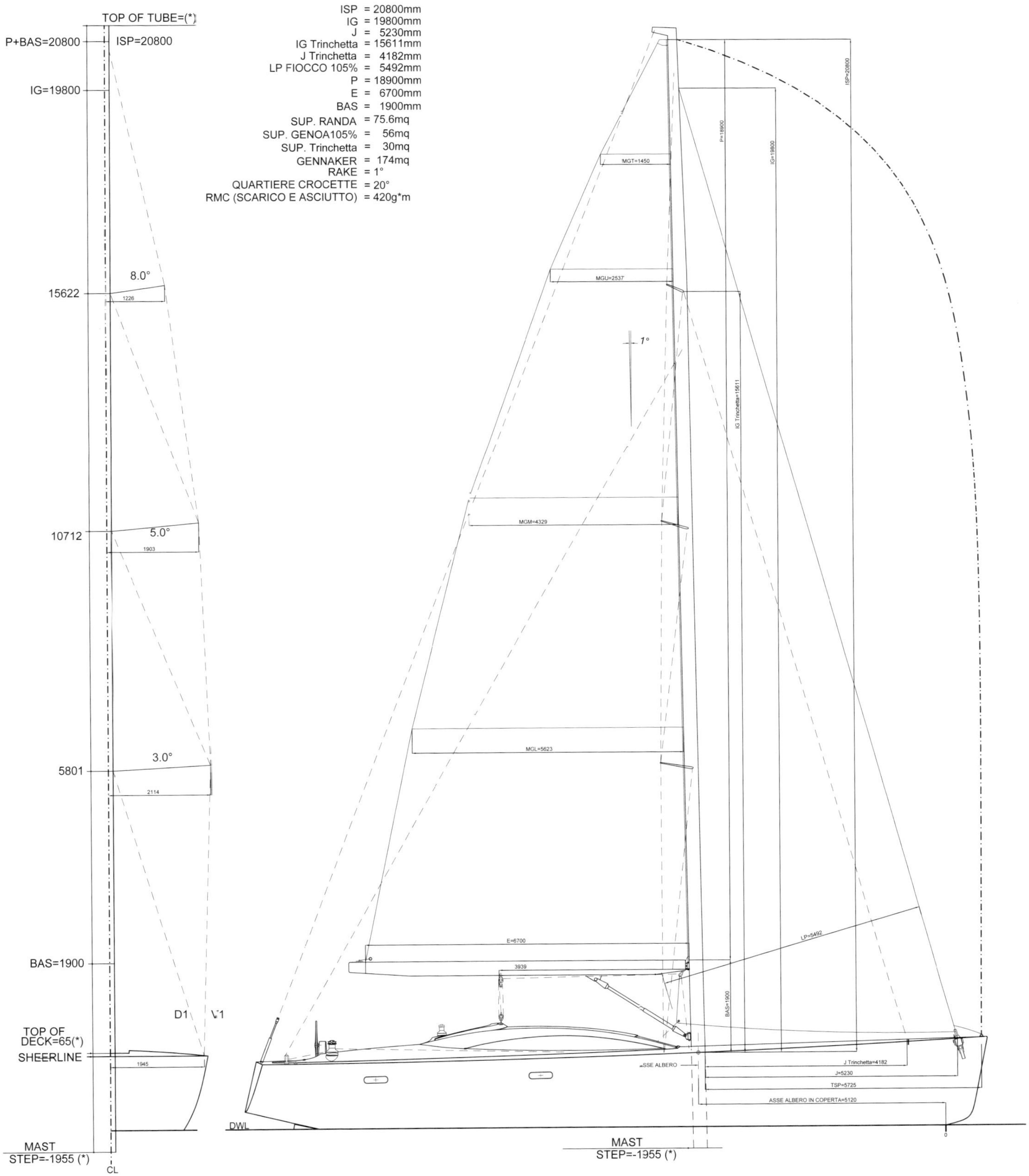
ISP = 20800mm
IG = 19800mm
J = 5230mm
IG Trinchetta = 15611mm
J Trinchetta = 4182mm
LP FIOCCO 105% = 5492mm
P = 18900mm
E = 6700mm
BAS = 1900mm
SUP. RANDA = 75.6mq
SUP. GENOA105% = 56mq
SUP. Trinchetta = 30mq
GENNAKER = 174mq
RAKE = 1°
QUARTIERE CROCETTE = 20°
RMC (SCARICO E ASCIUTTO) = 420g*m
TOP OF TUBE=(*)
P+BAS=20800
ISP=20800
IG=19800
8.0°
15622
5.0°
10712
3.0°
5801
BAS=1900
D1
TOP OF DECK=65(*)
SHEERLINE
MAST STEP=-1955 (*)
CL
1°
DWL
MAST STEP=-1955 (*)

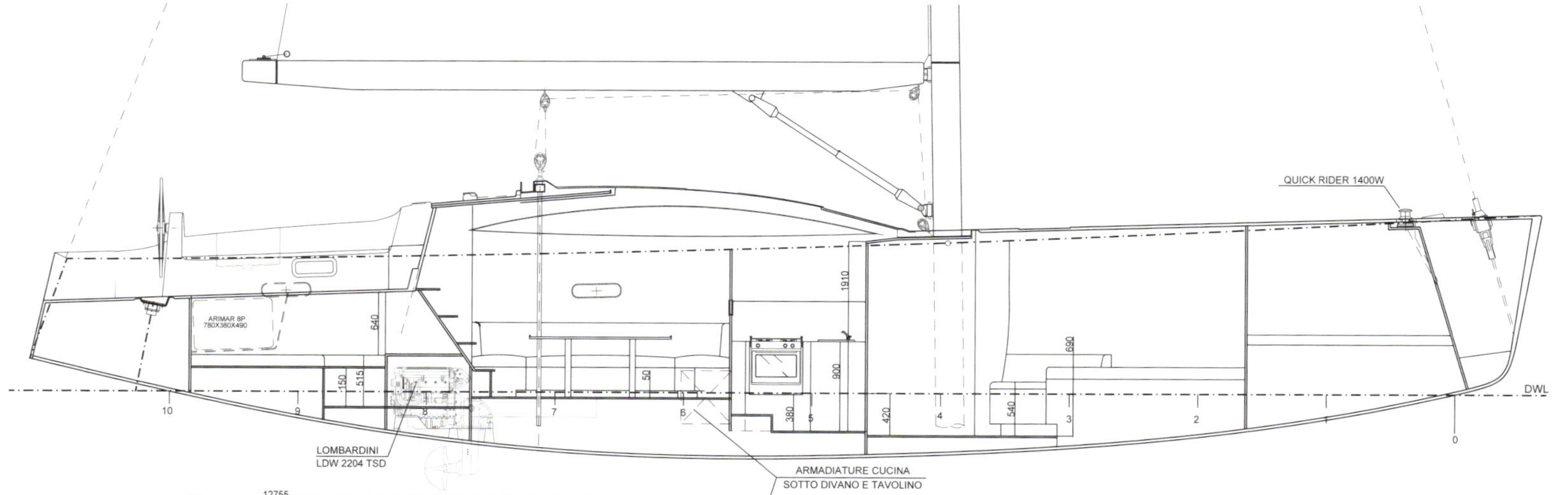
QUICK RIDER 1400W
ARIMAR 8P
780X380X490
LOMBARDINI
LDW 2204 TSD
ARMADIATURE CUCINA
SOTTO DIVANO E TAVOLINO
DWL

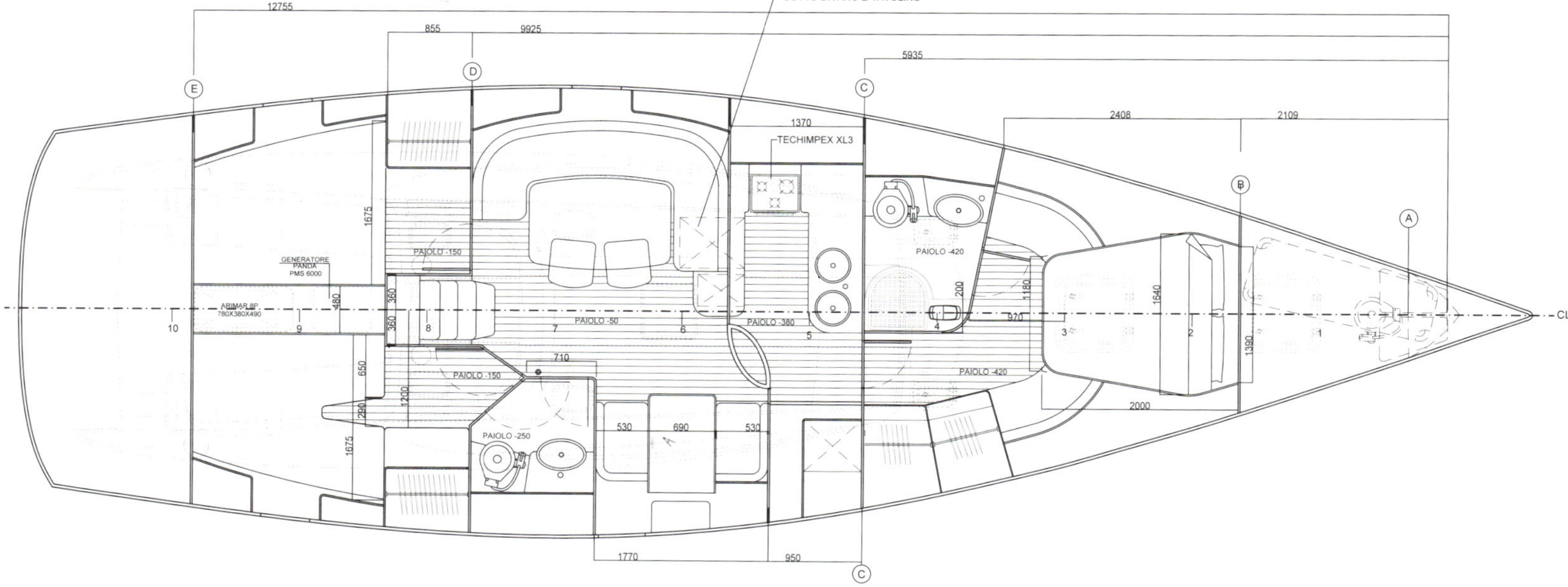
TECHIMPEX XL3
GENERATORE
PANDA
PMS 6000
ARIMAR 8P
780X380X490
PAIOLO -150
PAIOLO -50
PAIOLO -380
PAIOLO -420
PAIOLO -250
CL

General arrangement.

Galley and saloon, navigation station, owner's cabin.

Details of the keel bolts and cast iron and lead bulbs.

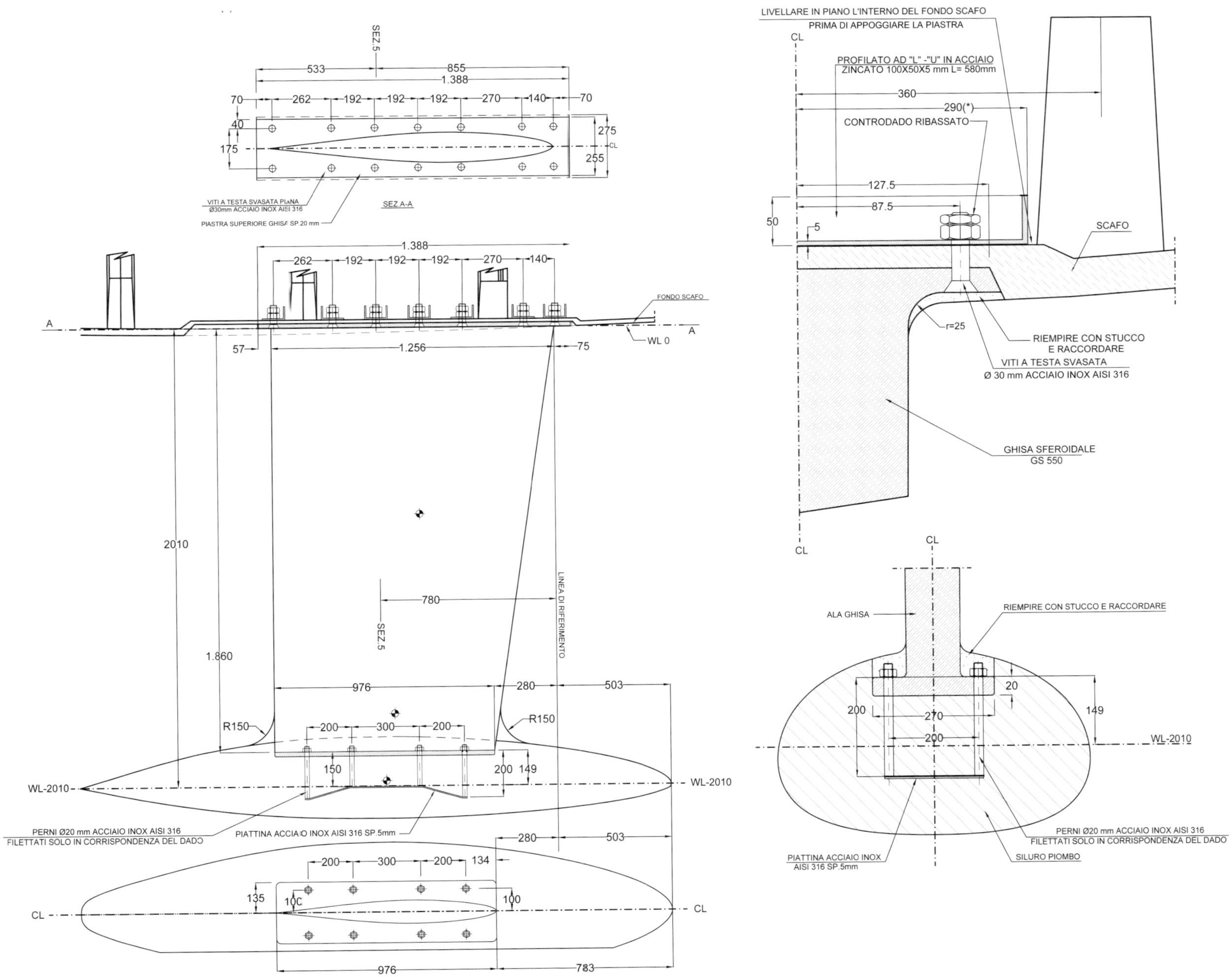

MOTORYACHTS

Studio Vallicelli began motoryacht sector operations indirectly in the 1990s by taking on design work (mainly engineering) for small military units on behalf of Montedison's Intermarine shipyard. The Studio made its official entry into the motoryacht sector via the French group Rodriguez, which at that time owned the ISA shipyard in Ancona. From 2003 onwards, the Studio designed the ISA 120 and ISA 140 models, two mass-produced molded fiberglass yachts offering rather innovative functional and aesthetic solutions. These models stood out for their organic architectural forms, expressing harmony between hull and superstructure quite rare for this type of boat. The Studio went on to design several megayachts in steel and aluminum: *Kolaha*, *Liberty*, *Aziza* and the award-winning *Okto*. All of these designs are characterized by highly sculptured superstructures and hulls with generously-proportioned overhangs that show stylistic affinities with sailing boats.

ISA 120

This was the first motor yacht that we designed for series production of fiberglass boats in a mold. It was characterized by a 36-meter planing hull, capable of exceeding 30 knots, and by a general two deck layout featuring five cabins, including the owner's cabin, a saloon, and crew quarters located in the after area of the lower deck. From this area, which also includes the galley, there was direct access to the tender garage and the engine room. The level of the bridge was set higher than the main deck. The style of the boat was characterized by a slender profile with essential features and by solutions regarding the creation of volumes that gave it an almost sculptural expression. The deckhouse and the bridge were characterized by bold lateral framing running longitudinally, giving a strong dynamic tension to the whole. The two external stairs, connecting the beach and the sun deck, were visible and acquired an architectural value strengthened by the two lateral ribs connecting the frame beams with the bridge.
The bulwark was deliberately left without handrails and pulpits, yet integrated with the yacht's topsides to give a more classic and more seagoing balance to the relationship between the masses of the hull and those of the superstructure. The main saloon had a continuous lateral window, ending with two semicircular sliding doors, which, when opened, provided a living area integrated with the cockpit.
The beach area, created for water games and bathing, was easily accessible from the cockpit and the sun deck by means of continuous, shallow steps.

Year 2002
Builder ISA Yachts Ancona
Construction GRP
LOA 36.45 m
B MAX 7.4 m
GT 173 t
Engine 3 X MTU 16V 2000 M93

Construction of the wooden model for the fiberglass superstructure mold.

ISA 120 at speed.

Motoryacht profile and superstructure deck plan.

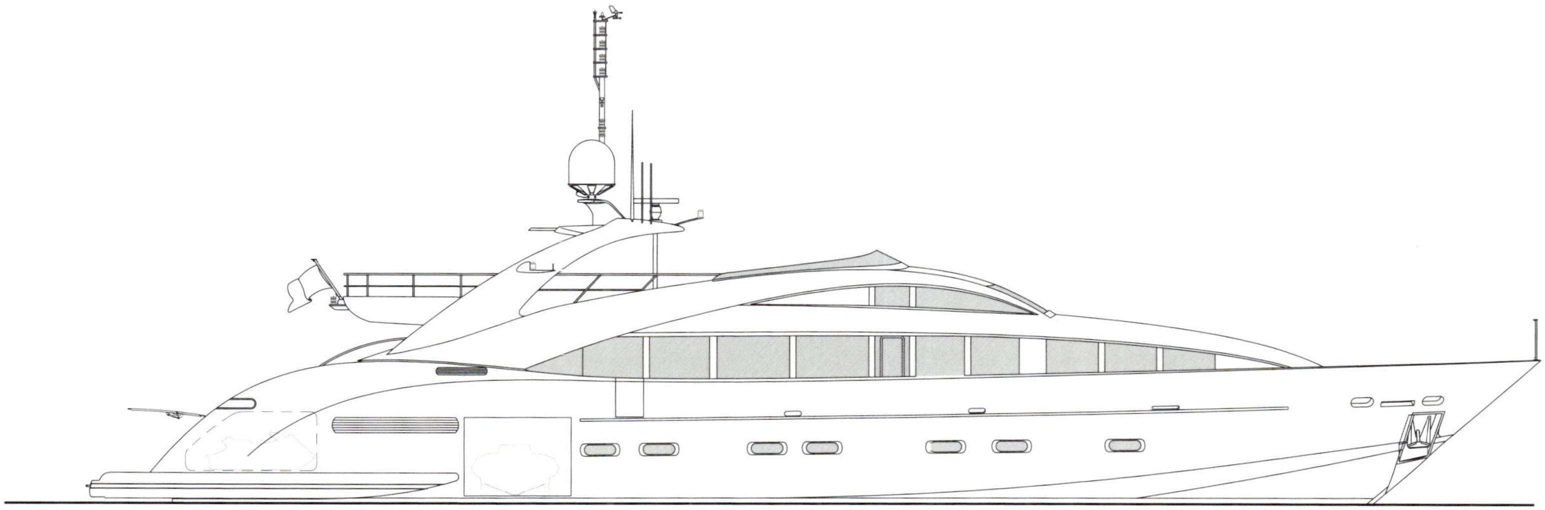

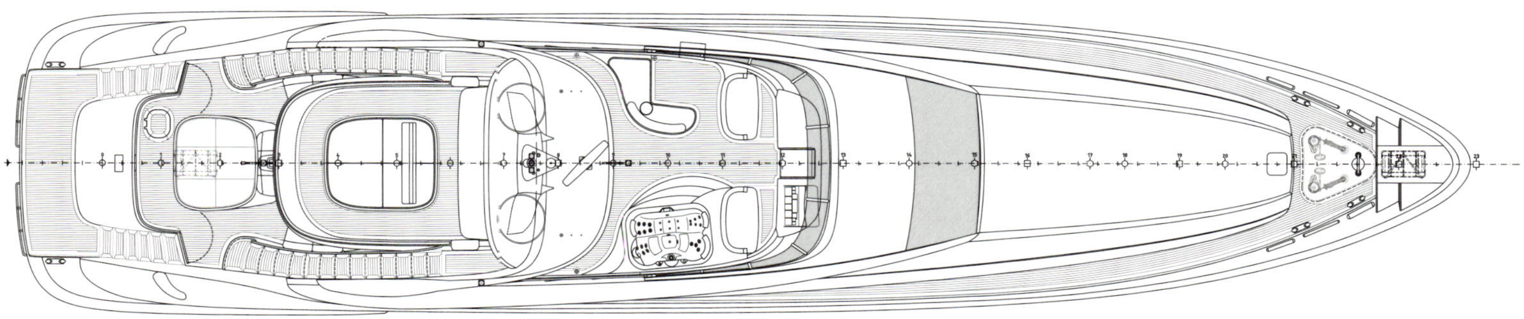

Silver Wind

This was our second project for a fast production fiberglass motoryacht and was born as a natural evolution of the ISA 120, both in terms of size and stylistic references, reinforcing a line that gained considerable interest from the market and international awards. In 2014 Silver Wind was the first vessel to emerge from the molds that had been unused for eight years. The profile of the boat was made slender not only by the proportions of the hull and the superstructure but also by the use of a continuous bulwark without stanchions, which reduced the visual impact of the deckhouse above the boat's topsides. The design was characterized by very expressive volumetric forms: the deckhouse volumes were marked by lateral framing running longitudinally, giving the whole a very accentuated dynamic appearance. The deck windows were continuous from bow to stern and ended with two sliding doors, which, when open, extended the main saloon into the cockpit, creating a very large covered living area. The main external deck, like the upper one, ran from bow to stern with wide, level areas. The interiors were on two decks and offered four guest cabins, the owner's suite, a saloon, as well as the crew's quarters, technical rooms, and bridge.

Year 2006-2014
Builder ISA Yachts Ancona
Construction GRP
LOA 43.66 m
B MAX 8.3 m
GT 230 t
Engine 2 X MTU 16V4000 M93L

Construction of the fiberglass superstructure.

Silver Wind at speed.

Motoryacht profile and deck plan.

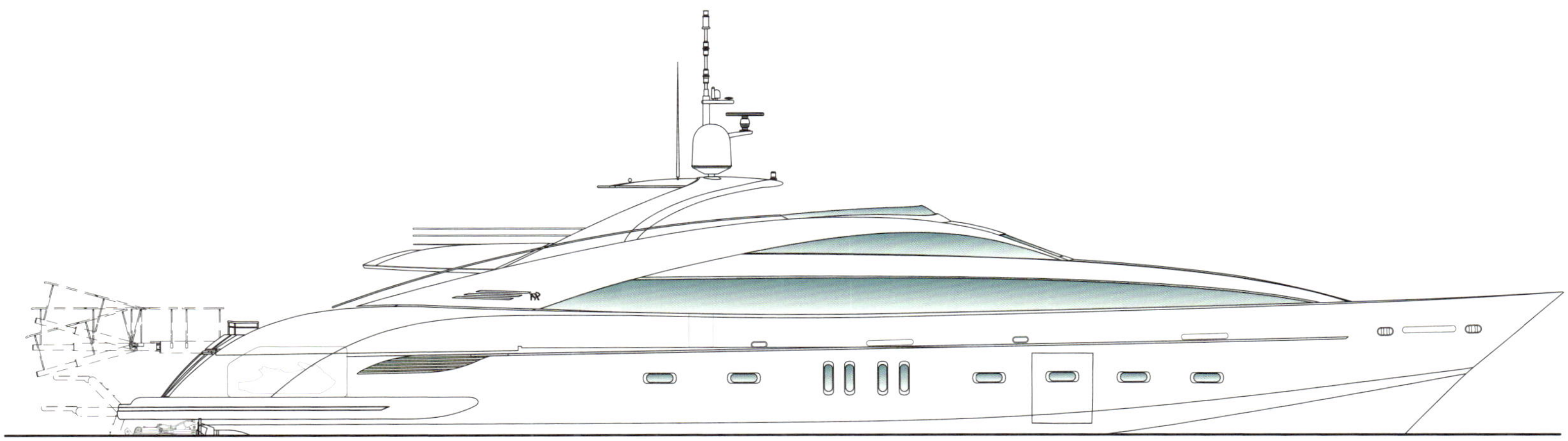

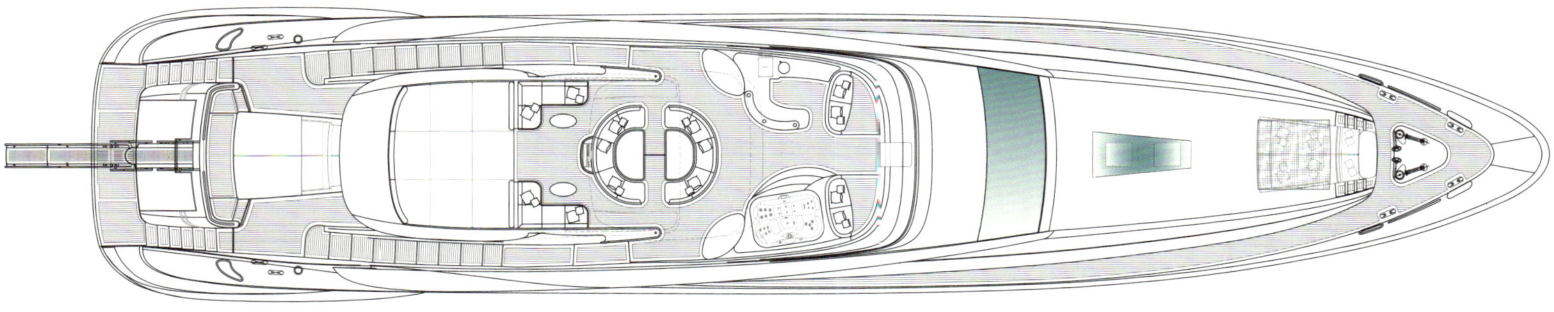

Kolaha

Kolaha was a yacht with a 63-meter displacement hull we designed with essential stylistic features, characterized by very soft lines. She was built with three interior decks, a steel hull, and a light aluminum alloy superstructure. The interiors offered five guest cabins, including the owner's cabin, two saloons located at different levels, the crew's quarters, and technical rooms. The steep angle of the transom, reminiscent of a racing sailboat, and the raked stem, gave the hull volumes a sweeping line. The sheer line traced an inverted curve in the bow area and gently descended towards the stern, giving the whole profile a dynamic appearance, underscored by the elongated lozenge shape of the large full-beam window. The shape of the superstructure was characterized by the bold volumes of the two roofs covering the deckhouse: the first located forward on the main deck, where the owner's cabin was located; the second on deck B, above the wheelhouse. The main deck boasted very wide lateral passages, while the main saloon had a continuous side window ending with two sliding doors giving onto the aft cockpit. Another peculiar element of the project concerned the creation of a large beach area. The sun deck is divided as follows: in the forward part there was a large jacuzzi with adjacent sunbathing areas; further aft there were sofas arranged in front of the bar area equipped with fridge, sinks and a dumb waiter connecting it to the main galley.

Year 2010
Builder ISA Yachts Ancona
Construction Steel – Aluminum
LOA 63.12 m
B MAX 12 m
GT 1180 t
Engine 2 X CAT 3512C

Construction at the ISA yard in Ancona.

Kolaha at cruising speed.

Motoryacht profile and deck plan.

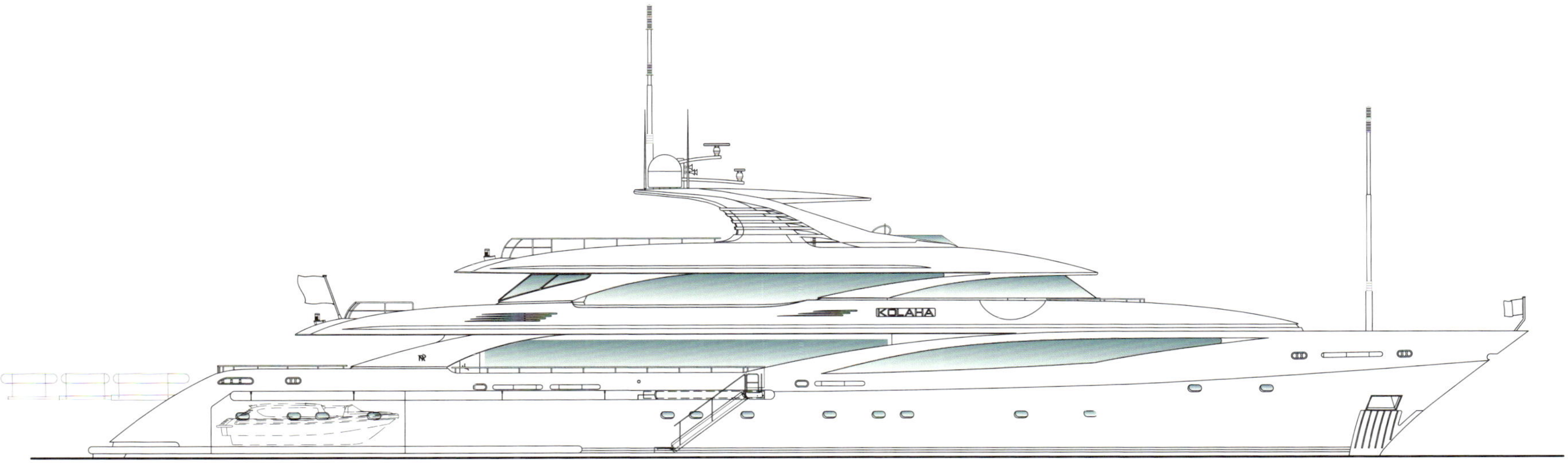

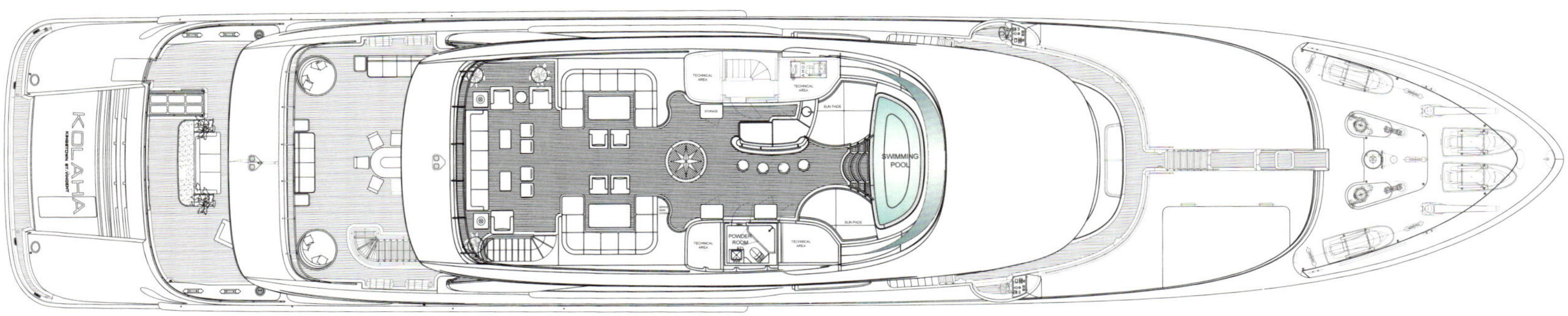

Liberty

Liberty was a 50-meter displacement motor yacht, built with a steel hull and a light alloy superstructure. The interiors, arranged on three decks, included four guest cabins, one owner's cabin, two saloons and an area for the crew. On top of this were the technical spaces, such as the bridge, engine room and tender garage. *Liberty's* style was characterized by continuous and essential lines. The hull had an accentuated bow and a sloping transom, without external stairs but equipped with a large opening hatch giving access to the beach area. Within this space, the open deck was at the same level as the beach deck, so that when the hatch was open, the area intended for water games could become very large. The transom was designed by tracing a curve that descends gently from bow to stern, giving *Liberty* a dynamic and elegant external profile. The convex shape of the gunwale running across the full beam was emphasized by the shape of the lateral windows. The deck forward of the wheelhouse was furnished with a sofa, allowing guests to enjoy this area during a cruise. The main deck had very wide lateral passageways, and the main saloon was surrounded by a continuous window ending with two sliding glass doors. These opened completely towards the cockpit, creating an extended living area between the interior and exterior. On the sun deck, there were sunbathing areas, a Jacuzzi and a dining table shaded by a rigid canopy inserted in the roll bar.

Year 2011
Builder ISA Yachts Ancona
Construction Steel – Aluminum
LOA 49.95 m
B MAX 9 m
GT 490 t
Engine 2 X MTU 12 V 4000 M71

Launch day.

Liberty at cruising speed.

Motoryacht profile and deck plan.

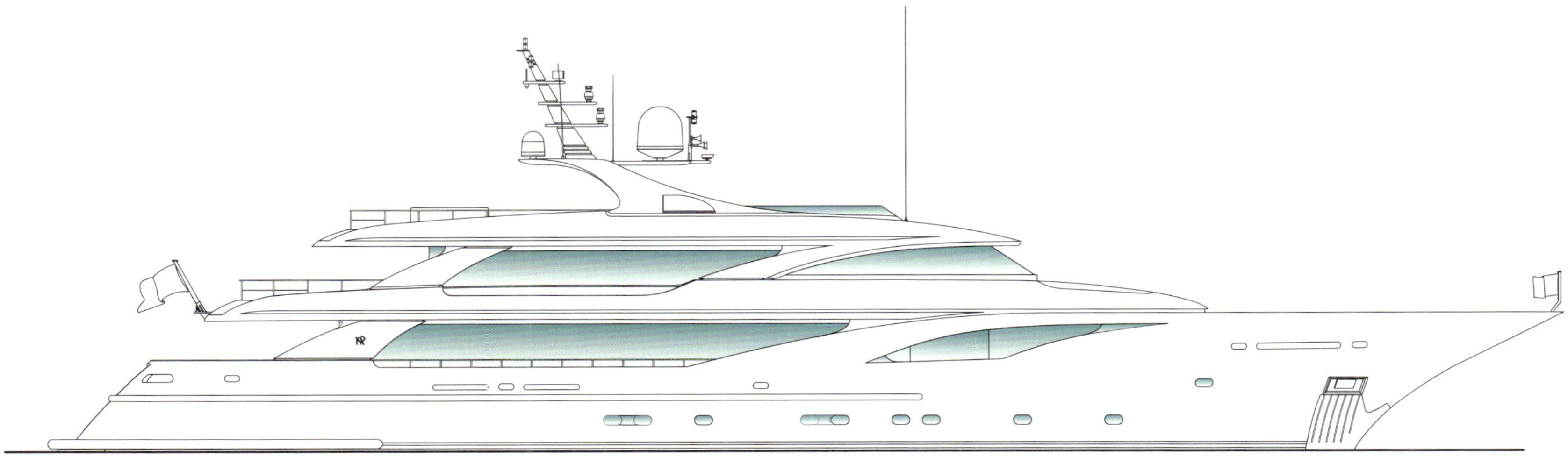

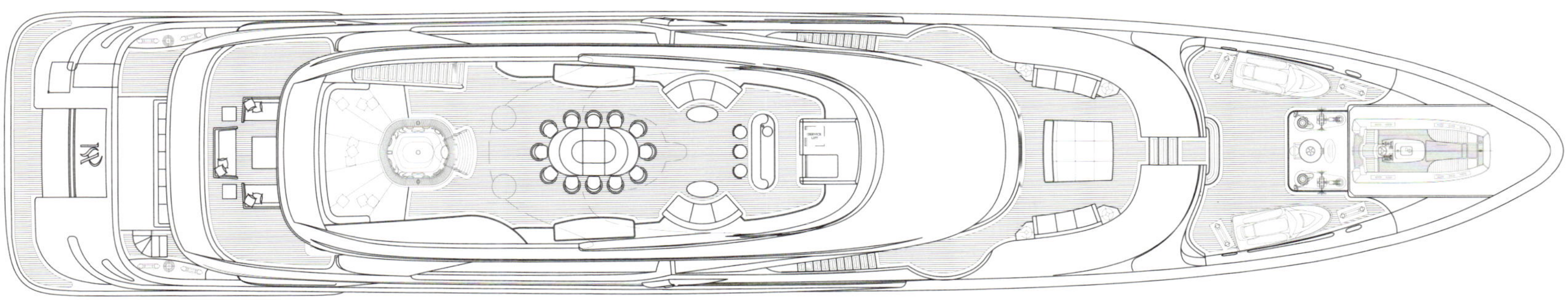

Okto

Okto was a 66-meter displacement vessel with three covered decks (plus the upper deck dedicated to the enclosed gym), built with a steel hull and light aluminum alloy superstructure. Below decks were four guest cabins, one owner's cabin, two saloons (main and upper) as well as the crew's quarters and technical rooms. Her layout was stylistically characterized by a morphological and chromatic contrast between the organic lines of the various decks and the rigid diamond geometry of the transom and the deckhouse on the sun deck. The lateral profile of the superstructure was designed around a framing element connecting the roof of the bridge with the aft handrail on the main deck, giving the whole a strongly dynamic effect and recalling some of the recurring styles in our earlier motoryacht projects. The sheer line affected an upside-down curve in the bow area and gently descended towards the stern, giving *Okto* a streamlined and dynamic appearance, even though she is a displacement yacht. This trend in the profile was underscored even more by the shape of the lateral windows in the full-beam area. The forward elevation's volume narrowed as it moved upward, marked by the very sculptural volume of the roof of the wheelhouse. The very slender hull lines included a plumb bow, maximizing waterline length and improving performance both in terms of consumption and speed, and increasing the volumes available for service rooms and technical spaces. The bow deck was completely flush and included a helicopter pad. The mooring area and garage for a tender were located beneath this deck. The main saloon featured a continuous window ending at the stern with two sliding doors that give onto an area, including a swimming pool measuring about 6 meters in length, around which descended the stairs giving external access to the beach area.

Year 2014
Builder ISA Yachts Ancona
Construction Steel – Aluminum
LOA 66.4 m
B MAX 10.5 m
DSPL 1150 t
Engine 2 X CAT 3512C

Construction at the ISA yard in Ancona.

Okto at slow speed.

Motoryacht profile and deck plan.

65M-GT-H

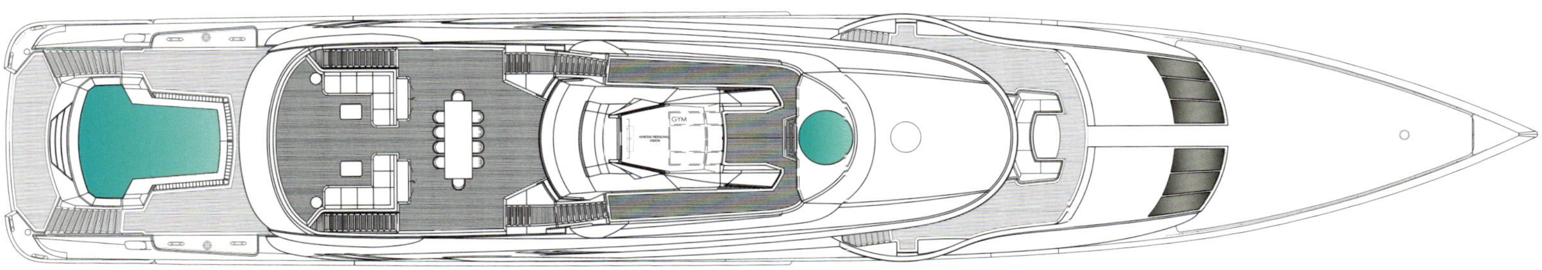
GYM

Genesia

Genesia was a refit and transformation project for a 40-meter supply vessel (construction had begun in 2005 at the Termoli Shipyard but was never completed) with the aim of creating a motoryacht which retained some of the characteristics of the original workboat. In addition to improving hull shape to optimize fuel consumption, the most important architectural modifications concerned lengthening the stem, the installation of a new wheelhouse and roll-bar, and the extension of the main-deck superstructure towards the stern. The latter modification allowed us to create a large upper-deck for the two large tenders and their davits. The naval architecture tests regarding the improved hull shape were performed at the INSEAN (the Maritime Research Institute of Rome) facilities, initially with CFD programs and then by tank testing with a 1:8 scale model. For the interior layout, our intention was to keep the main bulkheads in their original position as far as possible and to create a new layout that included four guest cabins, a full-beam owner's apartment, an upper-deck saloon with media room, and a main-deck saloon that opens aft onto a large cockpit. The interior decor was conceived to tie very sober, minimal shapes to details that harmonize with the original "metallic" soul of the supply vessel. The refit was carried out by the Cantieri Navali di Chioggia CNC, with the interiors by Tecno-Legno Yacht of Carbonera (Treviso). In 2018 *Genesia* won the Boat International Award for the best rebuild and the International Super Yacht Society award for the best refit.

Year 2005-2017
Builder Cantieri Navali di Termoli (CB) 2005 / Cantieri CNC di Chioggia (VE)
Construction Steel
LOA 40.3 m
B MAX 10.25 m
DSPL 580 t
Engine 2 X CAT C32 1.300 HP

The unfinished vessel at Termoli.
Tank testing at INSEAN in Rome.
Adding the bulbous bow.
During the refit at the CNC yard in Chioggia.

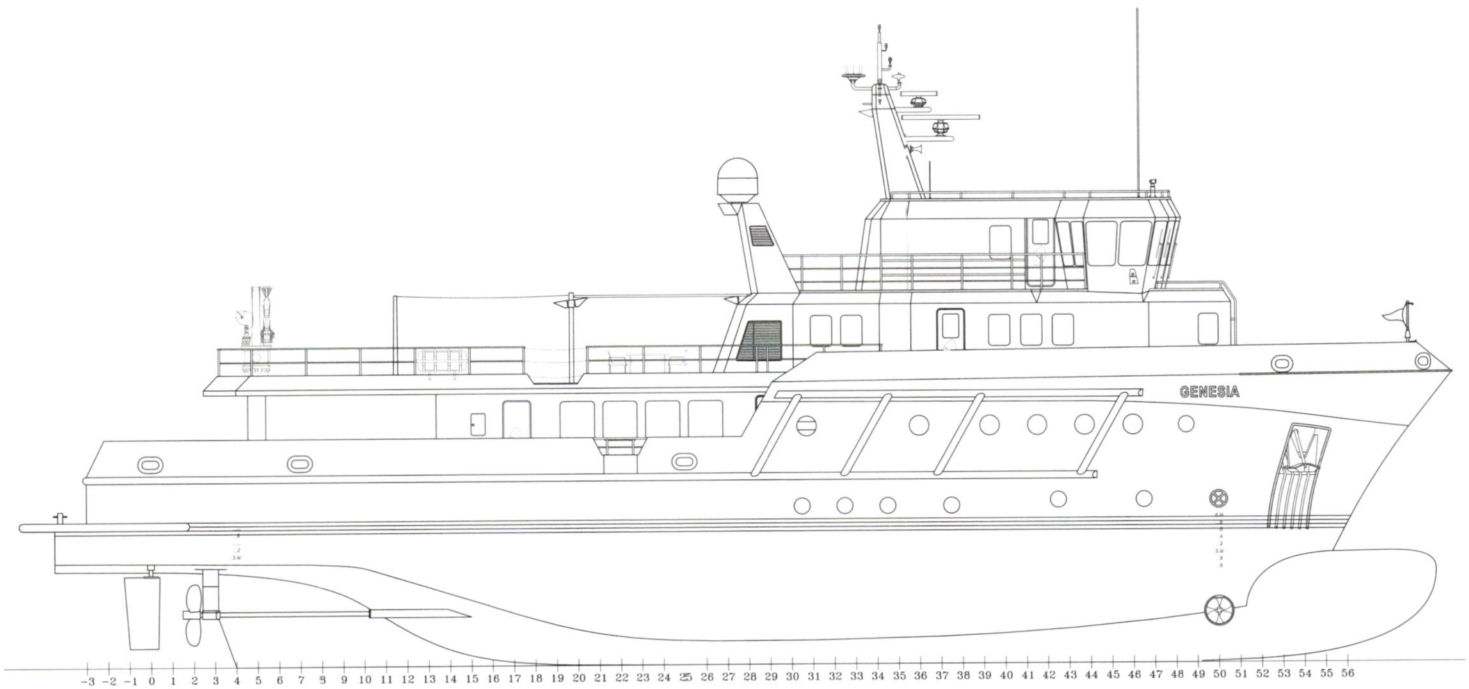

Genesia underway.

The main saloon.

Profile drawing.

Regest

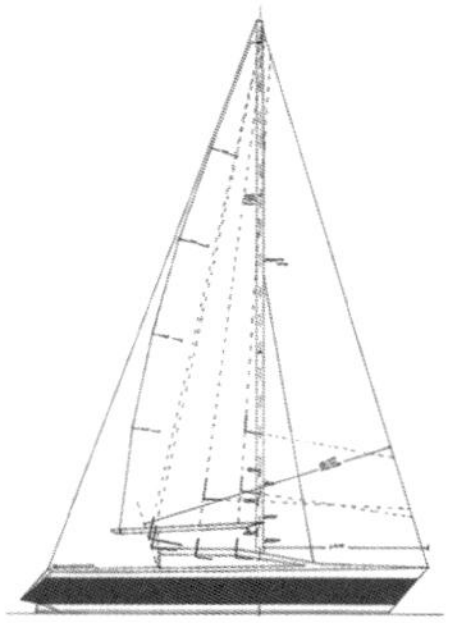

ZIGGURAT

Year 1975
Builder Alaver Verona – Tecnomar Roma
Construction GRP
LOA 9.15 m
B MAX 2.95 m
D 1.72 m
DSPL 3.10 t
Mainsail 16 mq
Headsail 30 mq
Spi 65 mq

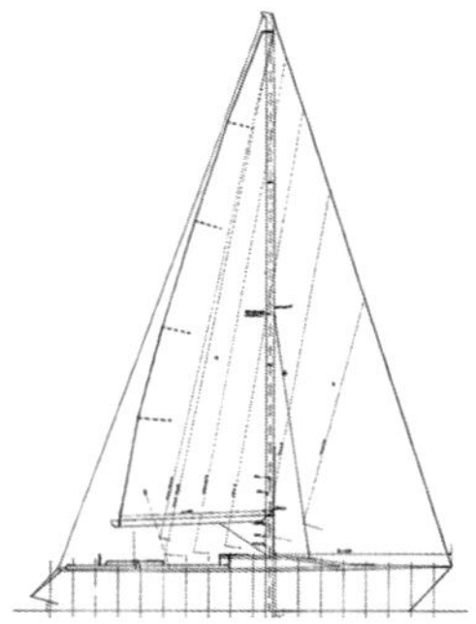

GATTONE

Year 1975
Builder Artmare Roma
Construction GRP
LOA 11.26 m
B MAX 3.50 m
D 1.90 m
DSPL 5.60 t
Mainsail 24 mq
Headsail 56 mq
Spi 120 mq

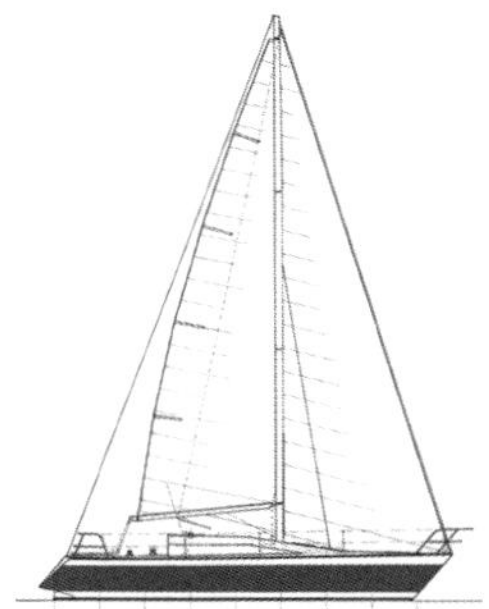

CACTUS

Year 1976
Builder Biasi Roma
Construction GRP
LOA 9.14 m
B MAX 3 m
D 1.70 m
DSPL 3.20 t
Mainsail 18
Headsail 36
Spi 78

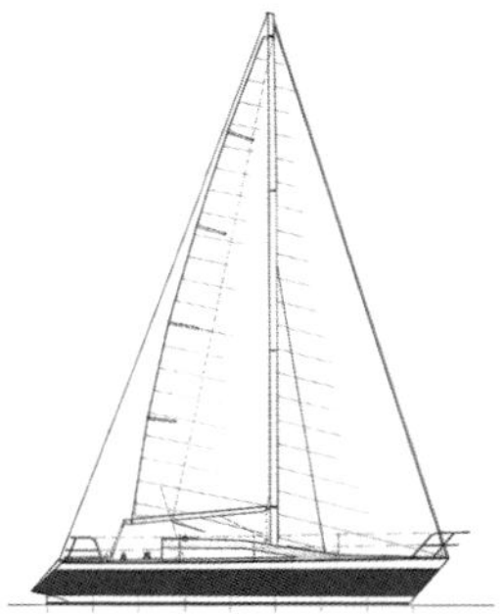

ZIGGURAT 916

Year 1976
Builder CPR Fiumicino (RM)
Construction GRP
LOA 9.16 m
B MAX 3.02 m
D 1.70 m
DSPL 3.10 t
Mainsail 18 mq
Headsail 30 mq
Spi 75

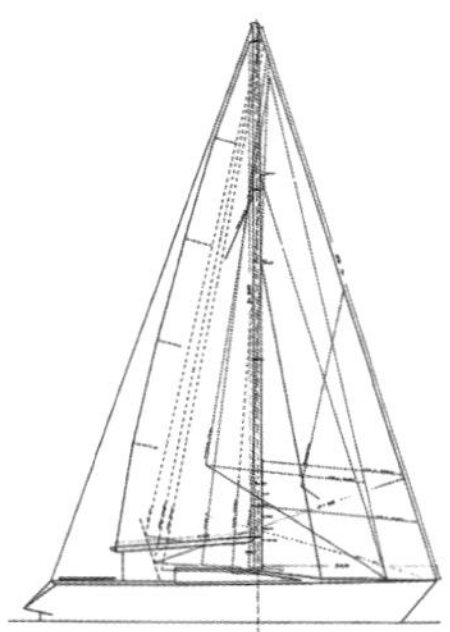

NOSTRA SIGNORA DEI TURCHI

Year 1976
Builder Biasi Roma
Construction GRP
LOA 11.08 m
B MAX 3.47 m
D 1.95 m
DSPL 6 t
Mainsail 25 mq
Headsail 44 mq
Spi 125 mq

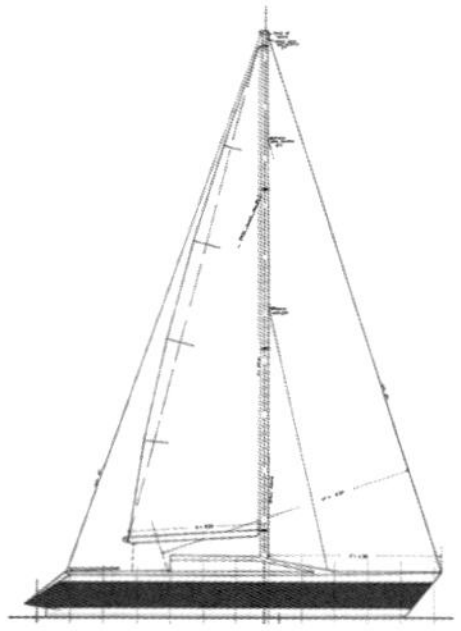

ARGENTO VIVO

Year 1976
Builder De Cesari Cervia (RA)
Construction Cold molded wood
LOA 10.15 m
B MAX 3.21 m
D 1.80 m
DSPL 3.90 t
Mainsail 20 mq
Headsail 42 mq
Spi 85 mq

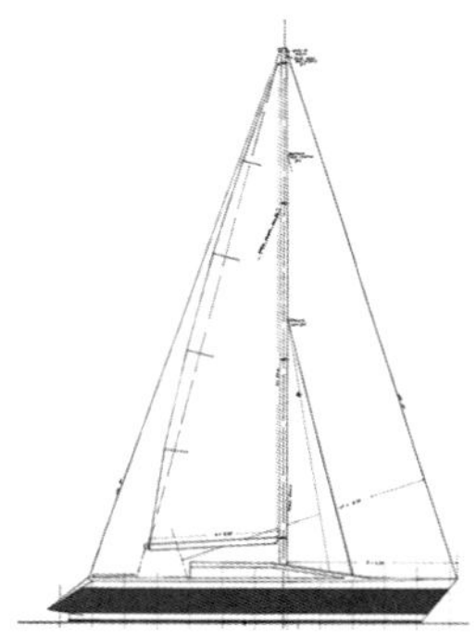

CANADOS 33

Year 1976
Builder Canados Ostia (RM)
Construction GRP
LOA 10.08 m
B MAX 3.20 m
D 1.80 m
DSPL 4.90 t
Mainsail 25 mq
Headsail 42 mq
Spi 90 mq

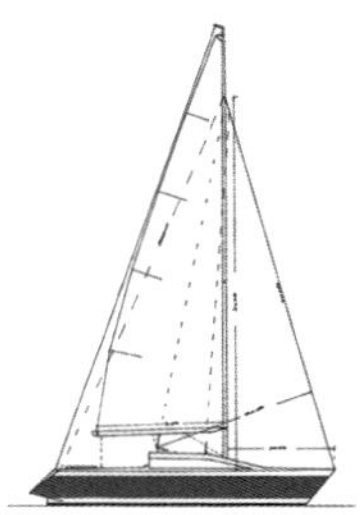

AVVENTURA 703

Year 1976
Builder Nauta S.n.c. Milano
Construction GRP
LOA 7.03 m
B MAX 2.48 m
D 1.45 m
DSPL 0.85 t
Mainsail 12 mq
Headsail 18 mq
Spi 45 mq

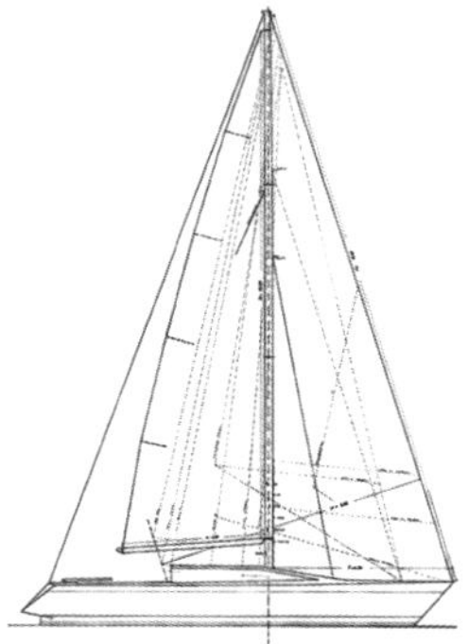

EL CID

Year 1976
Builder Crisman & Giraldi Trieste
Construction Cold molded wood
LOA 11.29 m
B MAX 3.46 m
D 1.95 m
DSPL 6.30 t
Mainsail 28 mq
Headsail 48 mq
Spi 110 mq

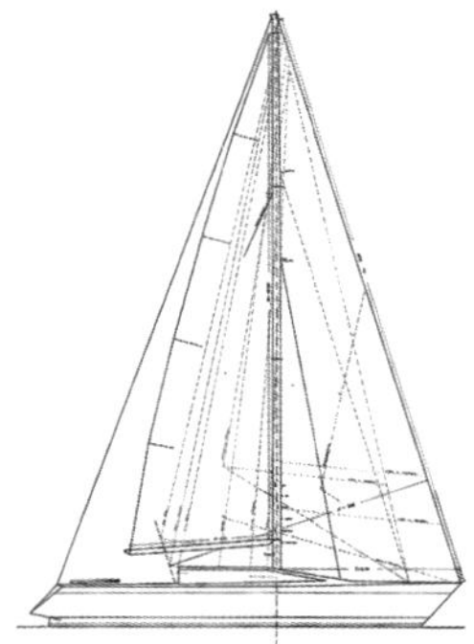

ALMIC

Year 1976
Builder Morri & Para Rimini
Construction Cold molded wood
LOA 11.20 m
B MAX 3.47 m
D 1.97 m
DSPL 6.30 t
Mainsail 28 mq
Headsail 48 mq
Spi 90 mq

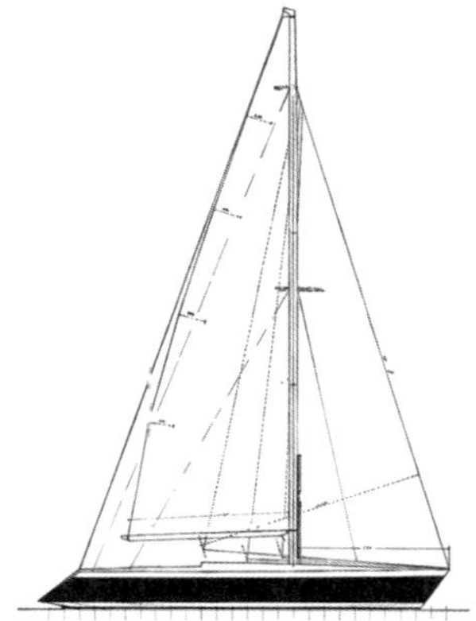

CACCIA ALLA VOLPE

Year 1977
Builder Bani Monte Argentario (GR)
Construction Marine plywood
LOA 13.40 m
B MAX 4.10 m
D 2.29 m
DSPL 6.72 t
Mainsail 48 mq
Headsail 58 mq
Spi 120 mq

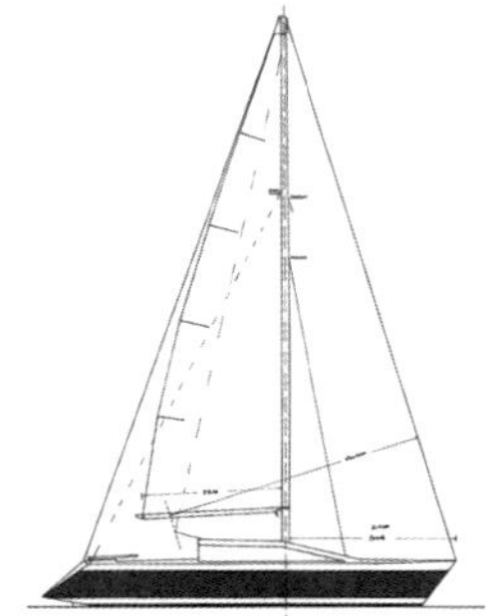

ATAHUALPA

Year 1978
Builder Bani Monte Argentario (GR)
Construction Cold molded wood
LOA 9 m
B MAX 2.92 m
D 1.75 m
DSPL 3.50 t
Mainsail 18 mq
Headsail 32 mq
Spi 75 mq

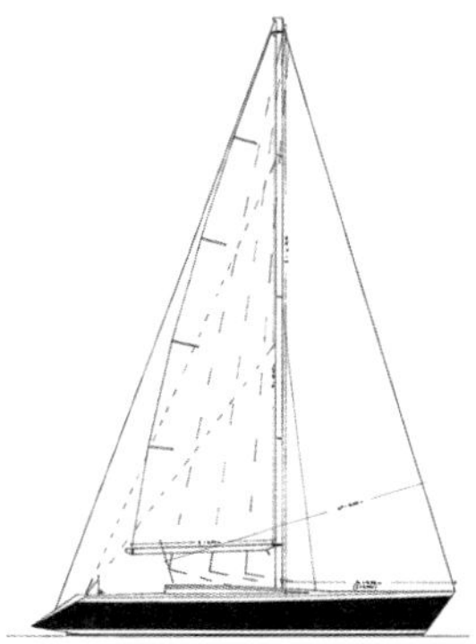

MANUA

Year 1978
Builder Petronio & Pecarich Trieste
Construction Cold molded wood
LOA 12.77 m
B MAX 3.82 m
D 2.10 m
DSPL 8 t
Mainsail 36 mq
Headsail 68 mq
Spi 135 mq

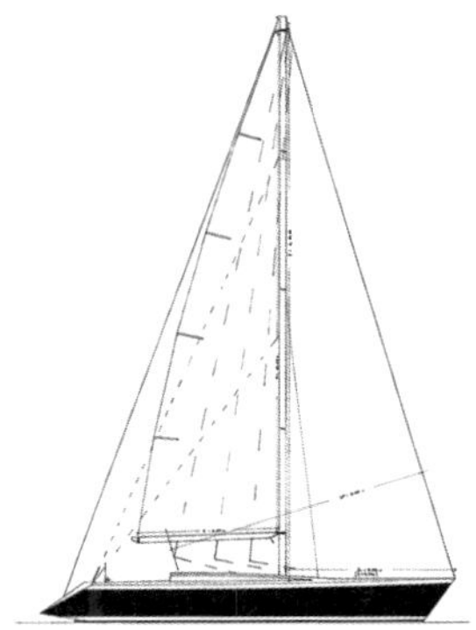

KOALA V

Year 1978
Builder De Cesari Cervia (RA)
Construction Cold molded wood
LOA 12.76 m
B MAX 3.80 m
D 2.20 m
DSPL 8 t
Mainsail 31 mq
Headsail 70 mq
Spi 141 mq

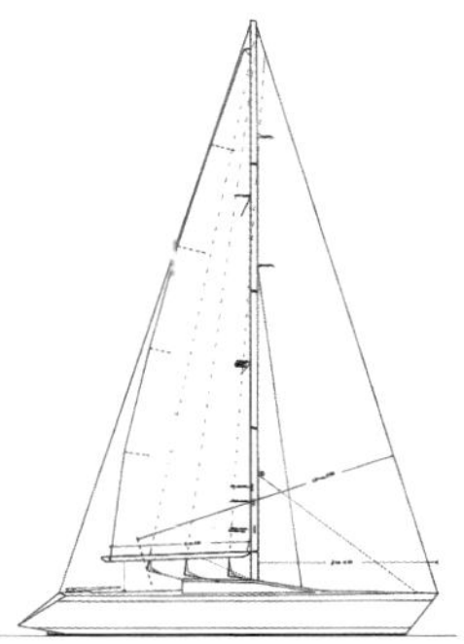

SOGNO BAGNATO

Year 1978
Builder Cant. Nav. Perotto Casarza Ligure (GE)
Construction GRP
LOA 11.01 m
B MAX 3.46 m
D 1.90 m
DSPL 5.25 t
Mainsail 28 mq
Headsail 45 mq
Spi 97 mq

BOTTA A DRITTA

Year 1978
Builder Gallinari Anzio (RM)
Construction Cold molded wood
LOA 10.14 m
B MAX 3.19 m
D 1.80 m
DSPL 4 t
Mainsail 20 mq
Headsail 43 mq
Spi 95 mq

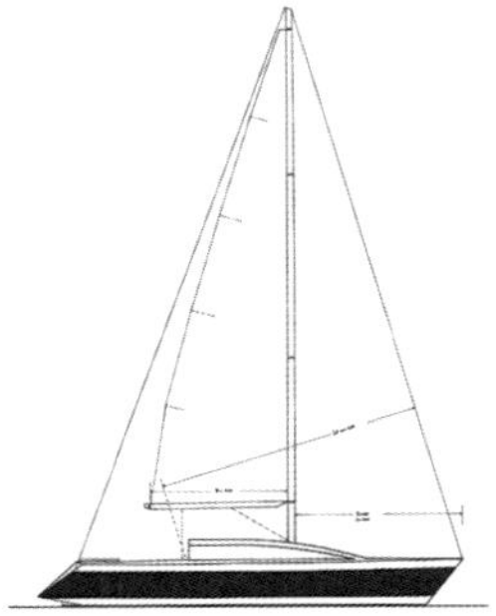

DRAGO AZZURRO

Year 1978
Builder Bani Monte Argentario (GR)
Construction Cold molded wood
LOA 9.68 m
B MAX 3.22 m
D 1.77 m
DSPL 3.70 t
Mainsail 18 mq
Headsail 38 mq
Spi 75 mq

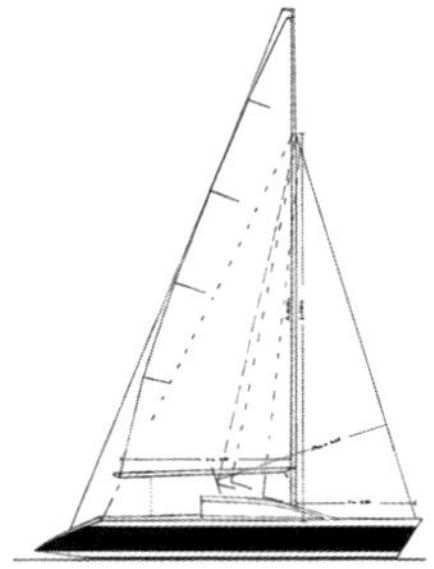

ASABRANCA

Year 1978
Builder CPR Fiumicino (RM)
Construction GRP
LOA 9.40 m
B MAX 3.20 m
D 1.80 m
DSPL 3.10 t
Mainsail 24 mq
Headsail 22 mq
Spi 48 mq

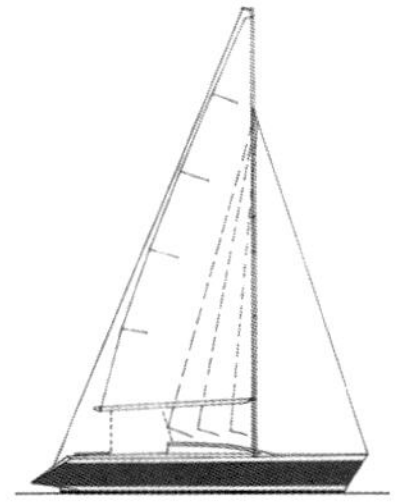

ZIGGURAT JR 8M

Year 1979
Builder CPR Fiumicino (RM)
Construction GRP
LOA 7.99 m
B MAX 2.90 m
D 1.50 m
DSPL 2.50 t
Mainsail 19 mq
Headsail 17 mq
Spi 35 mq

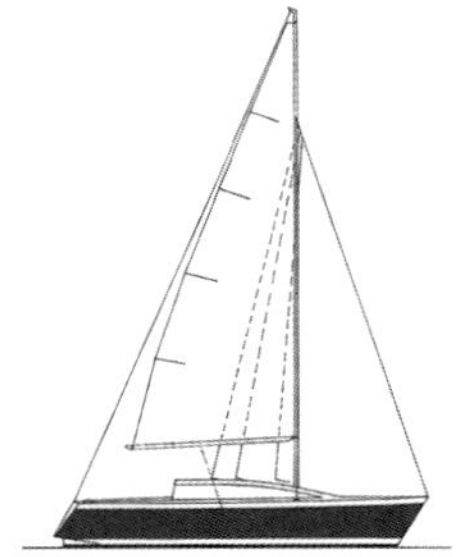

ZIGGURAT 812

Year 1979
Builder CPR Fiumicino (RM)
Construction GRP
LOA 8.13 m
B MAX 2.90 m
D 1.53 m
DSPL 3 t
Mainsail 17 mq
Headsail 16 mq
Spi 35 mq

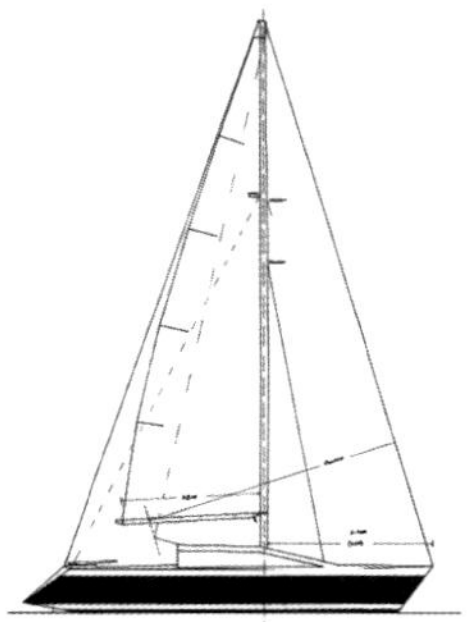

LEONELLA CELESTE

Year 1979
Builder De Cesari Cervia (RA)
Construction Cold molded wood
LOA 9.05 m
B MAX 3 m
D 1.79 m
DSPL 3 t
Mainsail 18 mq
Headsail 38 mq
Spi 75 mq

ORO FINO

Year 1979
Builder Morri & Para Rimini
Construction Cold molded wood
LOA 10.15 m
B MAX 3.34 m
D 1.93 m
DSPL 3.90 t
Mainsail 34 mq
Headsail 34 mq
Spi 68 mq

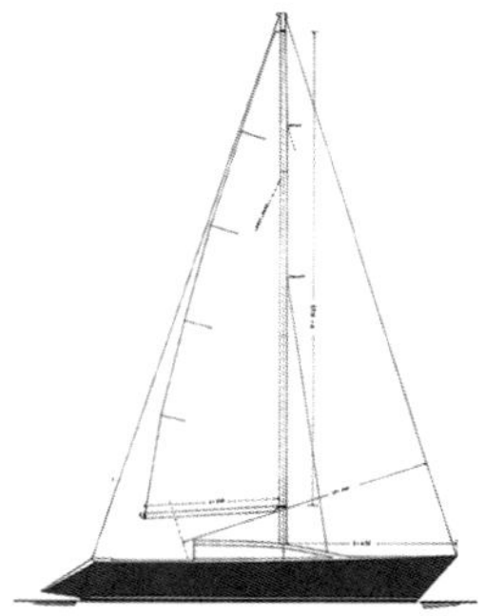

ASHANTI

Year 1979
Builder De Cesari Cervia (RA)
Construction Cold molded wood
LOA 10.30 m
B MAX 3.34 m
D 1.82 m
DSPL 3.80 t
Mainsail 22 mq
Headsail 40 mq
Spi 85 mq

NONNO GIGI

Year 1979
Builder Bani Monte Argentario (GR)
Construction Cold molded wood
LOA 12.06 m
B MAX 3.80 m
D 2.17 m
DSPL 6.50 t
Mainsail 49 mq
Headsail 51 mq
Spi 99 mq

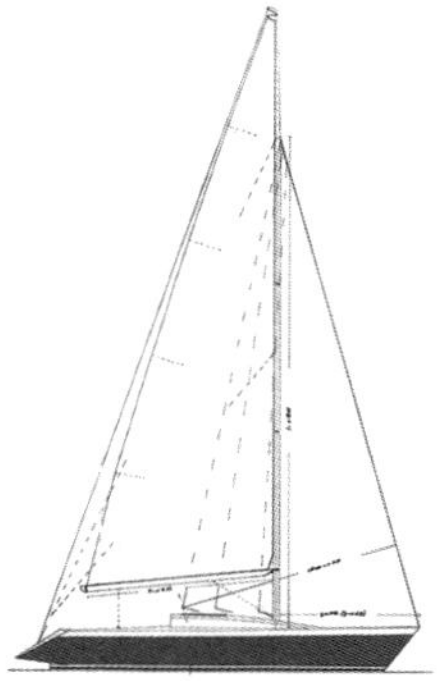

FILO DA TORCERE

Year 1979
Builder De Cesari Cervia (RA)
Construction Cold molded wood
LOA 11.29 m
B MAX 3.70 m
D 2.04 m
DSPL 5 t
Mainsail 43 mq
Headsail 43 mq
Spi 87 mq

BRAVA 13.40

Year 1979
Builder Minneford New York
Construction Aluminum
LOA 13.45 m
B MAX 4.15 m
D 2.57 m
DSPL 10 t
Mainsail 40 mq
Headsail 76 mq
Spi 153 mq

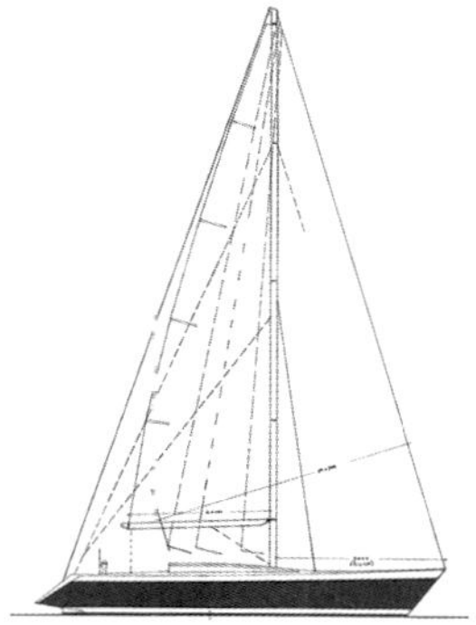

BLU SHOW

Year 1979
Builder Pezzini Viareggio (LU)
Construction Cold molded wood
LOA 12.49 m
B MAX 3.96 m
D 2.35 m
DSPL 7.70 t
Mainsail 35.50 mq
Headsail 69 mq
Spi 139 mq

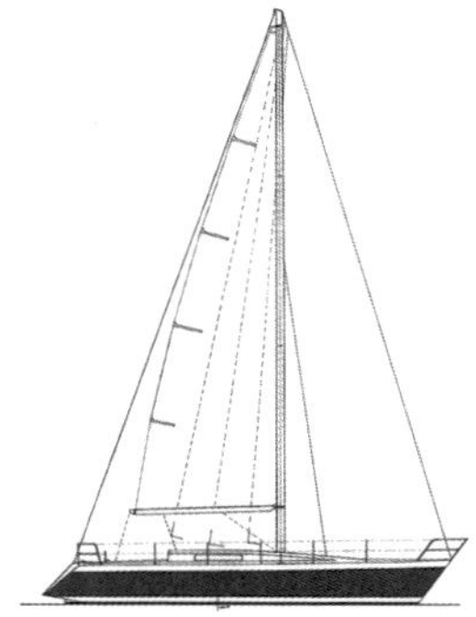

MARCHI 39

Year 1979
Builder Arredomar Campalto (VE)
Construction GRP
LOA 11.65 m
B MAX 3.62 m
D 2.02 m
DSPL 6.80 t
Mainsail 30.50 mq
Headsail 60 mq
Spi 120 mq

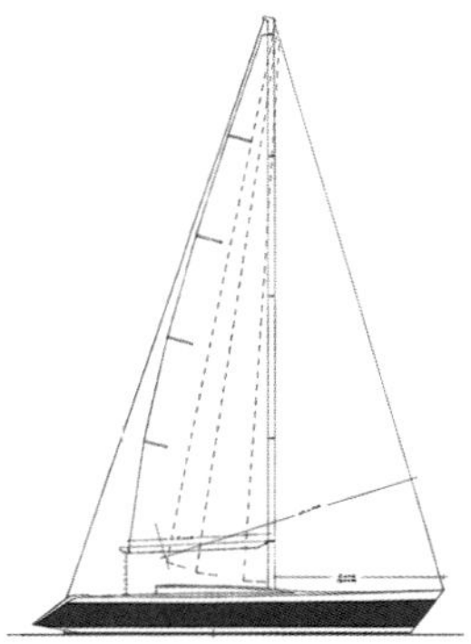

SLOOP 12.80

Year 1979
Builder Conavi Viareggio (LU)
Construction Aluminum
LOA 12.77 m
B MAX 3.88 m
D 2.19 m
DSPL 7 t
Mainsail 37 mq
Headsail 71 mq
Spi 145 mq

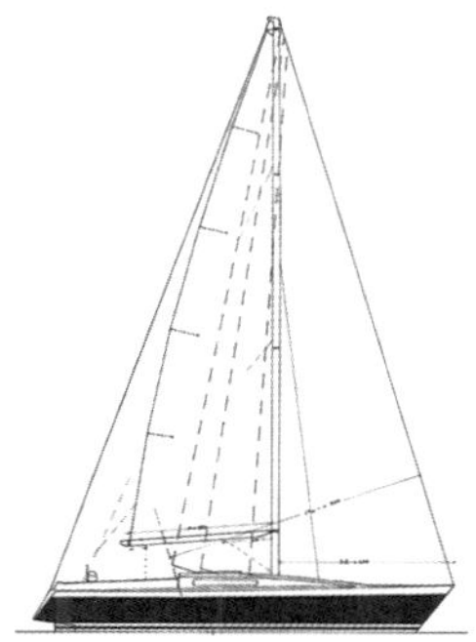

CANADOS 37

Year 1979
Builder Canados Ostia (RM)
Construction GRP
LOA 11.30 m
B MAX 3.63 m
D 1.98 m
DSPL 6.20 t
Mainsail 27 mq
Headsail 55 mq
Spi 110 mq

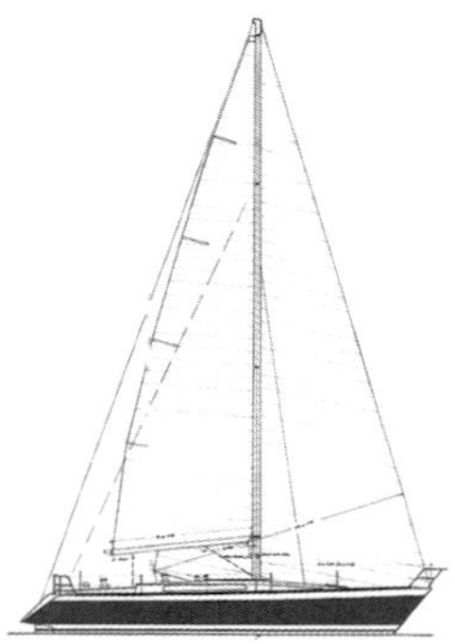

SHOW 42

Year 1979
Builder Barberis La Spezia
Construction GRP
LOA 12.65 m
B MAX 3.96 m
D 2.10 m
DSPL 8.40 t
Mainsail 36 mq
Headsail 70 mq
Spi 160 mq

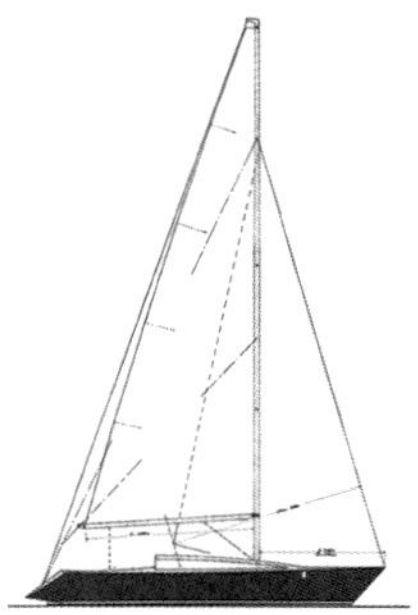

BIRBONA

Year 1979
Builder Petronio & Pecarich Trieste
Construction Cold molded wood
LOA 10.30 m
B MAX 3.39 m
D 1.89 m
DSPL 4.80 t
Mainsail 36 mq
Headsail 35 mq
Spi 70 mq

LSD OT

Year 1979
Builder Galetti Desenzano del Garda (BS)
Construction Cold molded wood
LOA 11.42 m
B MAX 3.70 m
D 2 m
DSPL 5.10 t
Mainsail 45 mq
Headsail 40 mq
Spi 85 mq

CUNEGONDA

Year 1979
Builder Bani Monte Argentario (GR)
Construction Cold molded wood
LOA 11.20 m
B MAX 3.67 m
D 1.94 m
DSPL 5 t
Mainsail 30 mq
Headsail 52 mq
Spi 107 mq

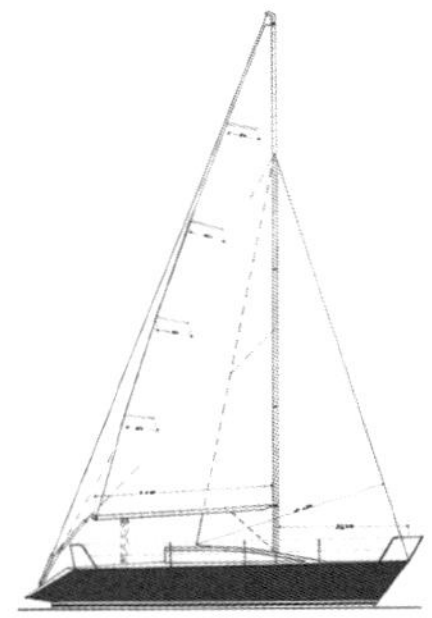

HIPAZIA

Year 1980
Builder Galetti Desenzano del Garda (BS)
Construction Cold molded wood
LOA 9.36 m
B MAX 3.12 m
D 1.89 m
DSPL 3.50 t
Mainsail 28 mq
Headsail 25 mq
Spi 50 mq

ZIGGURAT 995

Year 1980
Builder CPR Roma
Construction GRP
LOA 9.95 m
B MAX 3.30 m
D 1.70 m
DSPL 3.85 t
Mainsail 25 mq
Headsail 35 mq
Spi 55 mq

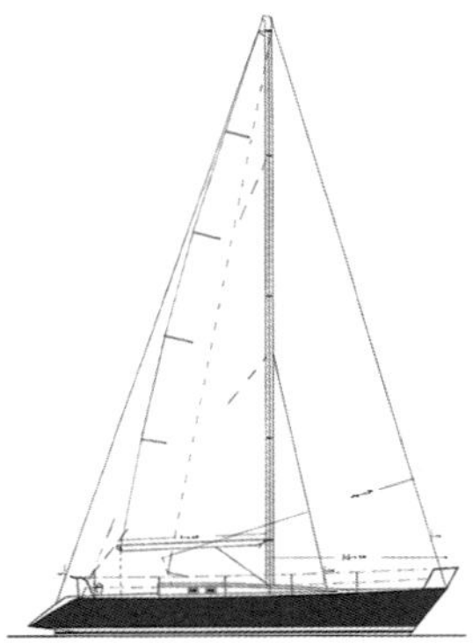

VOGLIA MATTA

Year 1980
Builder De Cesari Cervia (RA)
Construction Cold molded wood
LOA 12.63 m
B MAX 3.96 m
D 2.31 m
DSPL 8.50 t
Mainsail 35 mq
Headsail 68 mq
Spi 137 mq

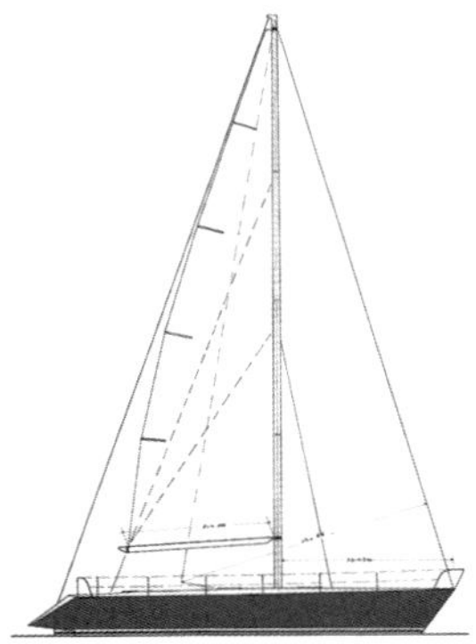

FANTASQUE

Year 1980
Builder Pouvreau La Rochelle (France)
Construction Aluminum
LOA 13.50 m
B MAX 4.13 m
D 2.55 m
DSPL 9 t
Mainsail 41 mq
Headsail 76 mq
Spi 155 mq

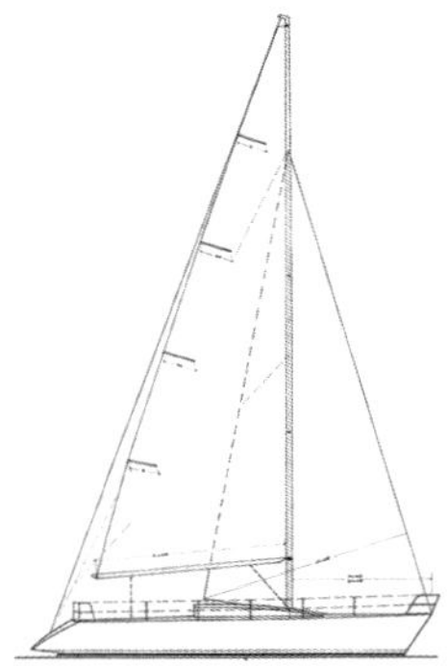

PRIMADONNA

Year 1980
Builder Mannini Sanremo (IM)
Construction Composite
LOA 12.66 m
B MAX 4.05 m
D 2.31 m
DSPL 7.10 t
Mainsail 50 mq
Headsail 53 mq
Spi 115 mq

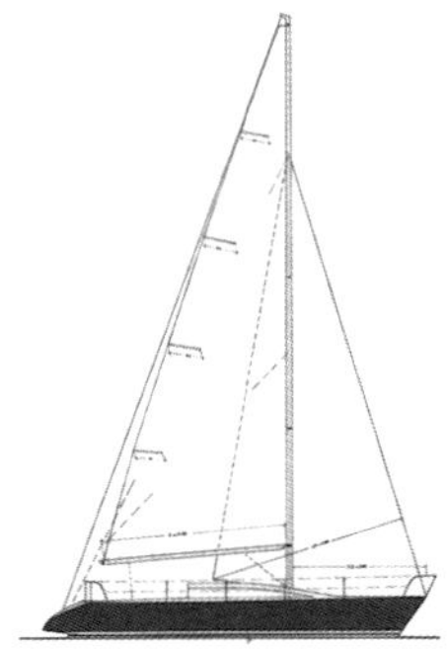

I VEZZI MIEI

Year 1981
Builder Maccianti Chiavari (GE)
Construction Cold molded wood
LOA 11.27 m
B MAX 3.74 m
D 2.07 m
DSPL 6.50 t
Mainsail 43 mq
Headsail 38 mq
Spi 78 mq

LSD TQT

Year 1982
Builder Galetti Desenzano del Garda (BS)
Construction Cold molded wood
LOA 10.18 m
B MAX 3.46 m
D 1.85 m
DSPL 3.70 t
Mainsail 33 mq
Headsail 32 mq
Spi 66 mq

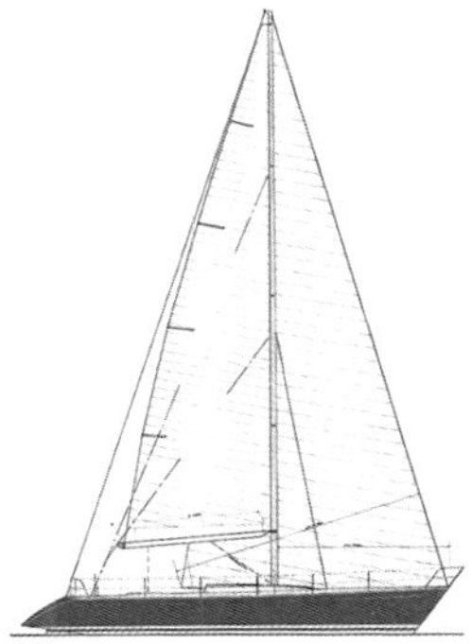

GEMINI

Year 1982
Builder De Cesari Cervia (RA)
Construction Cold molded wood
LOA 13.73 m
B MAX 4.11 m
D 2.60 m
DSPL 9.20 t
Mainsail 78 mq
Headsail 41 mq
Spi 155 mq

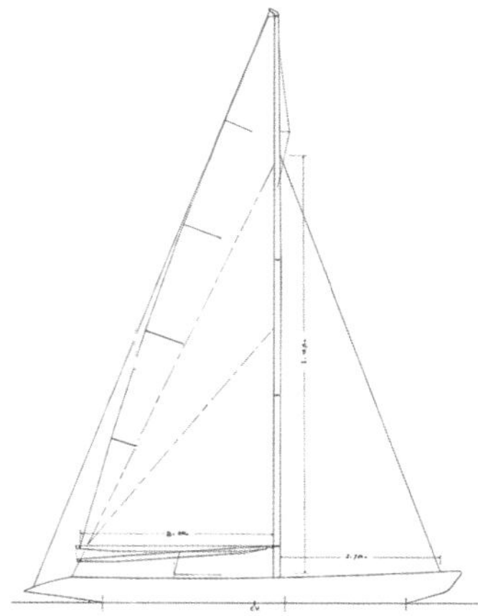

AZZURRA

Year 1982
Builder Yacht Officine Pesaro
Construction Aluminium
LOA 19.68 m
B MAX 3.73 m
D 2.70 m
DSPL 24.90 t
Mainsail 104 mq
Headsail 115 mq
Spi 280 mq

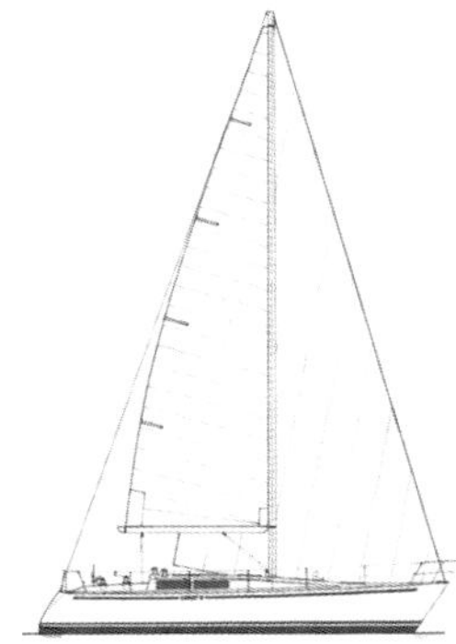

COMET 12

Year 1983
Builder Comar Forlì
Construction GRP
LOA 12.62 m
B MAX 3.90 m
D 2.50 m
DSPL 8.50 t
Mainsail 31 mq
Headsail 60 mq
Spi 135 mq

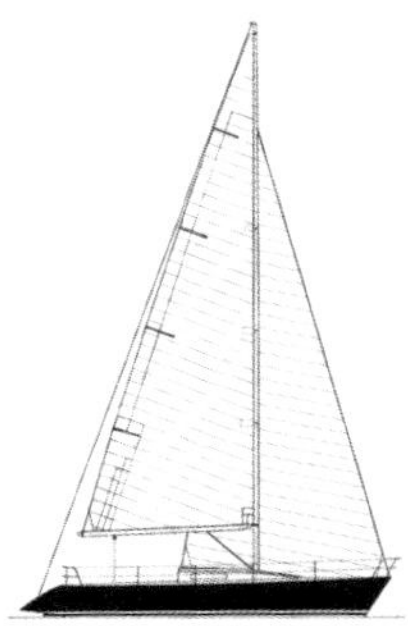

BRAVA 30.5

Year 1983
Builder Morri & Para Rimini
Construction Cold molded wood – Composite deck
LOA 12.19 m
B MAX 3.91 m
D 2.35 m
DSPL 6 t
Mainsail 49 mq
Headsail 48 mq
Spi 97 mq

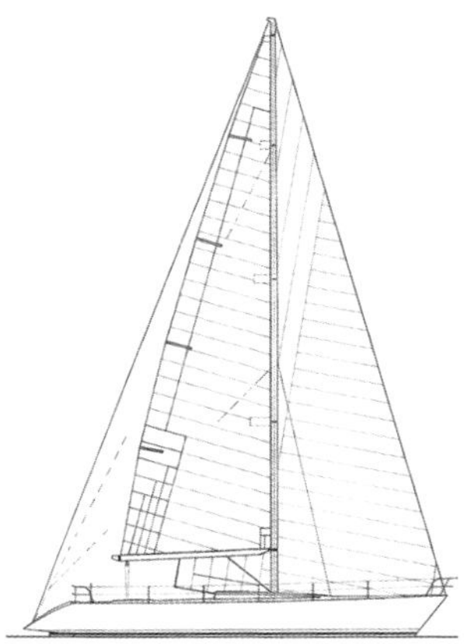

ENTREPRISE

Year 1983
Builder CMN Cherbourg (France)
Construction Aluminum – Composite deck
LOA 15.40 m
B MAX 4.50 m
D 2.80 m
DSPL 13.50 t
Mainsail 52 mq
Headsail 62 mq
Spi 150 mq

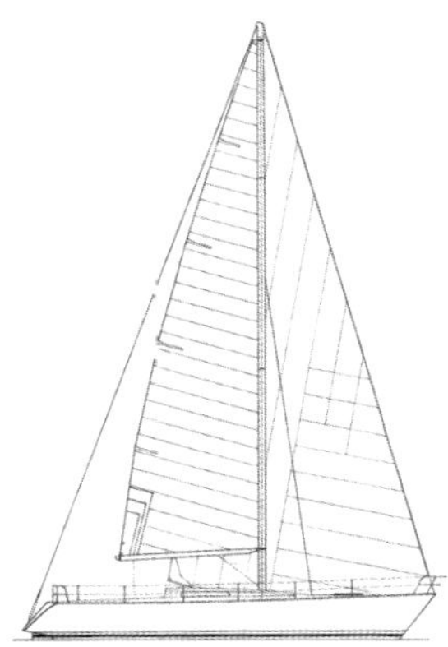

SLOOP 14.50

Year 1984
Builder Valdettaro La Spezia
Construction Cold molded wood
LOA 14.50 m
B MAX 4.28 m
D 2.70 m
DSPL 12 t
Mainsail 47 mq
Headsail 91 mq
Spi 184 mq

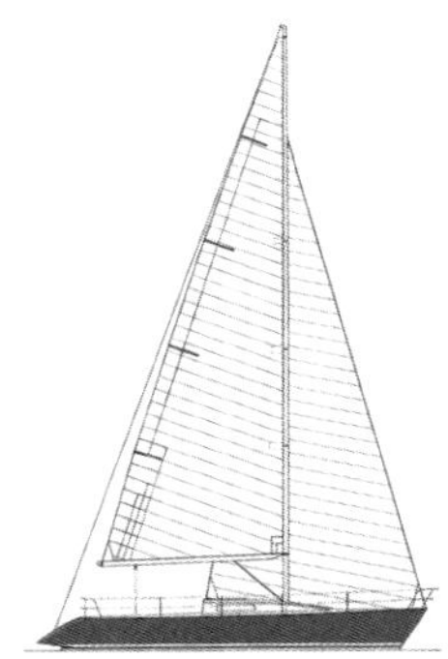

NUOVA

Year 1984
Builder Griggio Padova
Construction Aluminum
LOA 12.18 m
B MAX 3.93 m
D 2.30 m
DSPL 5.50 t
Mainsail 47 mq
Headsail 42 mq
Spi 95 mq

HT MM SERIE CASTORE E POLLUCE
Year 1984
Builder De Cesari Cervia (RA)
Construction Cold molded wood
LOA 9.21 m
B MAX 3.18 m
D 1.80 m
DSPL 2.60 t
Mainsail 27 mq
Headsail 28 mq
Spi 56 mq

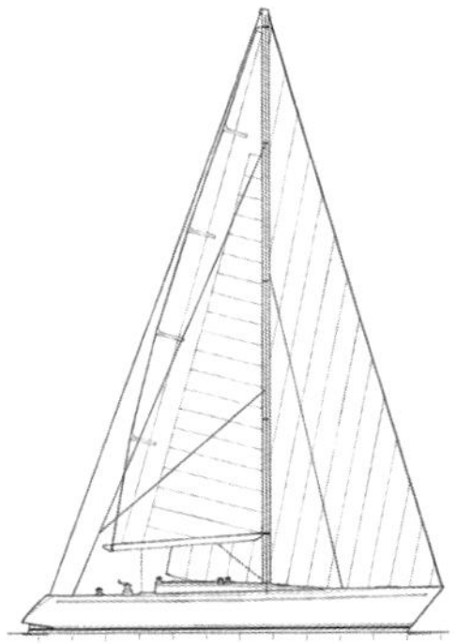

DIOSCURA

Year 1985
Builder Bani Monte Argentario (GR)
Construction Cold molded wood
LOA 12.60 m
B MAX 3.90 m
D 2.40 m
DSPL 8.90 t
Mainsail 37 mq
Headsail 68 mq
Spi 137 mq

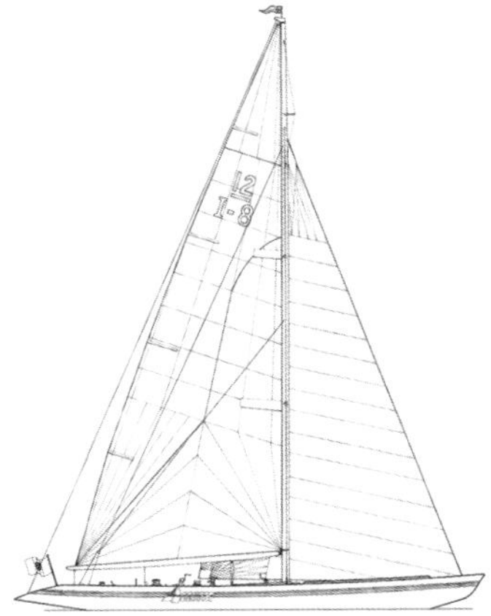

AZZURRA II

Year 1985
Builder Scardellato Treviso
Construction Aluminum
LOA 19.87 m
B MAX 3.75 m
D 2.70 m
DSPL 25 t
Mainsail 105 mq
Headsail 115 mq
Spi 280 mq

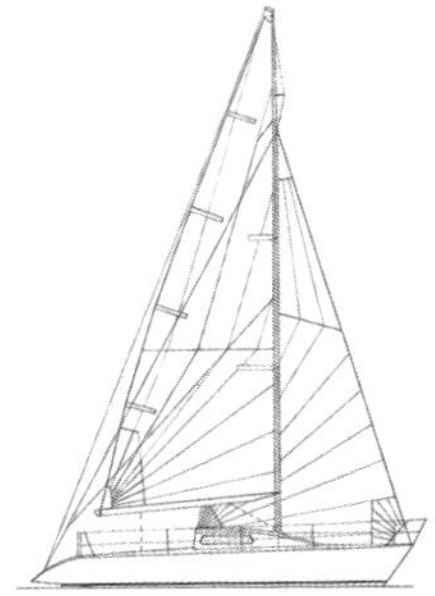

SPQR

Year 1985
Builder CPR – British Marine
Fiumicino (RM)
Construction Composite
LOA 9.28 m
B MAX 3.20 m
D 1.76 m
DSPL 3 t
Mainsail 27 mq
Headsail 25 mq
Spi 50 mq

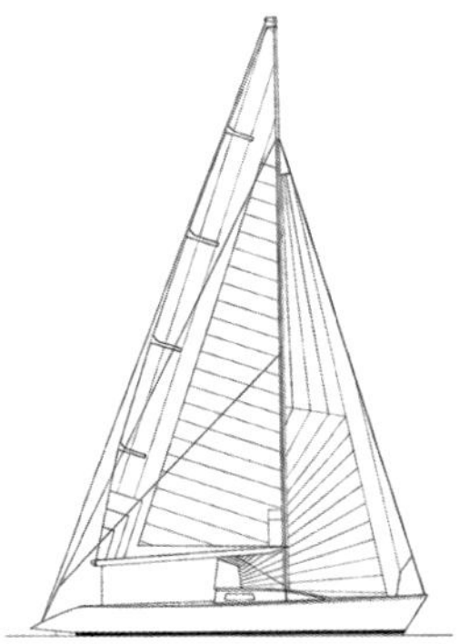

CAPRICORNO

Year 1985
Builder Galetti Desenzano
del Garda (BS)
Construction Composite
LOA 12.94 m
B MAX 4.15 m
D 2.35 m
DSPL 6.40 t
Mainsail 60 mq
Headsail 54 mq
Spi 109 mq

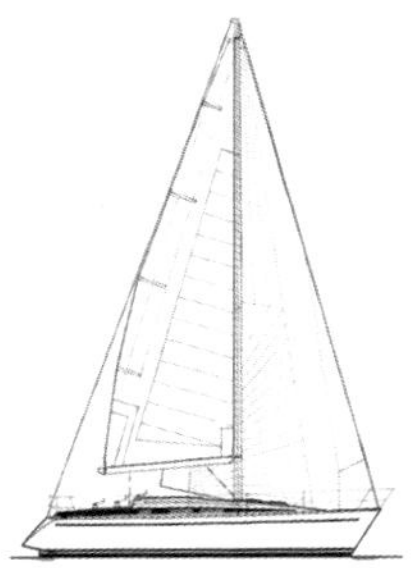

COMET 333

Year 1986
Builder Comar Forlì
Construction GRP
LOA 9.87 m
B MAX 3.35 m
D 1.50 m
DSPL 5.10 t
Mainsail 23.50 mq
Headsail 38 mq
Spi 85 mq

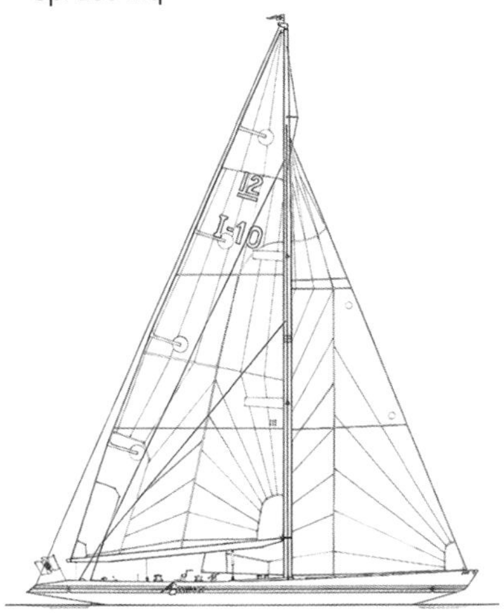

AZZURRA III

Year 1986
Builder SAI Ambrosini Passignano (PG)
Construction Aluminum
LOA 19.45 m
B MAX 3.70 m
D 2.70 m
DSPL 23.50 t
Mainsail 100 mq
Headsail 110 mq
Spi 270 mq

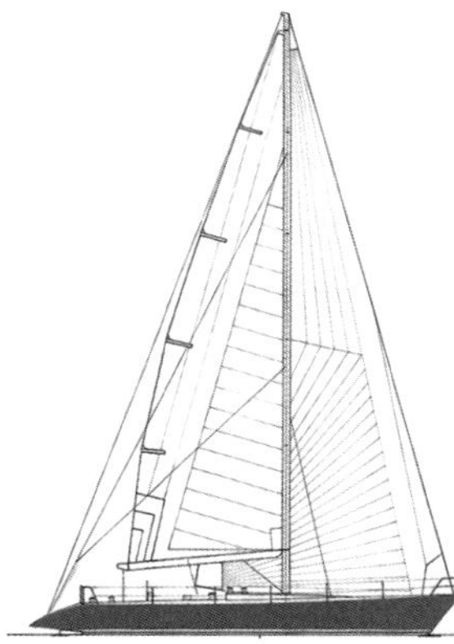

SPRINGBOK

Year 1986
Builder Geraghty Marine San Diego
(California)
Construction Composite
LOA 15.08 m
B MAX 4.54 m
D 2.70 m
DSPL 11.40 t
Mainsail 55 mq
Headsail 95 mq
Spi 190 mq

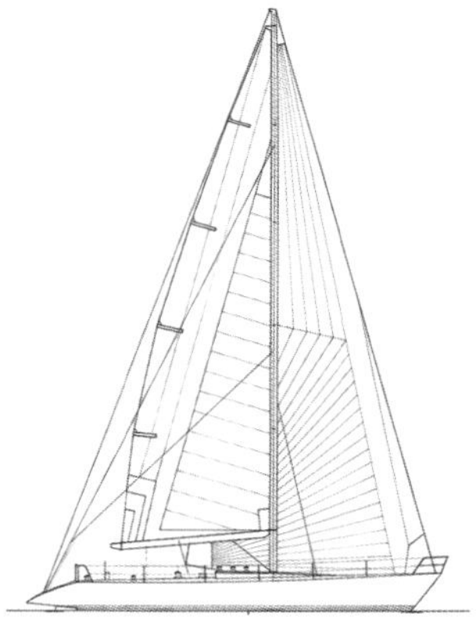

SIRENA

Year 1986
Builder Regat Ltd Istambul
Construction Composite
LOA 15.20 m
B MAX 4.25 m
D 2.76 m
DSPL 11.40 t
Mainsail 55 mq
Headsail 95 mq
Spi 190 mq

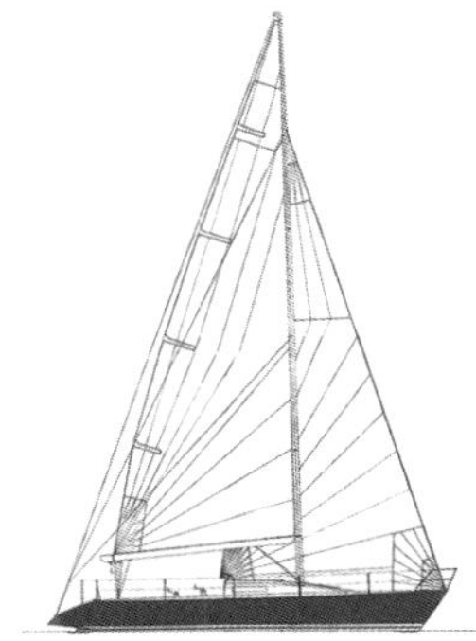

BRAVA LES COPAINS

Year 1986
Builder Green Marine Southampton (UK)
Construction Composite
LOA 12.07 m
B MAX 3.85 m
D 2.31 m
DSPL 5 t
Mainsail 48 mq
Headsail 47 mq
Spi 95 mq

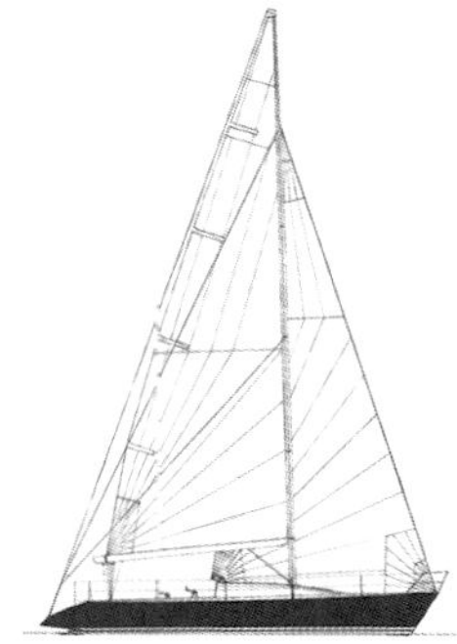

MEROPE

Year 1986
Builder Nautica C4 Treviso
Construction GRP
LOA 12.06 m
B MAX 3.93 m
D 2.16 m
DSPL 5.60 t
Mainsail 48 mq
Headsail 46 mq
Spi 93 mq

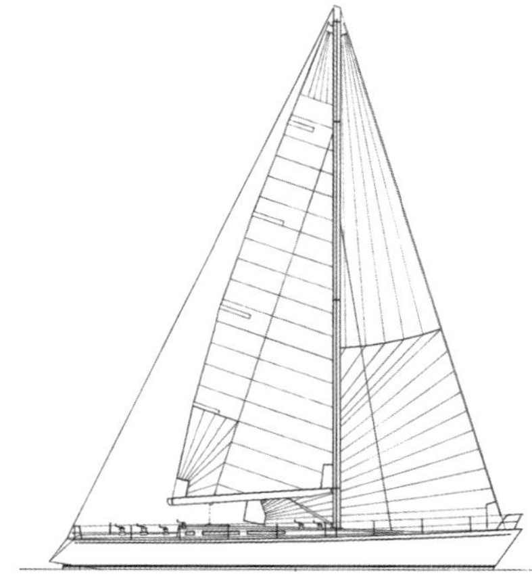

ULDB 65

Year 1987
Builder VR Yacht Orio al Serio (BG)
Construction GRP
LOA 19.70 m
B MAX 4.90 m
D 2.80 m
DSPL 15 t
Mainsail 70 mq
Headsail 114 mq
Spi 224 mq

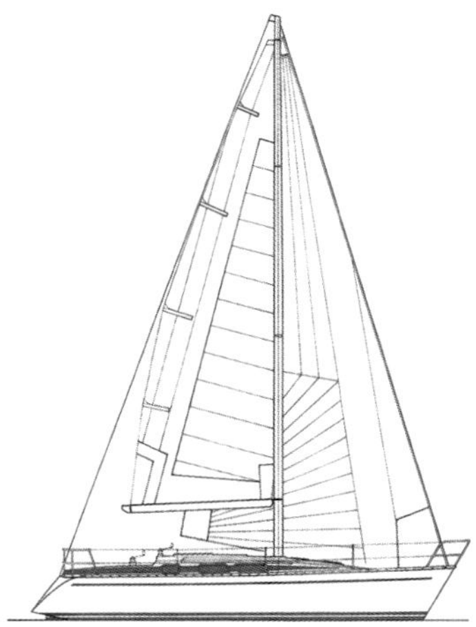

COMET 303

Year 1987
Builder Comar Forlì
Construction GRP
LOA 9.13 m
B MAX 3.15 m
D 1.85 m
DSPL 3.60 t
Mainsail 19 mq
Headsail 35 mq
Spi 73 mq

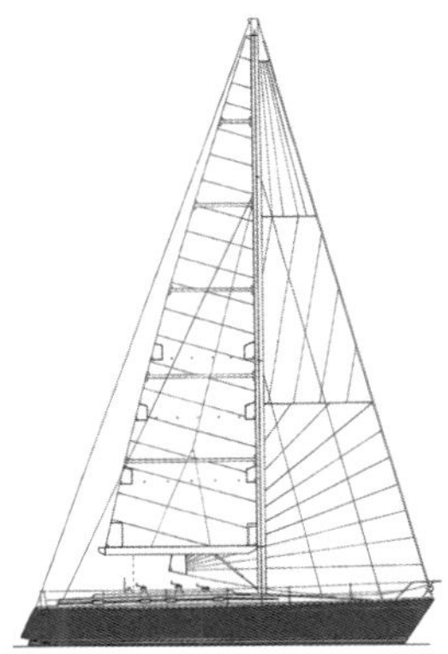

KOALA VI

Year 1988
Builder De Cesari Cervia (RA)
Construction Cold molded wood
LOA 15.02 m
B MAX 4.25 m
D 2.60 m
DSPL 16.80 t
Mainsail 50 mq
Headsail 87 mq
Spi 175 mq

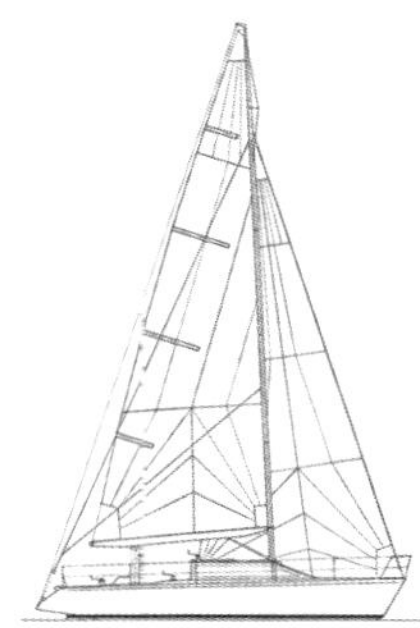

BRANCALEONE TQT 87

Year 1988
Builder Nautica C4 Treviso
Construction Composite
LOA 9.85 m
B MAX 3.31 m
D 1.85 m
DSPL 3.40 t
Mainsail 35 mq
Headsail 30 mq
Spi 67 mq

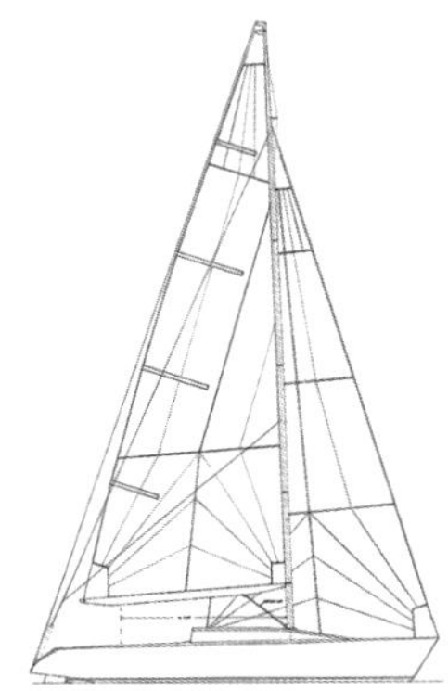

SPEEDWAVE

Year 1988
Builder Speedwave GMBH Monaco
Construction Cold molded wood
LOA 12.45 m
B MAX 3.77 m
D 2.40 m
DSPL 6 t
Mainsail 55 mq
Headsail 52 mq
Spi 105 mq

COMET 285

Year 1989
Builder Comar Forlì
Construction GRP
LOA 8.70 m
B MAX 2.95 m
D 1.65 m
DSPL 2.50 t
Mainsail 14 mq
Headsail 24 mq
Spi 48 mq

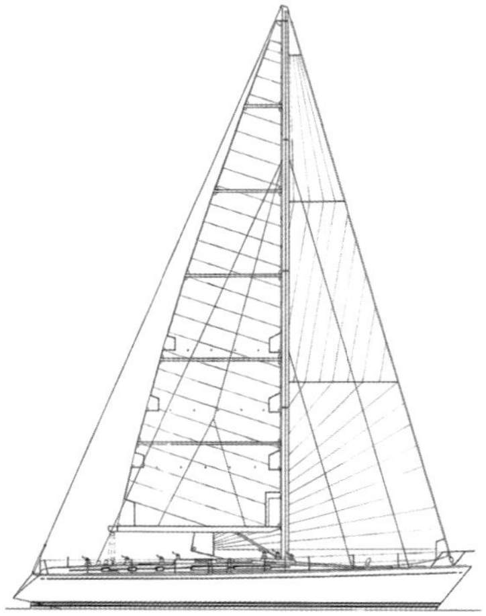

LA CARDINALA

Year 1989
Builder De Cesari Cervia (RA)
Construction Cold molded wood
LOA 21.60 m
B MAX 5.75 m
D 3 m
DSPL 35.50 t
Mainsail 100 mq
Headsail 150 mq
Spi 320 mq

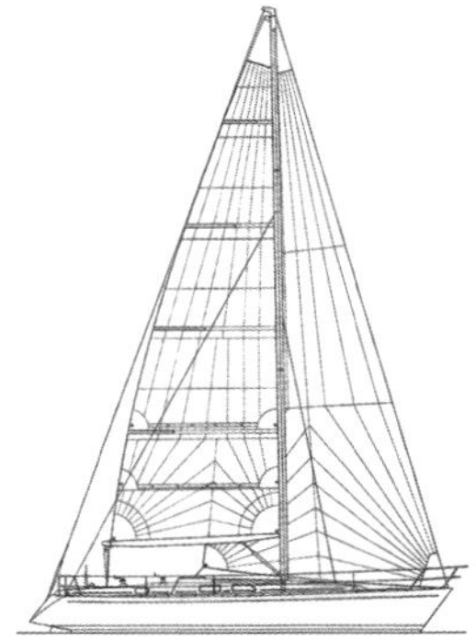

GENESI

Year 1989
Builder Comar Forlì
Construction GRP
LOA 13.10 m
B MAX 4.02 m
D 2.40 m
DSPL 9.10 t
Mainsail 43 mq
Headsail 57 mq
Spi 115 mq

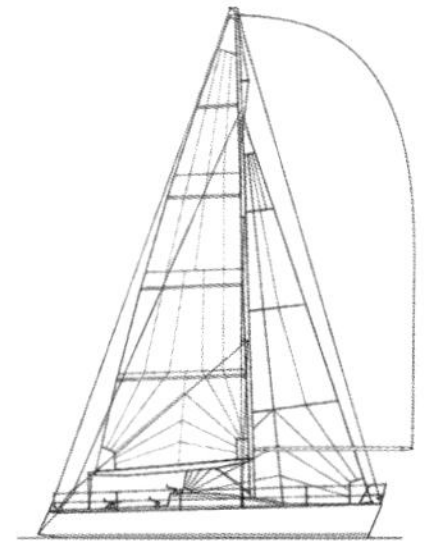

BRANCALEONE OPEN

Year 1990
Builder Morri & Para Rimini
Construction Cold molded wood
LOA 10.67 m
B MAX 3.24 m
D 2.10 m
DSPL 3.20 t
Mainsail 36 mq
Headsail 33 mq
Spi 72 mq

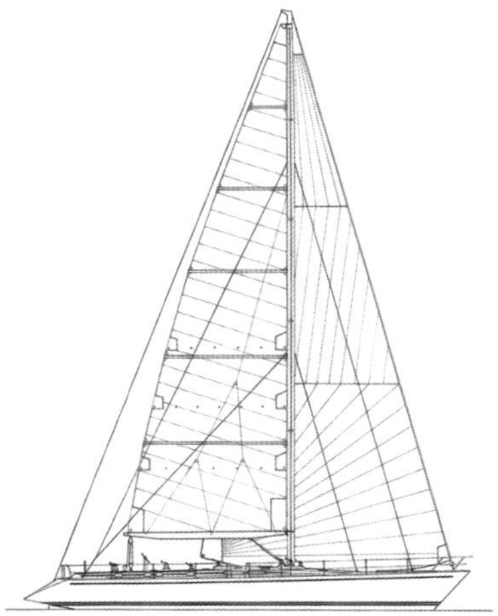

FETCH III

Year 1990
Builder De Cesari Cervia (RA)
Construction Cold molded wood
LOA 22.35 m
B MAX 5.87 m
D 3 m
DSPL 32.50 t
Mainsail 112 mq
Headsail 175 mq
Spi 350 mq

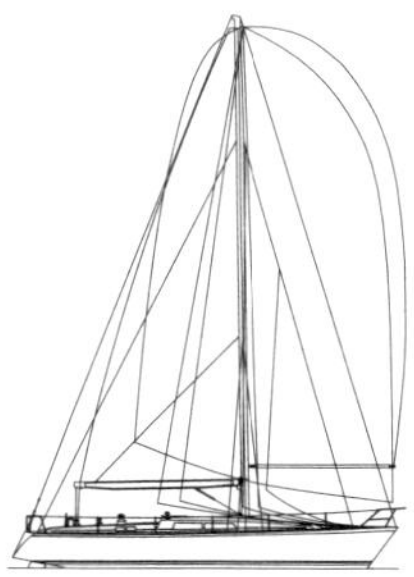

FENICIA

Year 1990
Builder C.N. Studio Altura Venezia
Construction Cold molded wood
LOA 13.70 m
B MAX 4.08 m
D 2.40 m
DSPL 8.50 t
Mainsail 47 mq
Headsail 69 mq
Spi 140 mq

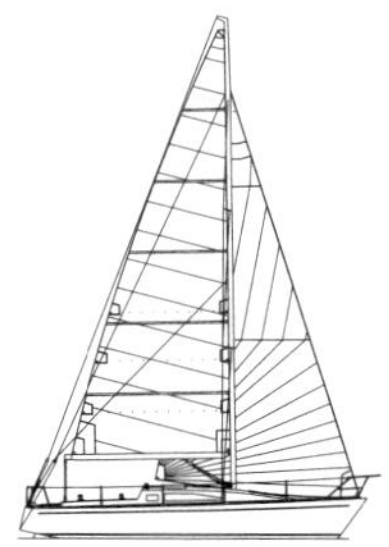

ESTE 35

Year 1991
Builder D'Este Fiumicino (RM)
Construction Composite
LOA 10.60 m
B MAX 3.27 m
D 1.90 m
DSPL 3.50 t
Mainsail 34 mq
Headsail 20 mq
Spi 90 mq

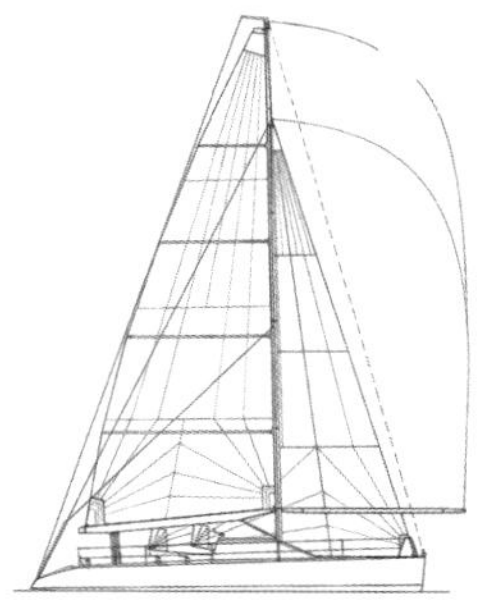

STRADIVARIA

Year 1992
Builder Soleri Ravenna
Construction Composite
LOA 12.55 m
B MAX 4.11 m
D 2.78 m
DSPL 2.95 t
Mainsail 49 mq
Headsail 47 mq
Spi 110 mq

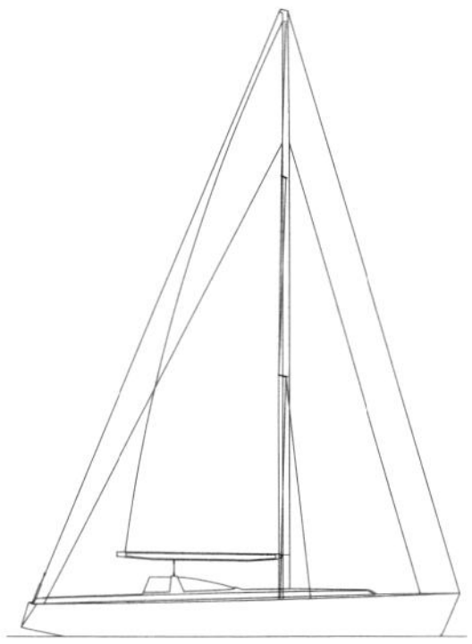

OCEANO 54

Year 1992
Builder VR Yacht Orio al Serio (BG)
Construction Composite
LOA 16.40 m
B MAX 4.92 m
D 2.60 m
DSPL 16.80 t
Mainsail 66 mq
Headsail 104 mq
Spi 160 mq

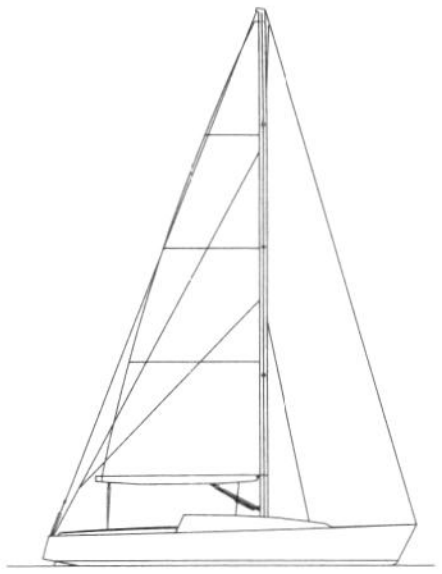

ESTE 39

Year 1992
Builder Cantieri D'este Fiumicino (RM)
Construction Composite
LOA 11.95 m
B MAX 3.75 m
D 2.40 m
DSPL 6.80 t
Mainsail 40 mq
Headsail 22 mq
Spi 125 mq

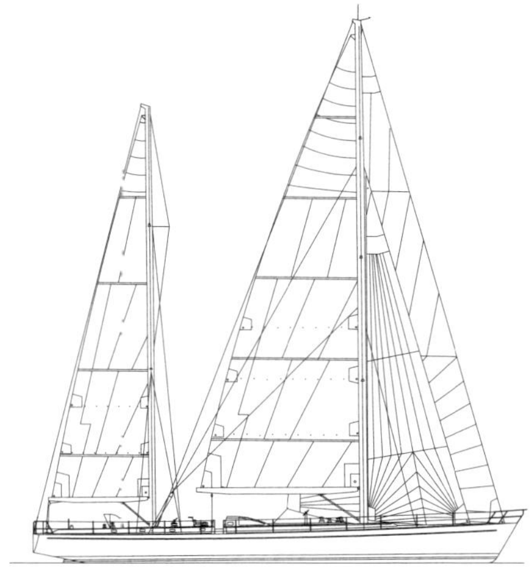

ORSA MAGGIORE

Year 1993
Builder Tencara Marghera (VE)
Construction Composite
LOA 28.30 m
B MAX 6.53 m
D 4.20 m
DSPL 75 t
Mainsail Maestra 140 mq Mezzana 68 mq
Headsail 180 mq
Spi 425 mq

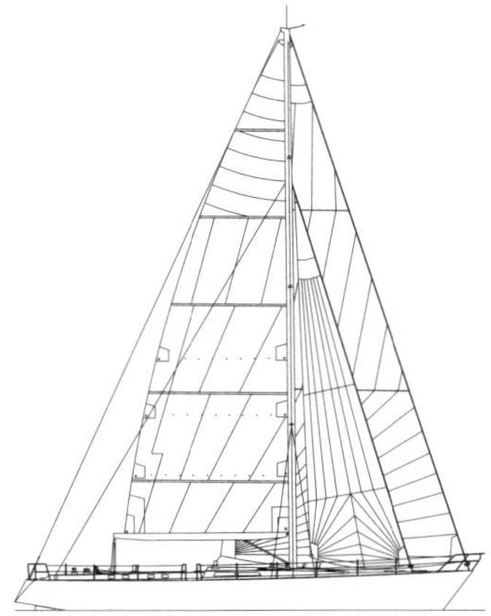

MIA

Year 1994
Builder Cassanelli – CNB Fano (PU)
Construction Aluminum
LOA 23.05 m
B MAX 5.90 m
D 2.94 m
DSPL 35 t
Mainsail 125 mq
Headsail 160 mq
Spi 305 mq

STAR CAT

Year 1995
Builder Latini Roma
Construction Cold molded wood
LOA 6.85 m
B MAX 2.45 m
D 1.20 m
DSPL 0.90 t
Mainsail 24 mq
Spi 18 mq

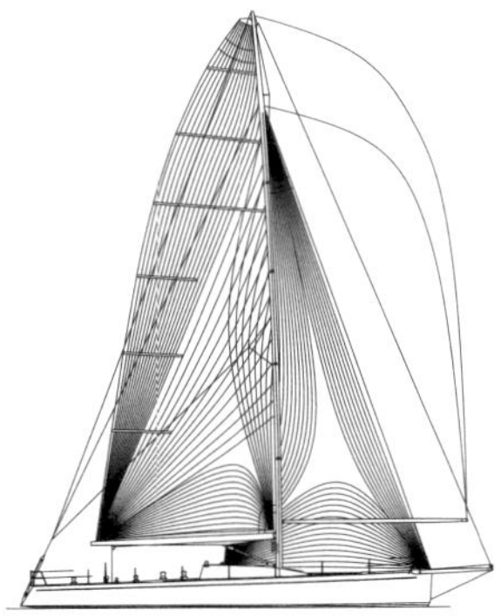

GAIA LEGEND

Year 1995
Builder Soleri Ravenna
Construction Composite
LOA 16.75 m
B MAX 4.76 m
D 3.50 m
DSPL 6.20 t
Mainsail 102 mq
Headsail 76 mq
Spi 240 mq

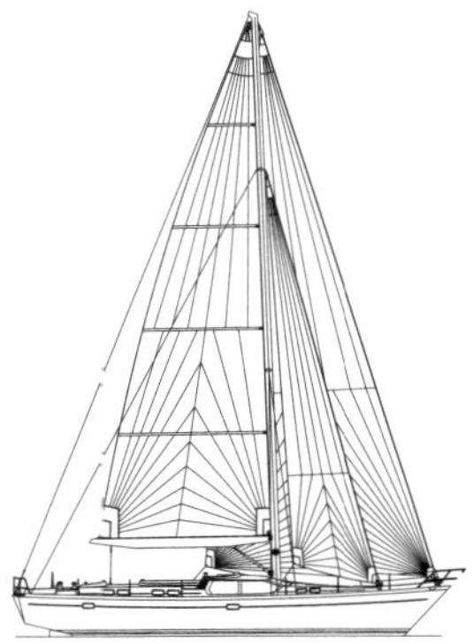

COMET 54

Year 1995
Builder Comar Fiumicino (RM)
Construction GRP
LOA 15.67 m
B MAX 4.59 m
D 2.50 m
DSPL 14 t
Mainsail 62 mq
Headsail 85 mq
Spi 180 mq

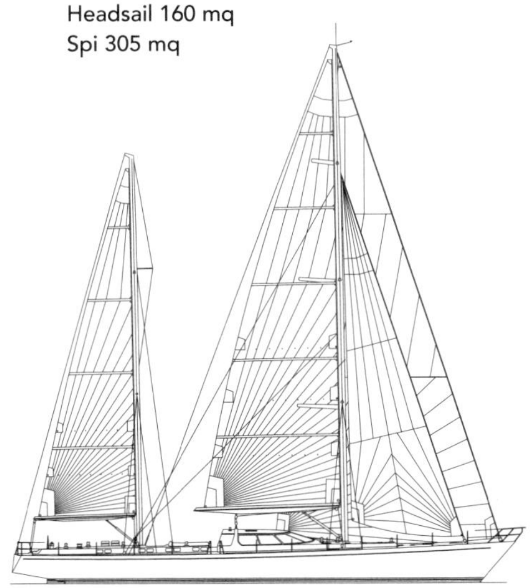

CRISTOFORO COLOMBO II

Year 1996
Builder SAI Ambrosini Passignano (PG)
Construction Composite
LOA 31.93 m
B MAX 6.69 m
D 3.52 m
DSPL 80 t
ANDA 75/135 mq
Headsail 215 mq
Spi 480 mq

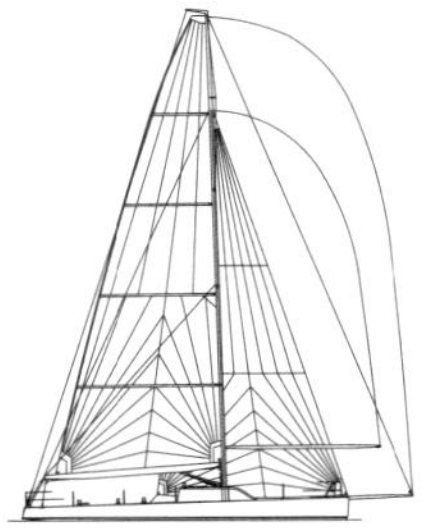

LISA

Year 1996
Builder Patuccelli Gargnano (BS)
Construction Composite
LOA 12.67 m
B MAX 4.09 m
D 2.97 m
DSPL 3 t
Mainsail 67 mq
Headsail 95 mq
Spi 240 mq

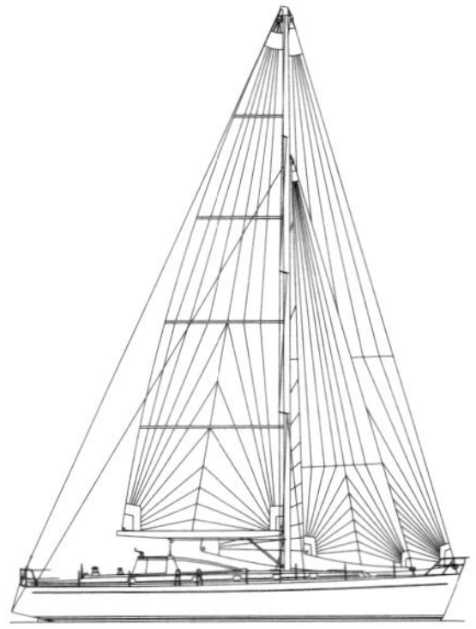

CONCHITA

Year 1996
Builder Carlini Rimini
Construction Cold molded wood
LOA 17.30 m
B MAX 5.27 m
D 2.95 m
DSPL 20.50 t
Mainsail 70 mq
Headsail 97 mq
Spi 210 mq

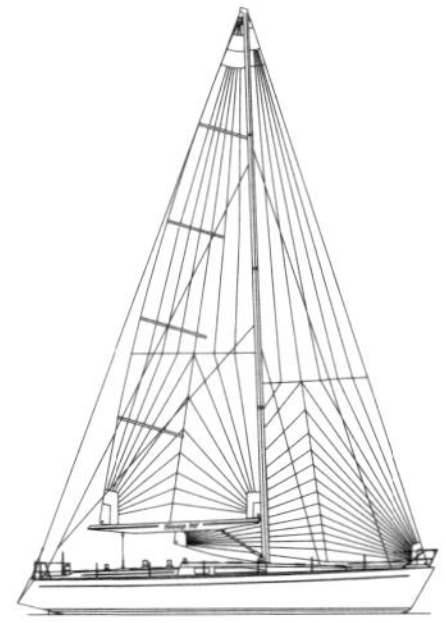

ATLANTIS

Year 1996
Builder Nautica C4 Treviso
Construction Cold molded wood
LOA 15.68 m
B MAX 4.58 m
D 2.80 m
DSPL 15 t
Mainsail 69 mq
Headsail 99 mq
Spi 199 mq

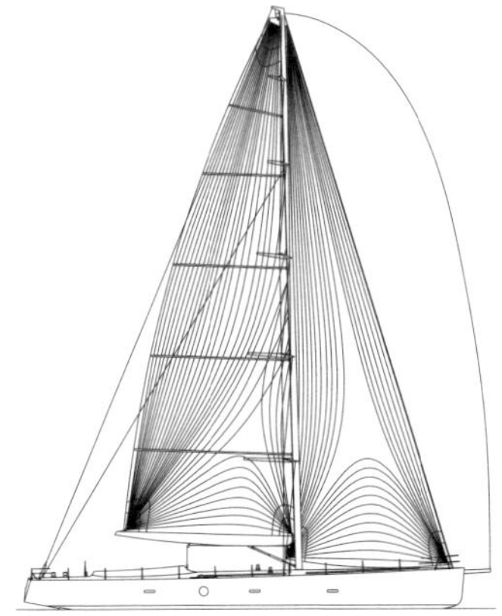

VIRTUELLE

Year 1997
Builder Tencara Marghera (VE)
Construction Composite
LOA 23.93 m
B MAX 6.20 m
D 3.92 m
DSPL 33 t
Mainsail 155 mq
Headsail 194 mq
Spi 410 mq

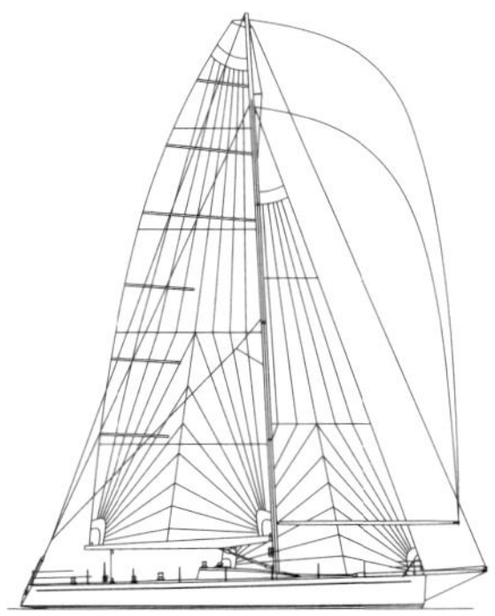

RIVIERA DI RIMINI

Year 1997
Builder Soleri Ravenna
Construction Composite
LOA 16.75 m
B MAX 4.76 m
D 4.2 m
DSPL 6.20 t
Mainsail 109 mq
Headsail 62 mq
Spi 235 mq

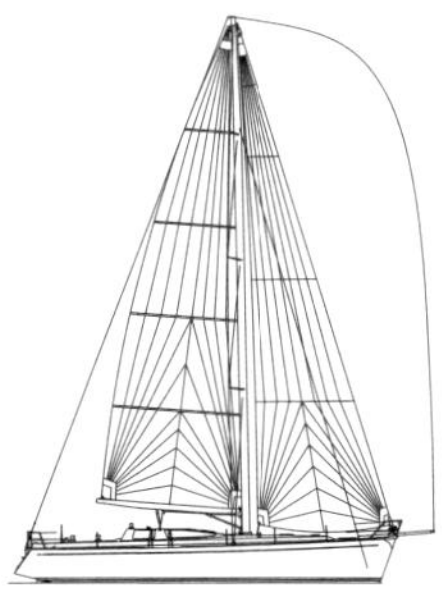

BELLAMORE

Year 1998
Builder Kalmar Ltda Brasile
Construction Cold molded wood
LOA 13.98 m
B MAX 4.45 m
D 2.27 m
DSPL 12.50 t
Mainsail 55 mq
Headsail 47 mq
Spi 180 mq

ILC 25

Year 1998
Builder Dolphin Capalbio (CR)
Construction Composite
LOA 7.60 m
B MAX 2.60 m
D 1.68 m
DSPL 1.20 t
Mainsail 25 mq
Headsail 20 mq
Spi 48 mq

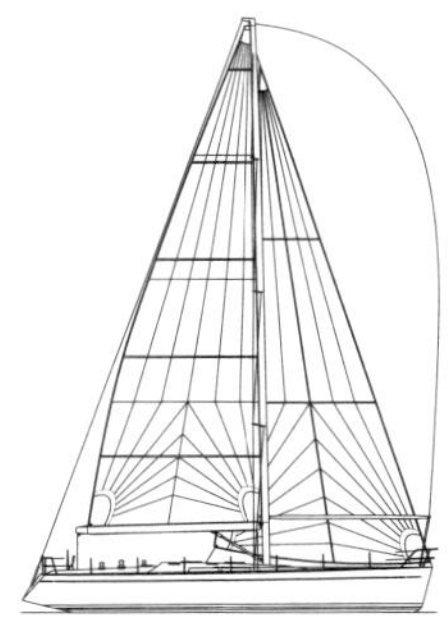

COMET 51

Year 2000
Builder Comar Fiumicino (RM)
Construction Composite
LOA 15.02 m
B MAX 4.69 m
D 3.04 m
DSPL 12 t
Mainsail 77 mq
Headsail 62 mq
Spi 168 mq

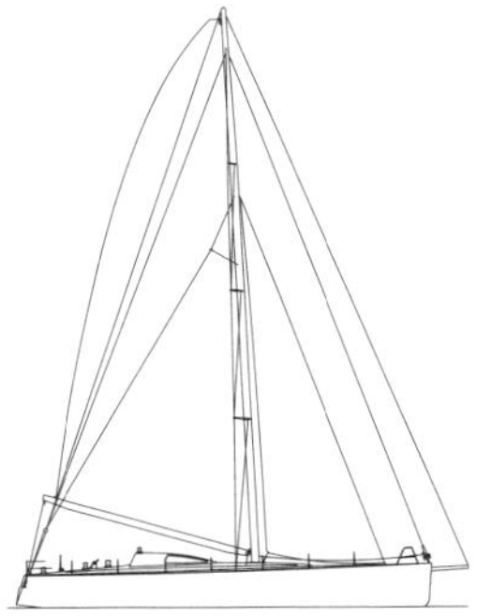

AMOR

Year 2000
Builder Serigi Aquileia (UD)
Construction Composite
LOA 16.34 m
B MAX 5.10 m
D 4 m
DSPL 9 t
Mainsail 105 mq
Headsail 63 mq
Spi 185 mq

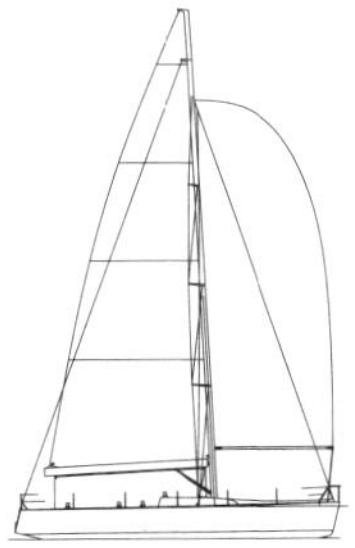

SHE DEVIL

Year 2001
Builder Latini Roma
Construction Composite
LOA 12.48 m
B MAX 3.38 m
D 2.66 m
DSPL 5.20 t
Mainsail 55 mq
Headsail 40 mq
Spi 110 mq

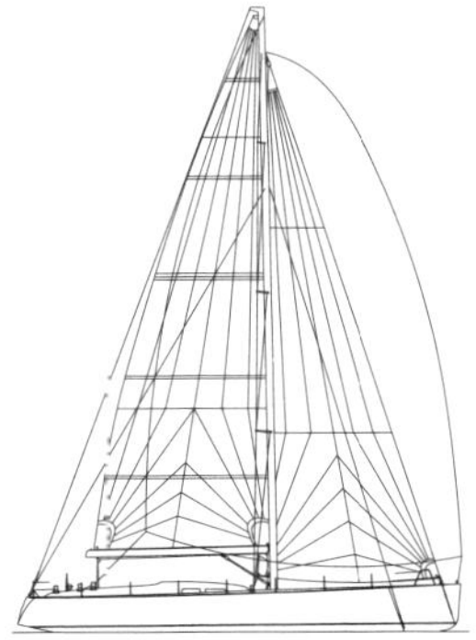

PENELOPE

Year 2001
Builder Alto Adriatico San Giorgio di Nogaro (UD)
Construction Composite
LOA 18.31 m
B MAX 5.25 m
D 3 m
DSPL 18 t
Mainsail 95 mq
Headsail 87 mq
Spi 240 mq

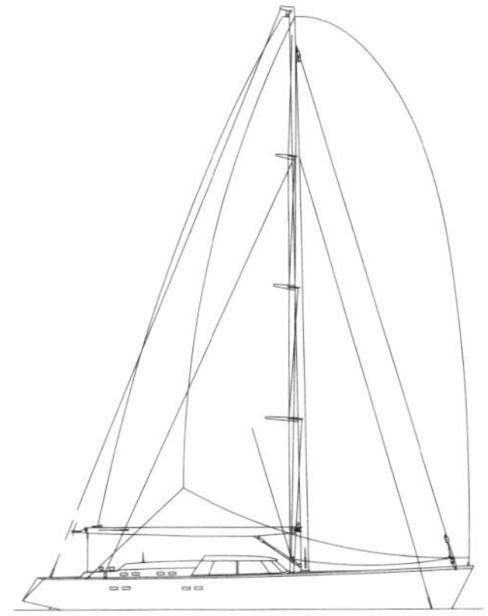

MARELLA

Year 2002
Builder De Cesari Cervia (RA)
Construction Cold molded wood
LOA 25.56 m
B MAX 6.40 m
D 3.40 m
DSPL 36 t
Mainsail 192 mq
Headsail 137 mq
Spi 425 mq

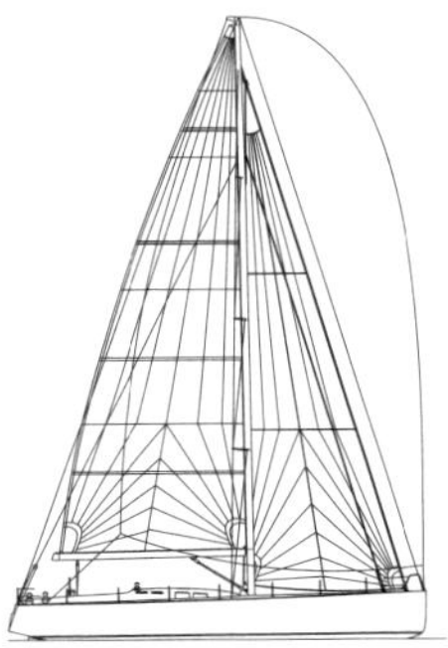

BLU SAIL 54

Year 2002
Builder Rimini Sail Rimini
Construction Composite
LOA 16.25 m
B MAX 5 m
D 4 m
DSPL 15 t
Mainsail 90 mq
Headsail 68 mq
Spi 260 mq

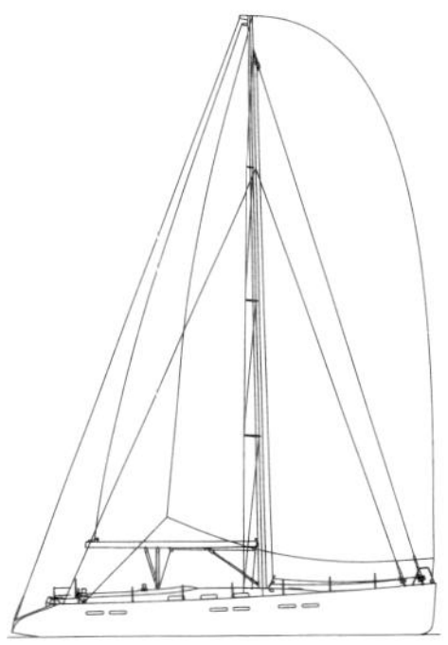

SLOOP 16,70M

Year 2003
Builder C.N. Yacht 2000 Salvirola (CR)
Construction GRP
LOA 16.82 m
B MAX 5.22 m
D 2.20 m
DSPL 16 t
Mainsail 77 mq
Headsail 62 mq
Spi 223 mq

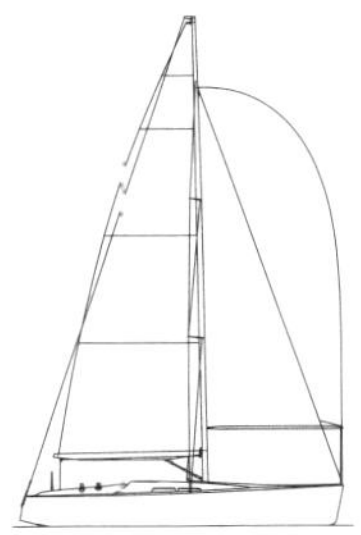

COMET 45

Year 2003
Builder Comar Fiumicino (RM)
Construction Composite
LOA 13.78 m
B MAX 4.20 m
D 2.63 m
DSPL 9.50 t
Mainsail 66 mq
Headsail 55 mq
Spi 146 mq

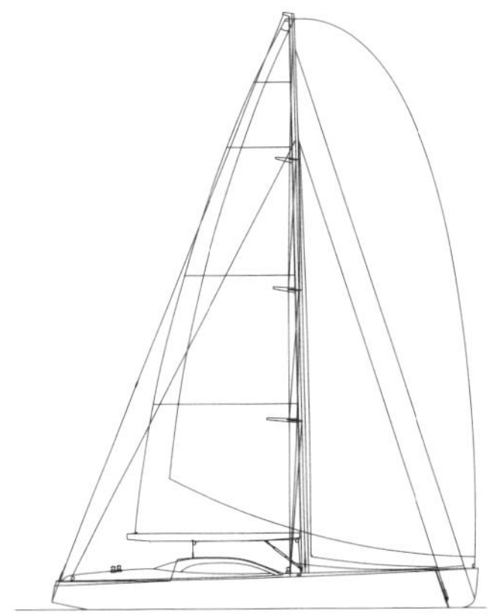

KAFRAD

Year 2003
Builder C.N. Yacht 2000 Salvirola (CR)
Construction Composite
LOA 23.10 m
B MAX 6 m
D 3.50 m
DSPL 34 t
Mainsail 150 mq
Headsail 130 mq
Spi 400 mq

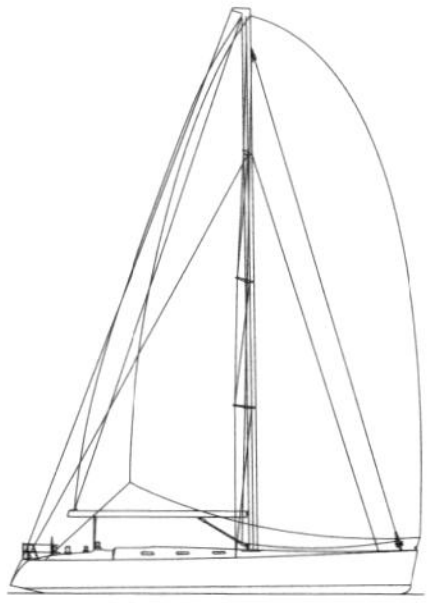

MARCHI 168

Year 2003
Builder Marchi Campalto (VE)
Construction Composite
LOA 16.79 m
B MAX 5.07 m
D 2.90 m
DSPL 16 t
Mainsail 86 mq
Headsail 68 mq
Spi 232 mq

COMET 41

Year 2004
Builder Comar Fiumicino (RM)
Construction Composite
LOA 12.45 m
B MAX 3.92 m
D 2.40 m
DSPL 7 t
Mainsail 55 mq
Headsail 54 mq
Spi 125 mq

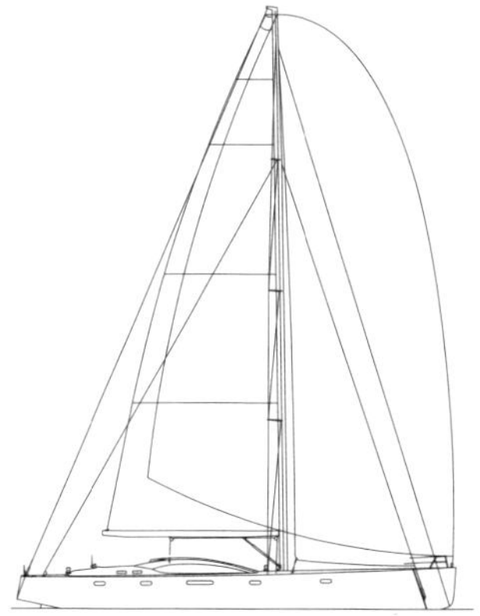

WIZARD

Year 2004
Builder C.N. Yacht 2000 Salvirola (CR)
Construction Composite
LOA 23.81 m
B MAX 6.01 m
D 3.27 m
DSPL 35 t
Mainsail 155 mq
Headsail 105 mq
Spi 400 mq

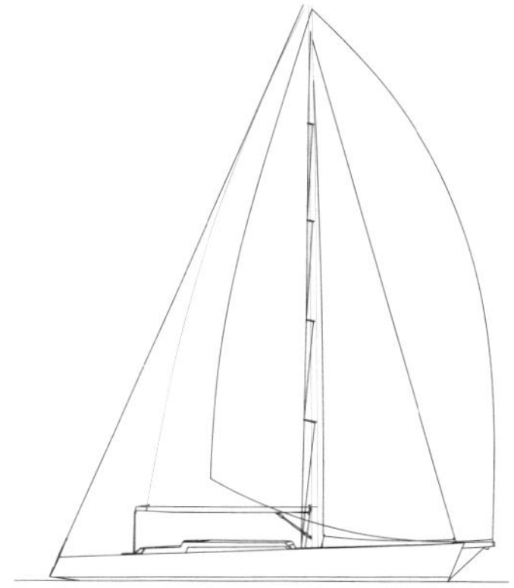

YOL

Year 2004
Builder C.N. Yacht 2000 Salvirola (CR)
Construction Composite
LOA 20.60 m
B MAX 4.90 m
D 2.80 m
DISPL 17 t
Mainsail 120 mq
Headsail 95 mq
Spi 500 mq

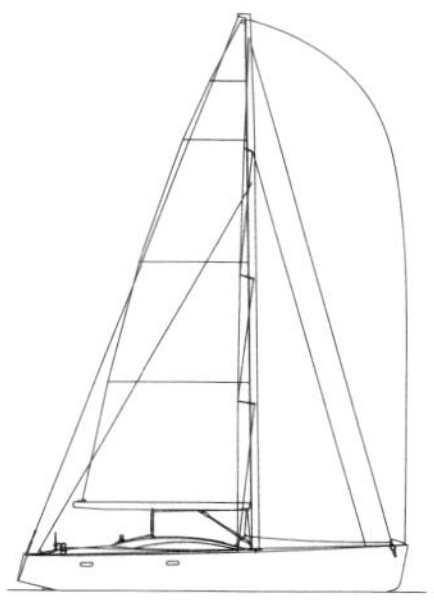

COMET 52 RS

Year 2005
Builder Comar Fiumicino (RM)
Construction Composite
LOA 15.30 m
B MAX 4.60 m
D 3.10 m
DSPL 13 t
Mainsail 75 mq
Headsail 55 mq
Spi 170 mq

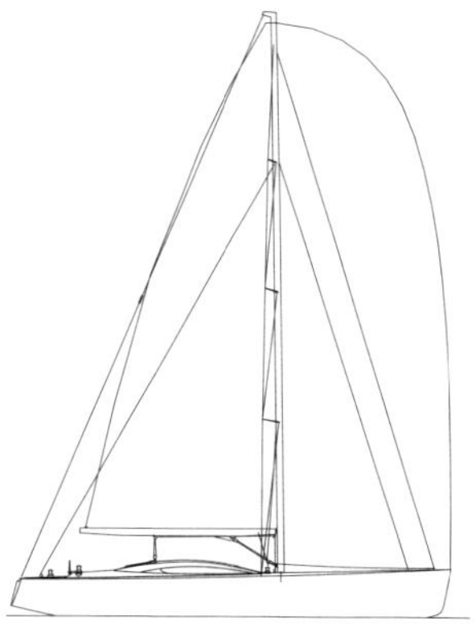

COMET 62 RS

Year 2006
Builder Comar Fiumicino (RM)
Construction Composite
LOA 19.05 m
B MAX 5.23 m
D 3.38 m
DSPL 20 t
Mainsail 108 mq
Headsail 77 mq
Spi 250 mq

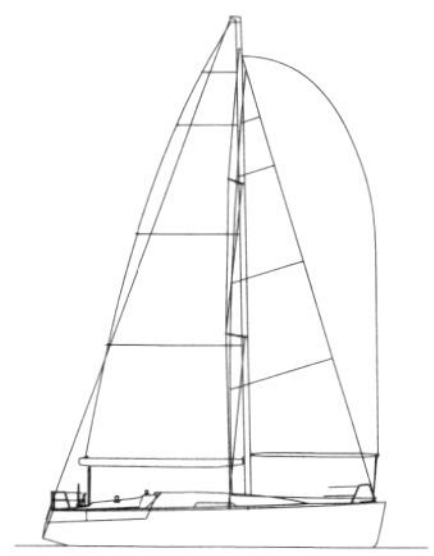

COMET 38

Year 2006
Builder Comar Fiumicino (RM)
Construction Composite
LOA 11.40 m
B MAX 3.75 m
D 2.10 m
DSPL 6.10 t
Mainsail 45 mq
Headsail 36 mq
Spi 92 mq

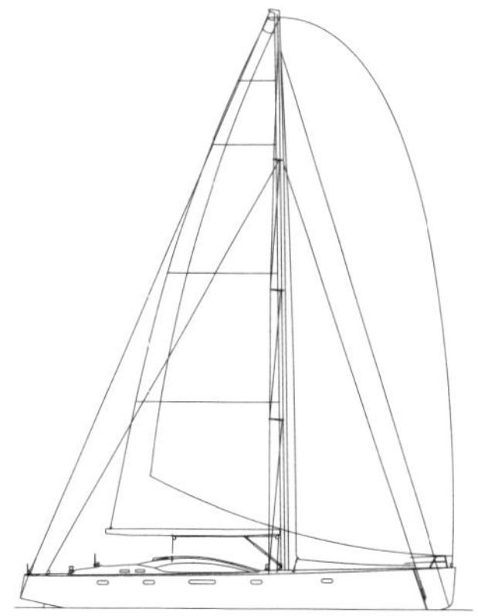

$H_2$0

Year 2006
Builder C.N. Yacht 2000 Salvirola (CR)
Construction Composite
LOA 23.81 m
B MAX 6.01 m
D 3.60 m
DSPL 35 t
Mainsail 160 mq
Headsail 100 mq
Spi 410 mq

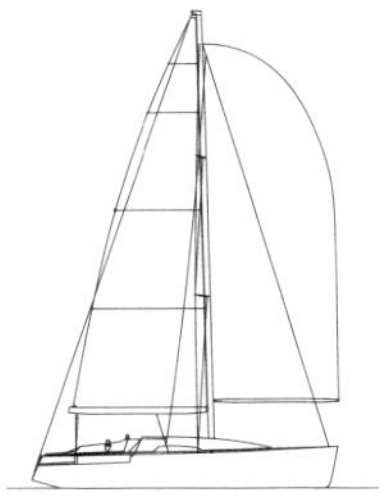

COMET 35

Year 2007
Builder Comar Fiumicino (RM)
Construction Composite
LOA 10.65 m
B MAX 3.64 m
D 2.09 m
DSPL 5.30 t
Mainsail 40 mq
Headsail 40 mq
Spi 93 mq

GALA

Year 2007
Builder Fanatek S.p.a. Roma
Construction Composite
LOA 7.57 m
B MAX 1.77 m
D 1.50 m
DSPL 0.75 t
Mainsail 15 mq
Headsail 7 mq
Spi 33 mq

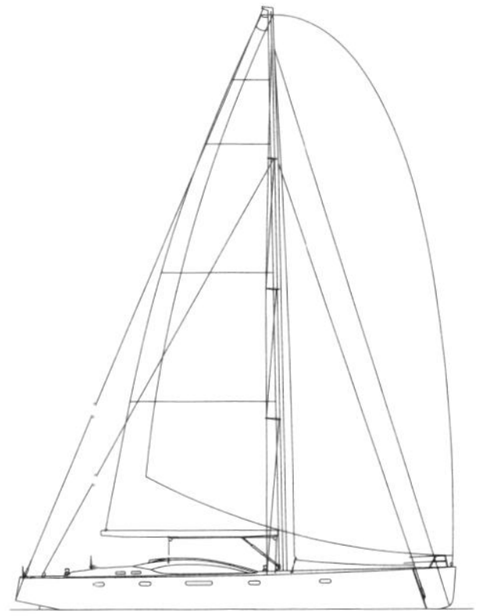

LADISEA

Year 2009
Builder C.N. Yacht 2000 Salvirola (CR)
Construction Composite
LOA 24.20 m
B MAX 6.01 m
D 3.25 m
DSPL 35 t
Mainsail 155 mq
Headsail 105 mq
Spi 400 mq

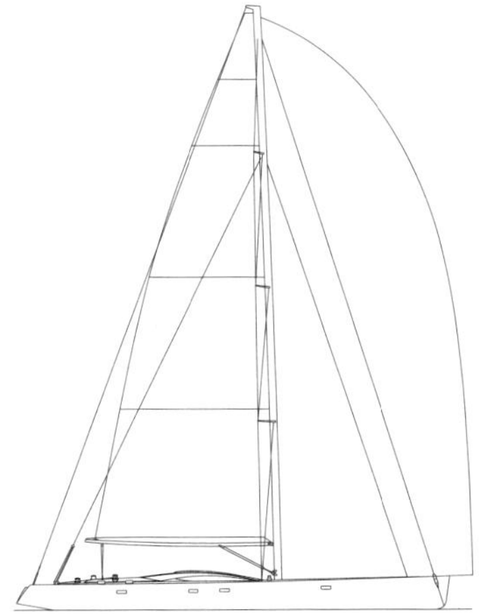

SHADOW

Year 2011
Builder Comar Fiumicino (RM)
Construction Composite
LOA 30.86 m
B MAX 6.70 m
D 4.80 m
DSPL 65 t
Mainsail 283 mq
Headsail 216 mq
Spi 703 mq

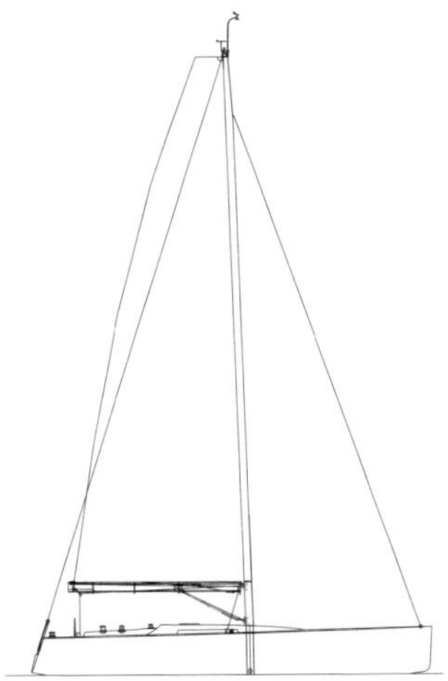

VERVE

Year 2011
Builder Comar Fiumicino (RM)
Construction Composite
LOA 15.10 m
B MAX 4.09 m
D 3.45 m
DSPL 9.50 t
Mainsail 80 mq
Headsail 69 mq
Spi 190 mq

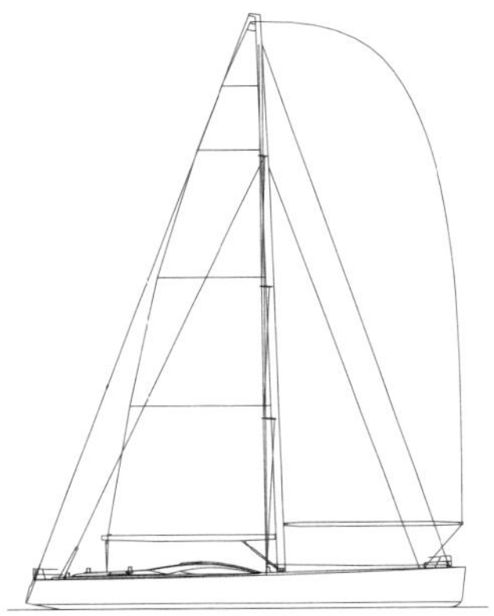

SCHOOL SHIP COAST GARD

Year 2012
Builder Comar Fiumicino (RM)
Construction Composite
LOA 25 m
B MAX 6.65 m
D 3.75 m
DISPL 47 t
Mainsail 172 mq
Headsail 157 mq
Spi 500 mq

ISA 120

Year 2002
Builder ISA Yachts Ancona
Construction GRP
LOA 36.45 m
B MAX 7.40 m
GT 173 t
Engine 3 X MTU 16V 2000 M93

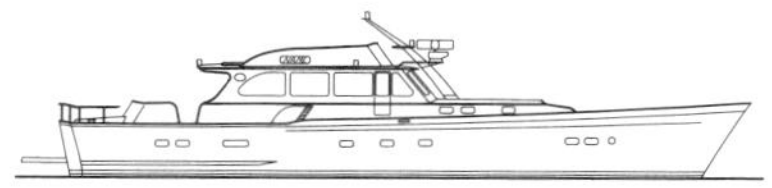

ARDIS MY 28M

Year 2003
Builder De Cesari Cervia (RA)
Construction Cold molded wood
LOA 28.72 m
B MAX 6.22 m
GT 72 t
Engine 2 X MTU 2000 M 70

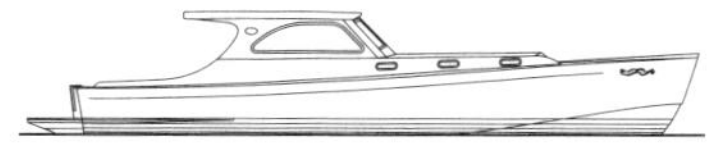

CLASSIC BOAT 16M

Year 2004
Builder De Cesari Cervia (RA)
Construction Cold molded wood
LOA 16.60 m
B MAX 4.55 m
GT 21 t
Engine 2 X CUMMINS QSM 11

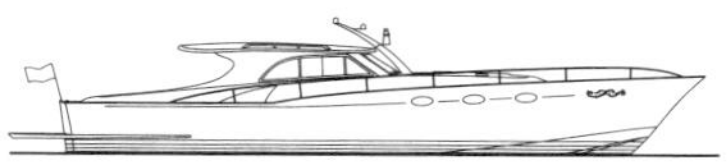

CLASSIC BOAT 18M

Year 2005
Builder De Cesari Cervia (RA)
Construction Cold molded wood
LOA 18.52 m
B MAX 5.07 m
GT 31 t
Engine 2 X MAN V8-900

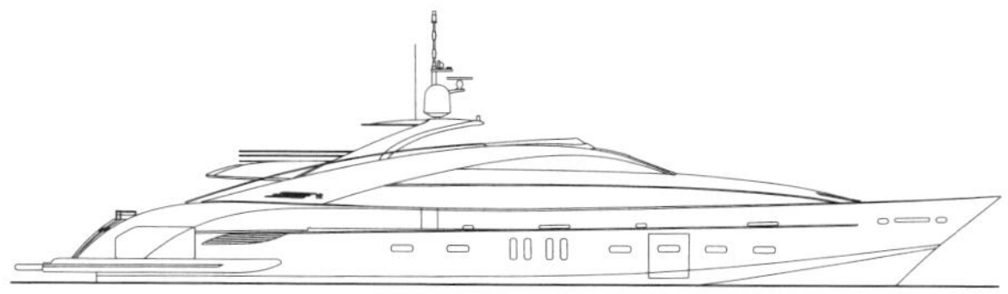

SILVER WIND

Year 2006-2014
Builder ISA Yachts Ancona
Construction GRP
LOA 43.66 m
B MAX 8.30 m
GT 230 t
Engine 2 X MTU 16V4000 M93L

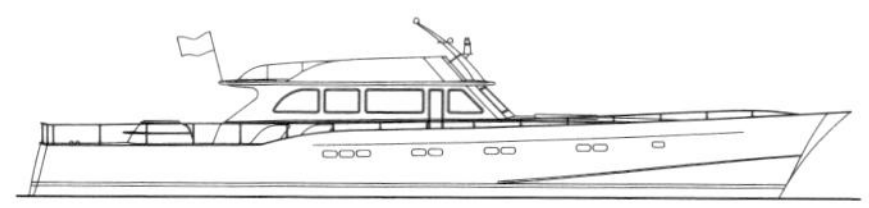

NYMPHA

Year 2006
Builder De Cesari Cervia (RA)
Construction Cold molded wood
LOA 31.80 m
B MAX 7.30 m
GT 90 t
Engine 2 X MTU 16 V 2000 M70

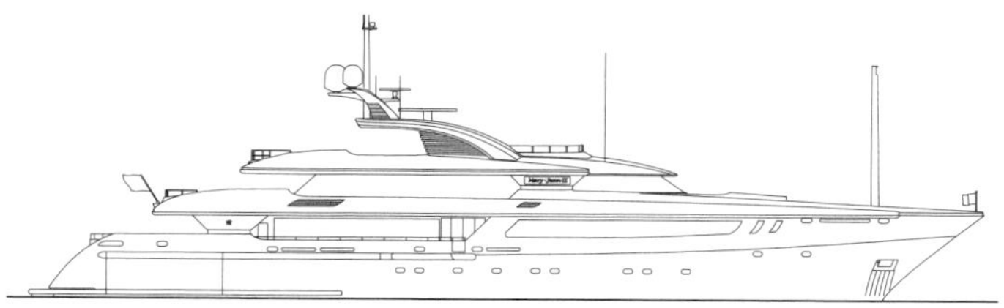

MARY JEAN II

Year 2010
Builder ISA Yachts Ancona
Construction Steel – Aluminum
LOA 60.50 m
B MAX 11.50 m
GT 980 t
Engine 2 X CAT 3512 C

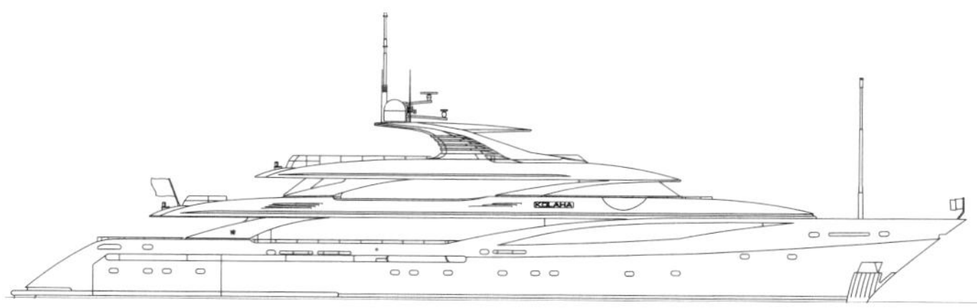

KOLAHA

Year 2010
Builder ISA Yachts Ancona
Construction Steel – Aluminum
LOA 63.12 m
B MAX 12 m
GT 1180 t
Engine 2 X CAT 3512C

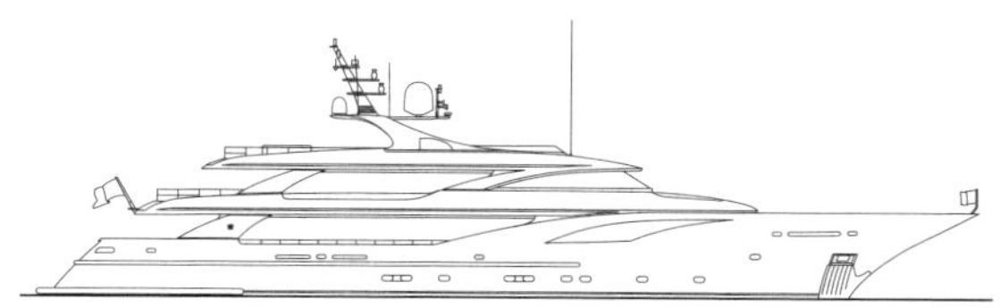

LIBERTY

Year 2011
Builder ISA Yachts Ancona
Construction Steel – Aluminum
LOA 49.95 m
B MAX 9 m
GT 490 t
Engine 2 X MTU 12 V 4000 M71

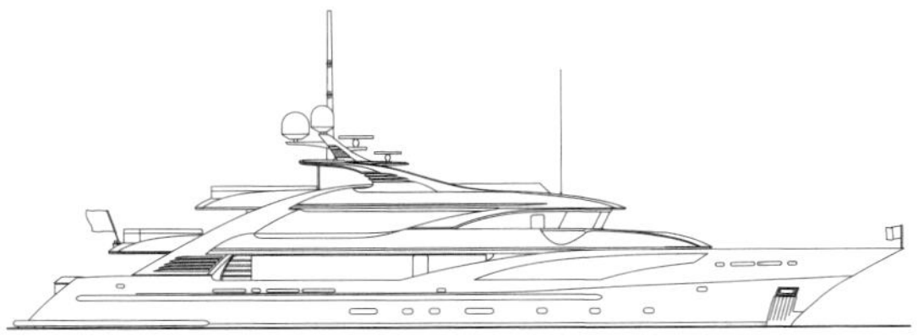

PAPI DU PAPI

Year 2011
Builder ISA Yachts Ancona
Construction Steel – Aluminum
LOA 50 m
B MAX 9 m
GT 500 t
Engine 2 X CAT 3512 C

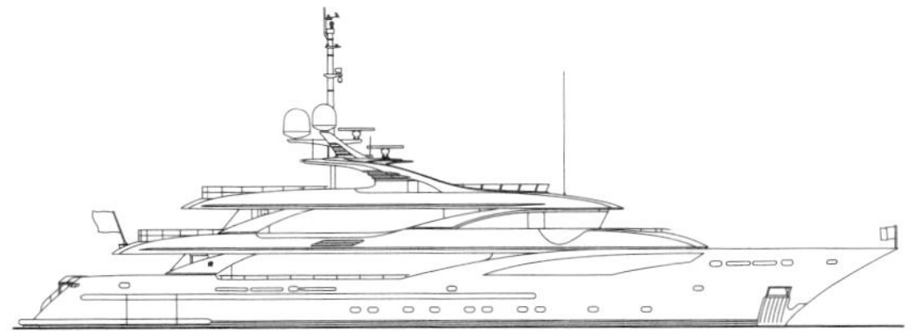

BELLE ANNA

Year 2011
Builder ISA Yachts Ancona
Construction Steel – Aluminum
LOA 49.95 m
B MAX 9 m
GT 495 t
Engine 2 X CAT 3512 C

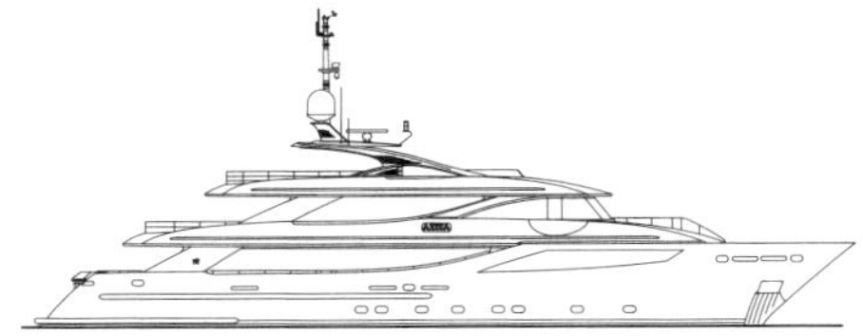

AZIZA

Year 2012
Builder ISA Yachts Ancona
Construction Steel – Aluminum
LOA 41 m
B MAX 8.40 m
GT 358 t
Engine 2 X CAT C18

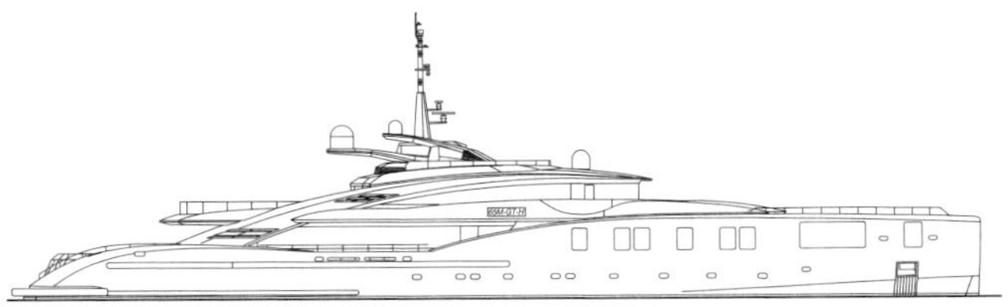

OKTO

Year 2014
Builder ISA Yachts Ancona
Construction Steel – Aluminum
LOA 66.40 m
B MAX 10.50 m
DSPL 1150 t
Engine 2 X CAT 3512C

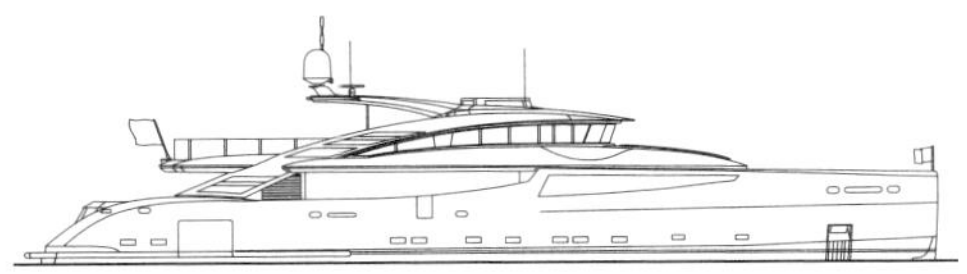

PHILMI

Year 2014
Builder ISA Yachts Ancona
Construction Steel – Aluminum
LOA 43.25 m
B MAX 8.60 m
DSPL 381 t
Engine 2 X CAT 3512C

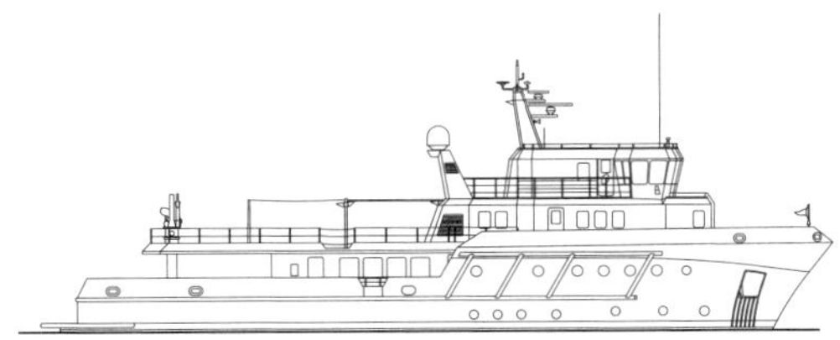

GENESIA

Year 2005-2017
Builder Cantieri Navali di Termoli (CB) 2005 / Cantieri CNC di Chioggia (VE)
Construction Steel
LOA 40.30 m
B MAX 10.25 m
DSPL 580 t
Engine 2 X CAT C32 1.300 HP

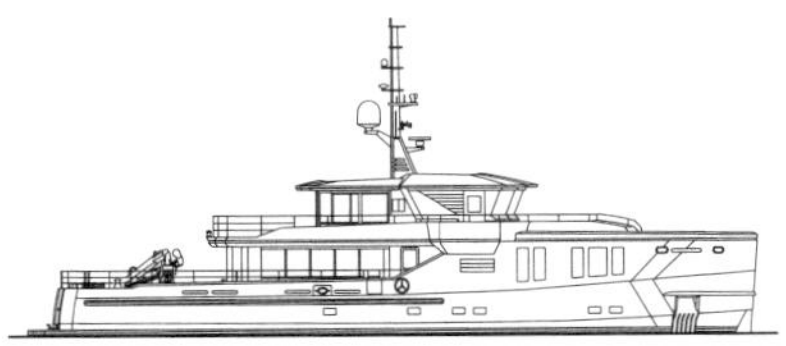

K-584

Year 2020
Builder CPN Ancona
Construction Steel – Aluminum
LOA 36 m
B MAX 7.60 m
GT 299 t
Engine 2 X MAN D2868 LE426 – 588 KW

Basic Glossary

by Carlo Bertorello

Angle of attack
The angle between the chord line of an airfoil and the axis of the body to which it is fitted. It is fixed and determined at design stage.

Aspect ratio
It is the ratio between the two main dimensions of a plane figure or the projection of a solid on a plane. It is the ratio between the span and the average chord of a wing or between the height and the mean width of a sail.

Bezier Curves
They are the most important type of polynomial curves. Widely used in computer graphics to model smooth curves. Defined by a set of control polygon points, the degree of the curve is given by the number of vertices less one of this polygon. The curve is contained in the convex set of its control points, and therefore the points can be displayed graphically and used to manipulate the curve intuitively.

Block Coefficient
It is the ratio between the volume of the immersed part of the hull and the one of a parallelepiped whose sides are length waterline, beam, and draft (of the hull). It is computed by the formula V/(L x B x T) and returns a quick estimate of the overall slenderness of the hull.

Bulwark
Part of the side of the hull that extends above the deck (with or without interruption at the deck level).

Bustle
After part of the hull that connects the hull forms to the leading edge of a rudder separated from the keel. Very common in the 60s and 70s yacht design, today almost never used.

Canard
An aeronautical term that indicates a layout of the trim plans forward of an aircraft wing. In naval architecture, it indicates the position of the main rudder and/or of an auxiliary rudder in front of the fin keel. Recently it is also used to indicate an auxiliary retractable daggerboard in front of a canting keel.

Canting keel
It is a ballast keel that can rotate in a transverse direction used to increase the stability of sailing boats by shifting the ballast weight to windward. Initially used only on racing boats, today, it is common also on large yachts for which the aim is to limit the heeling for comfort reasons.

Canoe body
It is the immersed part of the hull without the appendages (such as the fin keel and rudder). The definition has become common when the appendages, and particularly the fin keel, have become separated forms from the hull, and it has been necessary to distinguish the characteristics of the canoe body alone from those of a hull with appendages.

Center of the immersed lateral plan
Point of application of the resultant of the hydrodynamic forces on the immersed part of the hull and the appendages. It is conventionally placed in the geometric center of the hull and appendages underwater lateral plan.

Depths
Vertical distances at given measurement points from the bottom of the hull to the waterline aimed at conventionally defining the immersed forms of the hull. Mainly used in the measurement of IOR handicap system.

Flush deck
Deck configuration where no coachroofs and/or superstructures are placed. In some cases, the deck camber is particularly high to provide more height to the interior.

Foil
Immersed wing profile that generates a vertical force (lift) This is used as a hydrodynamic contribution to reduce Archimedes thrust and, therefore, the displacement, as well as to increase the righting moment of sailing boats.

Form Stability
Stability is given by the transversal position that the application point of the hydrostatic pressure assumes in the heeled hull (TCB). It is influenced by the transverse distribution of the hull volume, which can be assessed by the shape of the hull cross sections.

Frictional Coefficient
It is the non-dimensional value of the viscous resistance due to the motion of a body in a fluid, expressed by the formula $C_F = R_F/(0.5\ \rho\ v^2\ S)$, so the ratio between the viscous resistance of a hull and the product of the velocity squared by the wetted surface and by the density of the water divided by 2.

Gunwale
A line that identifies the upper part of the side planking. Once upon a time, the gunwale was the highest course of the side planking, often thicker.

Hull Measurement Girths
Measurement of the perimeter of the significant transversal sections, taken on the outside of the surface of the hull following the great circle line. Used to conventionally define the shapes of the hull and particularly, in the IOR handicap rule, the progression of the fore and aft overhangs.

Lines Plan
Conventional representation of the hull form through a set of longitudinal, horizontal, and transverse sections. It has been widely used in naval architecture. It is now frequently replaced by the three-dimensional representation of the hull surface.

Maximum Beam
Point of largest width of the hull, where, in the traditional wood construction, the longest beam was placed.

Metacentric radius
Distance between the center of the volume of the inclined hull and the point of intersection between the direction of the hydrostatic thrust of the inclined hull and the vertical direction of the upright hull, for an angle of inclination tending to zero.

NACA profiles
They are particular forms of wing profiles studied by the US National Advisory Committee for Aeronautics (NACA), for which the characteristics of lift and drag are known and available.

Power
Effective power (PE) Power required to tow the ship at a certain speed;
Shaft power (PD) Power to be supplied to the propeller to push the ship at a certain speed;
Brake power (PB) Rated power of the Engine adequate to reach the design speed.

Prismatic coefficient
It is the ratio between the volume of the hull (V), and that of a prism whose base is the transversal immersed section of maximum area (Am), and the height is the waterline length of the hull (L). It is computed by the formula V/(Am x L). It represents a more accurate indication of the slenderness of the hull than the one provided by the Block Coefficient. It is widely used for calculating the hull wave resistance with regressions based on experimental data and, in some cases, also for calculating the added resistance in waves.

Resistance Coefficient
It is the non-dimensional value of the total resistance of a hull, computed by the formula $C_T = R_T/(0.5\ \rho\ v^2\ S)$, so the ratio between the total resistance of the hull and the product of the speed squared by the wetted surface and by the density of the water divided by 2.

Sail plan center
Point of application of the resultant of the aerodynamic forces on the sails. It is conventionally placed in the geometric center of the sail plan.

Semi-angle of attack
It is the angle between the tangent at the forward end of the waterline and the centreline. It characterizes the slenderness of the bow, and in some cases, it was one of the parameters used in the calculation with regressions of the wave drag of the hull.

Skeg
Fixed appendage of the hull placed forward of the rudder and designed, as a function of the size and position of the rudder itself, to increase the directional stability and, in some cases, also the resistance to leeway at low speed.

Spline
It is a composite curve, made by joining adjacent segments of polynomial curves, imposing at the junction points the same tangent and curvature for both segments.

Tumblehome
Term corresponding to the widely used French "Fregatage." It represents the recessed shape of the higher part of the transversal sections of some ships and, in particular of sailing boats. The concept was used to reduce the resistance of the heeled hull, limiting its asymmetry.

Trim tab
Adjustable surface placed at the trailing edge of a keel with an airfoil profile designed to make the profile asymmetrical and thus increase its lift when necessary.

Underwater Lateral Plan (Lifting Surface)
Projection on the vertical longitudinal plan of the immersed part of the hull, of the wing surfaces of the appendages (fin keel, rudders, possible canards, and other bilgeboards) designed to counteract the leeway induced by the transverse component of the wind force.

Water ballast
It is fresh water or sea water ballast, contained in suitably arranged tanks, used to increase the transverse stability of sailing boats, or even to optimize the longitudinal trim of any boat. In the case of particularly light boats, the water ballast is also used to increase the displacement when sailing upwind in strong winds.

Weather–helm moment
It is the moment that generally arises when a sailboat heels transversally. When the boat heels, the center of buoyancy of the hull moves not only laterally, but due to the difference of the bow and stern volumes, also longitudinally, generally towards the stern. This produces a bow down longitudinal moment and a forward shift of the center of the lateral immersed plane that results in a luffing trend of the boat. Furthermore, the component (in the direction of motion) of the sail propulsive force, applied in the sail plan center generates a moment in the vertical plan with the component (in the opposite direction of the motion) of the hydrodynamic resistance produced by the immersed volumes of the hull. This moment is generating a further weather helm.

Winglet
Appendage placed at the tip of a wing. In the case of the sailing yacht at the bottom of the keel or of the rudder and designed to improve the efficiency by reducing the vortices due to the pressure difference between the two sides of the wing.

Acknowledgments
This book was published with funding from Co.Ge.Di. International S.p.A. – Gruppo Acque Minerali Uliveto e Rocchetta, and with support from the Department of Architecture at the Gabriele D'Annunzio University of Chieti-Pescara, which made its resources available for documentary research and graphic processing of design-related material.
Because images (drawings, illustrations, and photographs) play such a key role in telling this design-based story, I would first of all like to thank all of the authors who have so generously granted their use:
The photographers (listed in the credits), especially Carlo and Alessandra Borlenghi for their customary friendliness and patient helpfulness;
The editorial offices of "Giornale della Vela" and "Vela & Motore," which made their exhaustive archive materials available to us;
Giorgio Forattini and Massimo Bucchi for their cartoons *Buzzurra* and *Agnella*, Ugo Nespolo for his poster of the Italian Challenge in America's Cup, and Francois Chevalier and Jacques Taglang for drawings of boats of historical interest; a special thanks to Jessica Lagatta and Luca Mincarini, who did all of the highly delicate reworking of the graphic and photographic design material, and to Adriano Ciai with Virgilio Castellana for the research, selection, and classification of the drawings and images in our Studio's archive.
My thanks to everyone who meticulously reread the texts (in Italian or English), offering suggestions and flagging up inaccuracies: Elisabetta Strickland, Yula di Sambuy, Patrizia Ferri, Marco Profili, Giulia Vallicelli (my personal, daily editor, who also carried out much of the meticulous archival research); Vittorio Mariani, Alessandro Nazareth, and Nicola Sironi (for technical-specialist clarifications); James Robinson Taylor for the revision of the English translation, and Carlo Bertorello for editing the glossary.
Finally, my most heartfelt thanks naturally go to everyone who (albeit with different roles, at different times and for different periods) helped to produce the raw materials for this story – the drawings: Alessandro Nazareth (office partner), Patrizia Ferri, Nicola Sironi, Vittorio Mariani (founding members of the studio), Adriano Ciai, Virgilio Castellana, Massimo Di Nicolantonio, Camillo Giammarco (my former students and long-term collaborators), and starting from the very beginning, Paolo Bitetti, Carlo Bertorello, Antonio Di Stefano, Simone De Marinis, Paola Catella, Sira Lucchetti, Gianfranco Franciosi, Carlo Vallicelli, Giovanni Senatore, Tonino Matricini, Bacci Belgrano, Stefano Orlando, Marco Sironi, Stefano Carugno, Giorgio Cassetta, Giulia Vallicelli, Giulietto Ricci, Michelangelo Vallicelli, Fiorella Fino, Luigi Locuratolo, and finally Francesco Ricci, Leonardo Cozzi, Daniele Poli, Giulia Roncoroni, Michele Molino, Mario Biferali, Filippo Pernisco, Gianluca Roveri who, as interns, still offered valuable contributions to the project activities.

AV